D0075711

Food, Agriculture, and Rural Policy into the Twenty-First Century

Food, Agriculture, and Rural Policy into the Twenty-First Century

Issues and Trade-Offs

EDITED BY
Milton C. Hallberg,
Robert G. F. Spitze, and Daryll E. Ray

Westview Press
BOULDER • SAN FRANCISCO • OXFORD

Published in 1994 in the United States of America by Westview Press, Inc., 5500 Central Avenue, Boulder, Colorado 80301-2877, and in the United Kingdom by Westview Press, 36 Lonsdale Road, Summertown, Oxford OX2 7EW

Library of Congress Cataloging-in-Publication Data
Food, agriculture, and rural policy into the twenty-first century :
issues and trade-offs / edited by Milton C. Hallberg, Robert G.F.
Spitze, and Daryll E. Ray.
p. cm.
Includes bibliographical references and index.
ISBN 0-8133-8763-9
1. Food supply—Government policy—United States. 2. Agriculture and state—United States. 3. Nutrition policy—United States. 4. Agriculture—Economic aspects—United States. 5. Agriculture—Environmental aspects—United States. 6. United States—Rural conditions. 7. Agricultural laws and legislation—United States. I. Hallberg, Milton C. II. Spitze, R. G. F. III. Ray, Daryll E.
HD9006.F568 1994
338.1'973—dc20

93-50054
CIP

Printed and bound in the United States of America

The paper used in this publication meets the requirements of the American National Standard for Permanence of Paper for Printed Library Materials Z39.48-1984.

10 9 8 7 6 5 4 3 2 1

Contents

RURAL PROGRAMS, RESEARCH, AND EDUCATION

Preface

Current legislation for the U.S. food and agricultural sector is provided by the Food, Agriculture, Conservation, and Trade Act of 1990. The policies embodied in this Act encompass farm price and income support, food distribution, grain reserves, international trade, soil and water conservation, and rural development, as well as research and education programs vital to farm families, rural communities, agribusiness firms, consumers, and taxpayers. This legislation is due to expire in 1995. One option available to Congress and the President is to follow the 1990 Act with another that would provide more current, comprehensive policy for food and agriculture and again have a 5-year life. A second option is to extend existing legislation for a period beyond 1995, as appears to be the intent of Congress with the Omnibus Budget Reconciliation Act of 1993. In either case, informed decisions about the issues and policy choices must be made by all who would participate in the policy process.

Three political realities will likely overshadow the choices that will be made: (1) national concerns about the budget deficit will likely lead to reduction in or curtailment of public expenditures for many programs, including those focused on agriculture, (2) interest-group concerns about a variety of nonfarm issues, along with farm/nonfarm tensions, will continue to be influential in the formation of food and agricultural policy, and (3) environmental and natural resource consequences of agricultural activity are likely to continue, if not intensify, and will require more dialogue and cooperation before environmental/agricultural conflicts are resolved.

These political realities are by no means new in the history of policymaking for American agriculture, but they take on added significance in an era of (1) large budget deficits, (2) continuing urban expansion, leading to conflicts between urban and rural society, and (3) increasing concerns about human and animal health and the welfare of future generations. Thus, there are many policy issues still to be resolved and still much to be done through legislative initiative.

As 1995 policy decisions are crafted, attention will be directed toward trade-offs that exist between budget constraints and financial assistance for agriculture and rural residents, between international and national priorities for agriculture, between the costs and benefits of a diminished public involvement in agriculture, between protection of the environment and efficiency of agricultural production, between food costs and food safety, and between expansion and contraction of agricultural research and extension education programs. Public policy decisions are needed to accommodate these trade-offs in a manner that is economically sound and socially acceptable. All this will be widely debated before new legislation is crafted.

This book was designed to contribute significantly to 1995 policymaking by providing relevant and objective information to those who will participate as elected officials, policymaking advisors, industry spokespersons, lobbyists (for both agricultural and nonagricultural interests), educators, policy administrators, and concerned citizens. It offers recent information concerning a wide array of issues fundamentally important to the U.S. agricultural and food sector, reviews a diverse set of policy approaches for dealing with these issues, and assesses trade-offs among these alternative approaches. Each chapter of the book focuses on a current or emerging public concern that is likely to be a focus of attention in 1995 policy. The book should be a valuable policy guide for policymakers, agricultural extension educators, agricultural and food organization leaders, interest groups concerned about food and agriculture's future, as well as students enrolled in college courses on public policy.

This book was prepared as a major initiative of North Central Regional Research Project, NCR-151. The editors wish to extend a special thanks to Dr. Donald E. Anderson, University of North Dakota, who has served as administrative advisor to NCR-151 for the past several years and has contributed as a full partner in several projects of the committee as well as a strong supporter of the committee's overall efforts. Other agencies cooperating on this project include the Farm Foundation, the National Center for Food and Agricultural Policy, the Congressional Research Service, and Federal Extension Service. The Pennsylvania State University provided generous administrative and secretarial support.

As always, in efforts such as this, the authors of the various chapters of this book deserve a special commendation. They were selected to contribute because of their particular expertise and to reflect a truly "national" perspective. Their collective contributions have made possible a book that is informative and current, and that draws upon the expertise of professionals with a wide variety of perspectives.

In addition to chapter authors, the following professionals served as peer reviewers of the various chapters: Charles W. Abdalla, Resources for the Future and The Pennsylvania State University; Jay C. Anderson, Utah State University; Harry Ayre, University of Arizona; Del Banner,

University of Illinois; Peter Barry, University of Illinois; James T. Bonnen, Michigan State University; Thomas A. Brewer, The Pennsylvania State University; Arthur B. Daugherty, Economic Research Service, USDA; B. R. Eddleman, Texas A&M University; Barry Flinchbaugh, Kansas State University; Daryl Heasley, Director, Northeast Regional Center for Rural Development, The Pennsylvania State University; Dale Heien, University of California, Davis; Robert O. Herrmann, The Pennsylvania State University; Bengt Hyberg, Economic Research Service, USDA; Roger Johnson, North Dakota State University; Steve Koenig, Economic Research Service, USDA; Warren Lee, Ohio State University; Lynn H. Lutgen, University of Nebraska-Lincoln; Alden C. Manchester, Economic Research Service, USDA; William H. Meyers, Iowa State University; Louis H. Moore, The Pennsylvania State University; Wesley N. Musser, The Pennsylvania State University; Mechel S. Paggi, Senior Economist, American Farm Bureau Federation; James Pease, Virginia Polytechnic Institute and State University; David W. Price, Washington State University; Kenneth L. Robinson, Cornell University; Roland R. Robinson, Cooperative State Research Service, USDA; David Schweikhardt, Michigan State University; Larry V. Summers, Agricultural Marketing Service, USDA; Luther Tweeten, Ohio State University; Robert D. Weaver, The Pennsylvania State University; and Carl Zulauf, Ohio State University.

Glenn Bengtson served as technical editor for the book. He pleasantly and professionally edited the entire manuscript and provided invaluable guidance throughout the book's development. Finally, the senior editor's secretary, Jane Mease, was a most important contributor to the project. Jane established procedures, logged activity, typed copy, revised copy, worked with authors to develop appropriate graphic material, interpreted publisher guidelines and more. And she did it all graciously, efficiently, and professionally.

Milton C. Hallberg
Robert G. F. Spitze
Daryll E. Ray

About the Contributors

Walter J. Armbruster is managing director, Farm Foundation, Oak Brook, Illinois. He has provided leadership in organizing policy and professional symposiums on agricultural and food marketing issues; served as editor and contributing author to a number of marketing policy research and education publications; and focused his research on marketing efficiency and institutions, government programs, and marketing policy issues.

Harry S. Baumes is chief of the Western Hemisphere Branch, Agriculture and Trade Analysis Division, ERS. His primary research thrusts have been in the U.S. agricultural policy, supply/demand and price analysis, and international trade areas.

John R. Brake is the William I. Myers Professor of Agricultural Finance at Cornell University. His appointment includes teaching, research, and extension. He authored the report of the National Commission on Agricultural Finance and is co-editor of *Agricultural Finance Review.*

Roy R. Carriker is professor of food and resource economics at the University of Florida. He teaches and supports statewide extension education programs dealing with natural resource and environmental policy.

Julie A. Caswell is associate professor of resource economics, University of Massachusetts, Amherst. She conducts research and has written widely on the economics of food safety and nutrition, including analysis of public policy, the use of information policies such as food labeling, and the strategic response of food companies to changes in policy and consumer demand.

Robert L. Christensen is professor of resource economics and extension economist, University of Massachusetts. His teaching, research, and extension work has primarily focused on farm management and marketing. He served as director of the farm policy project for NASULGC during development of the 1990 agricultural and food policy.

Robert A. Cropp is professor of agricultural economics and Director of the Center for Cooperatives at the University of Wisconsin-Madison. His teaching, research and extension responsibilities focus on dairy marketing, dairy policy and agricultural cooperatives with emphasis in dairy cooperatives. He has authored numerous research and extension publications regarding these subjects.

Brady J. Deaton is professor and chair of the Department of Agricultural Economics at the University of Missouri and currently serves as Social Science Unit Leader in the College of Agriculture, Food and Natural Resources. His specialty is rural economic development and policy.

Michael R. Dicks is associate professor of agricultural economics and Director of the Great Plains Agricultural Policy Center. He teaches and conducts research in the areas of agricultural and natural resource policy. His research focus is on the land use implications of agricultural and natural resource policies.

Mark A. Edelman is professor of economics at Iowa State University and coordinator of the Iowa Public Policy Education Project. PPEP is a statewide nonpartisan citizen education project. In 1992, PPEP won national recognition by AAEA as the Distinguished Extension program. Edelman also researches and writes a media column series on a variety of agricultural and public issues.

Olan D. Forker is professor of agricultural economics at Cornell University. With a general interest in commodity markets and prices, his research focus is on measuring the economic impact of commodity checkoff programs. He is co-author of *Commodity Advertising: The Economics and Measurement of Generic Programs.*

Barry L. Flinchbaugh is professor and extension state leader at Kansas State University. He conducts public policy education programs on agricultural and economic policy and teaches a junior/senior level course in agricultural policy. Flinchbaugh administers an extension education program in farm management, marketing, policy and economic development, and lectures extensively throughout the country to farm organizations and agribusinesses.

A. L. (Roy) Frederick is professor and extension economist at the University of Nebraska-Lincoln. He specializes in public policy, with particular emphasis on the general economy, taxation, farm programs and international trade.

Bruce L. Gardner is professor of agricultural and resource economics at the University of Maryland. His research interests are primarily in agricultural policy analysis. During 1989-1992 he was Assistant Secretary for Economics in the U.S. Department of Agriculture, with responsibility for the Department's economic analysis and forecasting agencies.

Harold D. Guither is professor of agricultural policy, University of Illinois and serves as director of extension programs for the Department of Agricultural Economics. He has conducted research and authored reports and monographs on many public policy issues. His most recent research and writing deals with the animal rights and animal welfare movements.

Milton C. Hallberg is professor of agricultural economics at The Pennsylvania State University. He teaches and conducts research in agricultural policy and has authored or edited several reports and books in the field including *Policy for American Agriculture: Choices and Consequences.*

Thomas R. Harris is professor of agricultural economics and director, University Center for Economic Development at the University of Nevada, Reno. His research activities are focused on impact analyses of rural communities experiencing changes in economic structure or federal land management policies.

Dale E. Hathaway is director of the National Center for Food and Agricultural Policy in Washington, D.C. He has been making, researching, and writing about agricultural and trade policy in various positions for forty years.

Glenn A. Helmers is professor of agricultural economics at the University of Nebraska. His teaching and research involves several aspects of the economics of agriculture and the environment. He is currently analyzing agricultural policy and its impact on sustainable agriculture.

Dennis R. Henderson is chief, Marketing Economics, Economic Research Service, U.S. Department of Agriculture. His research has focused on industrial organization in the food sector, strategic behavior of multinational firms, determinants of domestic and international market competitiveness, and public policy that affects market performance. Prior to joining USDA, he was professor of agricultural economics at the Ohio State University.

Dana L. Hoag is associate professor of agricultural and resource economics at Colorado State University. His research and teaching is focused on production and policy issues addressing agriculture and the environment. He has accomplishments in sustainable agriculture, soil conservation, and groundwater contamination from pesticides.

Jean D. Kinsey is professor of agricultural and applied economics at the University of Minnesota. She teaches and conducts research on food marketing, food and consumer policy, household economics, and human capital. She is a co-author of *Food Trends and the Changing Consumer.*

Ronald D. Knutson, professor of agricultural economics and extension economist, is Director of the Agricultural and Food Policy Center at Texas A&M University. He works primarily in the area of policy and is the author or co-author of numerous publications including *Agricultural and Food Policy*, now in its third edition.

Carol S. Kramer is Associate Director of the Resources and Technology Division of the Economic Research Service, U.S. Department of

Agriculture. The division conducts an integrated program of research, policy analysis, and data development concerning the relationship between agriculture and the use of resources on the environment, food safety, and human health.

James A. Langley is senior policy analyst, Agricultural Stabilization and Conservation Service, USDA. As principal agricultural economist, he is responsible for the development of policy options and for the evaluation of program recommendations and legislative proposals on general commodity and program issues related to the Commodity Credit Corporation.

David R. Lee is associate professor of agricultural economics at Cornell University. His teaching and research is in agricultural and trade policy. During his recent leave in Washington, D.C., he served as visiting economist in the Economic Research Service of USDA and in the Office of the Administrator of USAID.

David A. Lins is professor of agricultural economics and adjunct professor of finance, University of Illinois. He teaches agricultural and corporate finance and conducts research on financial management issues for farms and agribusiness firms. He is a principal in the Center for Farm and Rural Business Finance.

Karl D. Meilke is a professor of agricultural economics and business at the University of Guelph, Ontario, Canada. He is currently serving on the Executive Committee of the International Agricultural Trade Research Consortium and his research is focused on agricultural commodity modeling and policy analysis.

William H. Meyers is professor of economics, Associate Director of the Center for Agricultural and Rural Development and Co-Director of the Food and Agricultural Policy Research Institute at Iowa State University. His research is in agricultural trade and policy, price analysis and transition economics, and he is North American Editor of *Agricultural Economics.*

John P. Nichols is professor of agricultural economics at Texas A&M University and teaches in the areas of food and agricultural marketing and policy. His research is focused on problems of marketing management and market development in the food system. Recent publications include an edited book, *Economic Effects of Generic Promotion Programs for Agricultural Exports.*

C. Tim Osborn is Leader, Resource and Commodity Policy Section, Economic Research Service, USDA, directed at the use, value, allocation, and quality of natural resources. His research has focused on the economic and distributional effects of Federal conservation policy, the economic impacts of soil erosion, effects of erosion on agricultural productivity, targeting of erosion control funding, and interactions between commodity and conservation programs.

Glen C. Pulver is professor emeritus of agricultural economics, University of Wisconsin-Madison. His research and consulting focuses on rural economic development policy. Among his publications are *Designing Development Strategies in Small Towns* and *Community Economic Development Strategies.*

Daryll E. Ray is professor of agricultural economics and rural sociology and Director, Agricultural Policy Analysis Center, and holds the Blasingame Chair of Excellence in Agricultural Policy, University of Tennessee. With research primarily on commodity programs, he has developed two well-known national models to estimate the economic impacts of agricultural policies and economic conditions.

C. Parr Rosson III is professor of agricultural economics and extension specialist-international trade and marketing at Texas A&M University. Extension, research, and teaching efforts focus on international trade, trade policy, and international marketing operations. He has recently co-authored the book titled *Introduction to Agricultural Economics.*

C. Ford Runge is professor, Center for International Food and Agricultural Policy, Department of Agricultural and Applied Economics, University of Minnesota. Recent publications include *Reforming Farm Policy: Toward a National Agenda,* with Willard W. Cochrane, and *Free Trade, Protected Environment: Balancing Trade Liberalization and Environmental Interests.*

Larry D. Sanders is professor of agricultural economics at Oklahoma State University. He teaches and does extension work in public policy related to agriculture, natural resources and environment, rural development, and international trade policy. He has written a variety of publications and chaired the National Public Policy Education Committee.

Jerry R. Skees is professor of agricultural economics, University of Kentucky. His research and teaching program in agricultural policy has resulted in a number of publications and special projects funded by USDA. He is co-author of the book, *Sacred Cows and Hot Potatoes,* Westview Press.

David M. Smallwood is the senior economist with the Food Economics Branch, Economic Research Service, U.S. Department of Agriculture. He conducts research and analysis on domestic food assistance and nutrition programs.

Daniel B. Smith is professor of agricultural and applied economics at Clemson University. He is coordinator of extension programs in agricultural economics. He conducts extension and applied research programs in farm management with an emphasis on policy and risk management alternatives.

Edward G. Smith is professor and extension specialist and holds the Distinguished Ray B. Davis Professorship in Agricultural Cooperation in the Department of Agricultural Economics at Texas A&M University. His extension and research activities focus on food and agricultural policy.

Mark E. Smith is leader of the Commodity Trade Programs Section in the Economic Research Service, U.S. Department of Agriculture. His work has included research on food aid, export credit, and export subsidy programs in the U.S. and other countries.

Robert G. F. Spitze is professor of agricultural economics, University of Illinois. Focusing his research and teaching on food and agricultural policy, primarily price, income, and human resource policy, his most recent publication is a co-authored book, *Food and Agricultural Policy: Economics and Politics.*

Ronald G. Trostle, chief of the Commodity and Trade Analysis Branch, Economic Research Service, USDA, is responsible for short- and long-term forecasting of foreign and U.S. agricultural trade. During 1990-92, he was ERS coordinator of economic analysis activities to support the Uruguay Round negotiations.

Fred C. White is the D. W. Brooks Distinguished Professor of Agricultural and Applied Economics at the University of Georgia. He teaches undergraduate and graduate courses in agricultural policy. His research focuses on the economic evaluation of policies affecting agriculture.

Norman K. Whittlesey is professor of agricultural economics at Washington State University. He has more than 30 years research and teaching experience in water resource management and public policy. His work has influenced federal agency methods for irrigation project evaluation and guided water allocation policy in the West.

Ronald C. Wimberley is professor of sociology at North Carolina State University where he specializes in rural, agricultural, and environmental issues. His recent presidential address for the Rural Sociological Society, "Policy Perspectives on Social, Agricultural and Rural Sustainability," is in the Spring, 1993 *Rural Sociology*.

Abner W. Womack, co-director of the Food and Agricultural Policy Research Institute, is professor of agricultural economics, University of Missouri. His primary research focus is price and policy analysis that relies heavily on the use of large scale econometric models of U.S. and global agriculture. A majority of his current work is associated with the assessment of U.S. and global issues under consideration by the U.S. Senate and House Agricultural Committees.

PART ONE

The Setting

1

The Economic Setting for U.S. Food and Agriculture

Daryll E. Ray and Roy Frederick

Agricultural and food policy in its broadest sense has been around since the beginning of the nation. Furthermore, nearly all other countries—developed and developing—have policies which are directed toward food and agriculture. The nature of these polices differs from country to country, but they all affect the availability and cost of food, as well as the level and stability of farm prices and incomes. Policy changes are influenced by a myriad of forces—some economic and some political. This chapter briefly reviews the sweeping changes which have occurred in U.S. agriculture and broadly summarizes the food system's supply, demand, and income conditions which followed recent legislation. Also, the projected farm economy during the debate on and expected duration of the 1995 food and agriculture legislation will be discussed.

Agriculture has changed dramatically since the mid-1800s. At that time 84 percent of the U.S. work force was in agriculture and produced 72 percent of the Gross National Product (Wallace 1986). The exodus of farm workers, and the reduction in agricultural input production and product processing on the farm has resulted in major changes in the structure of agriculture. The United States boasted 5.9 million farms and a farm work force of 11 million in 1945, but both have been declining since. By 1977, the number of U.S. farms had dropped to 2.7 million, and the farm work force dwindled to about 4 million workers (Emerson 1978). In the 12 years between 1988 and 2000, it is estimated that the number of farms will fall from about 2 million to 1.7 million (Martinez 1992).

World War II saw the genesis, in response to a wave of farm mechanization, of the continuing trend toward larger U.S. farms, as well as the shift of input production and product processing from the farm to nonfarm firms (Emerson 1978). Formation of the European Economic

Community (EEC) in the 1960s was accompanied by greater internationalization of agriculture; and permanent U.S. agricultural surpluses appeared (Wallace 1986). The United States reaped great benefits from this globalization of agriculture during the 1970s, because export demand sharply increased. In response, the United States and other countries boosted their own productive capacities, and export demand plunged, prompting the farm crisis of the 1980s.

During the late 1980s, net farm income rose sharply, though not for the reasons that policymakers expected. The early 1990s have seen declining commodity prices for crops, stable to increasing input prices, and flat net incomes to farmers.

The Economic Backdrop for Farm Legislation

The economic outlook for agriculture depends on the balance between the sector's expanding capacity to produce and the growth in market demand for its products (O'Brien 1992). The sections which follow discuss changes in resource utilization, input use, demand composition, prices, and net income since the late 1970s. Several of the graphic presentations are organized by the tenure of food and agricultural legislation:

- 1978-81 for the Food and Agriculture Act of 1977.
- 1982-85 for the Agriculture and Food Act of 1981.
- 1986-90 for the Food Security Act of 1985.
- 1991-95 for the Food, Agriculture, Conservation, and Trade Act of 1990 and projections.
- 1996-2000 for 1995 policy and projections.

For these graphs, the variable averages are indexed, with the 1982-85 average set to 100. Hence, variable averages for the years covered by the 1977 and 1985 acts and selected other periods are shown as percentages of the average for the 1981 act period. These delineations of act tenures also will be used throughout the discussion for convenience, although it should be noted that simply because an event is described as occurring during an act period does not imply that the act is responsible for the event. Projections for the 1990 act and the 1995 policy, unless otherwise noted, are from the Food and Agricultural Policy Research Institute (FAPRI 1993) baseline published in April 1993 or the Congressional Budget Office (CBO 1993) baseline published in February 1993.

FAPRI and CBO projections are not predictions of what will happen during the projection period; instead, they are intended to show how U.S. agriculture could look assuming no dramatic changes in a myriad of policy and macroeconomic conditions. For example, both the CBO and FAPRI projections assume Conservation Reserve Program contracts will be allowed to expire as scheduled beginning in 1996. Both also assume new food and agricultural legislation will be enacted in 1995 and it will have a 5-year life. The FAPRI projection further assumes stable long-term U.S. interest rates, annual increases of 2.8 percent in real U.S. Gross Domestic Product (GDP), and no significant agricultural policy changes in most other countries. FAPRI also assumes no conclusion to the General Agreement on Trade and Tariffs (GATT). For a full discussion of these assumptions, see FAPRI (1993) and CBO (1993).

Land Use: Planted and Diverted Cropland

Crop Acreages

Planted area for the 15 major crops[1] has declined considerably since the 1977 act (Figure 1.1). In part, this reduction was a response to reduced prices, but largely it was due to land diversion under federal farm programs, including the Conservation Reserve Program (CRP). During the period covered by the 1977 act, acreage planted to the 15 major crops averaged 297 million acres; 40 million acres were idled or diverted from production. For the periods covered by the 1981 and 1985 acts, the corresponding numbers are 290 million acres and 37 million acres, and 260 million acres and 62 million acres, respectively. The large increase in diverted acreage from 1982 to 1990 is due to the CRP. FAPRI projects acreage planted to the 15 major crops will increase slightly to an average of 262 million acres for the years covered by the 1990 farm act, with an average diverted acreage for the period of 60 million acres.

Assuming CRP contracts are allowed to expire, acreage diverted under annual programs and the CRP are projected to decline during the span of the 1995 policy to 42 million acres. Also during this time, planted area for the 15 major crops increases to 275 million acres. The sum of diverted acreages and acreages planted to the 15 crops during the tenure of the 1995 policy is projected to be 5.5 million acres less than during the 1990 act period, suggesting that a significant portion of the acreage coming out of the CRP is projected to remain in grass.

A similar pattern is evident for acres planted to the 7 major crops (see footnote 1). Acreage planted to corn, wheat, and soybeans generally

---

[1]The 15 major crops include 7 major program crops—corn, wheat, barley, grain sorghum, oats, soybeans, and cotton—as well as rice, sugar, sunflowers, peanuts, edible beans, tobacco, rye, and flaxseed.

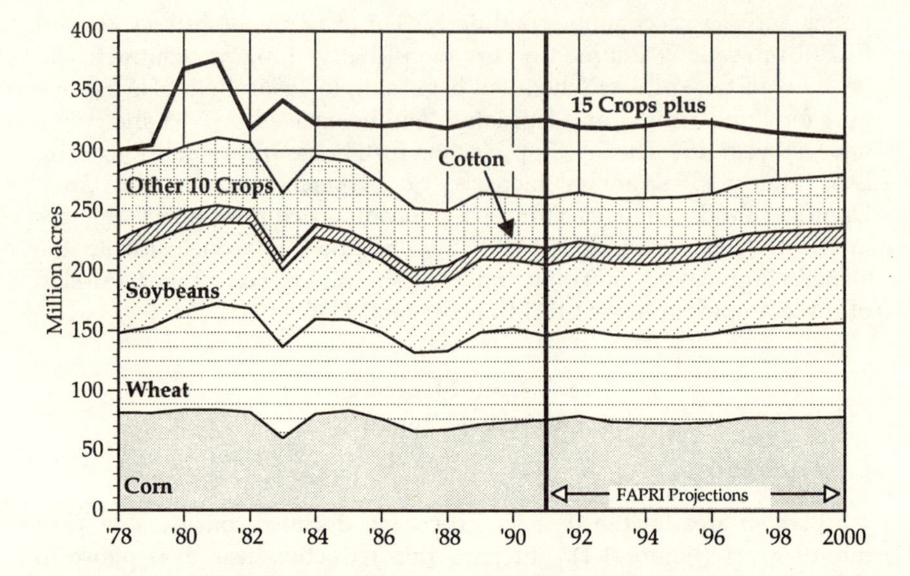

FIGURE 1.1 Changes in U.S. Planted Acreage, Total Acreage Selected for 15 Crops, 15-Crop Total Acreage, and 15-Crop Total Acreage Plus Idled, 1978-2000.
Sources: USDA (as compiled in Ray et al., 1993 and FAPRI 1993).

clined between 6 percent and 11 percent between the 1981 and 1985 act period averages. The reasons for these declines were that grain prices were pushed down by lackluster export demand and efforts to reduce grain inventories. The price declines led to increased use of short- and long-term land withdrawal programs. Much of the acreage taken out of wheat and soybeans was enrolled in the CRP, while most of the acreage previously in corn went to annual land diversion programs. However, cotton planted acreage increased 9 percent, as natural fibers and denim jeans became increasingly popular.

FAPRI projects increases in planted acreage for corn, wheat, and soybeans during the 1990s. Average wheat planted acreage is projected by FAPRI to increase about half a million acres during the 1990 act, compared to the previous 5 years, and another 4.7 million acres during the years to be covered by the 1995 policy. Cotton acreage also is projected to increase during the 1990s.

Input Use

With fewer acres and lower crop prices during the late 1980s, farmers needed fewer inputs and could not afford as much machinery and other depreciable assets as they could during the 1970s and early 1980s. Figure 1.2 shows these input usage changes, with total hours of farm work showing the sharpest drop at 15 percent between the 1981 act and 1985 act averages. Nonpurchased inputs, which include operator and family labor, operator-owned real estate, and other capital items, decreased 9 percent between the two periods.

The use of mechanical power and machinery declined 13 percent. Use of nitrogen fertilizer and agricultural chemicals held about constant between the two periods but, with fewer acres harvested, the application rate per acre increased. The combined quantities of feed, seed, and livestock purchased increased somewhat as did the use of miscellaneous inputs. Projections on input quantities are not available for the 1990s. However, production expenditure projections are discussed in a later section.

Production and Productivity Measures

Corn yields during the period of the 1985 act averaged 7 bushels above that of the prior 4 years, continuing a long-standing upward trend. Part of this increase may be due to a stepped-up diversion of less productive corn acreage under federal programs in the latter period. Average wheat yield actually decreased 6 percent between the 1981 and 1985 act periods (Figure 1.3). Unfavorable weather and lower prices probably are largely responsible for the lower average. The index of crop production per acre increased 6 percent between the two periods, and livestock production per breeding unit increased 14 percent.

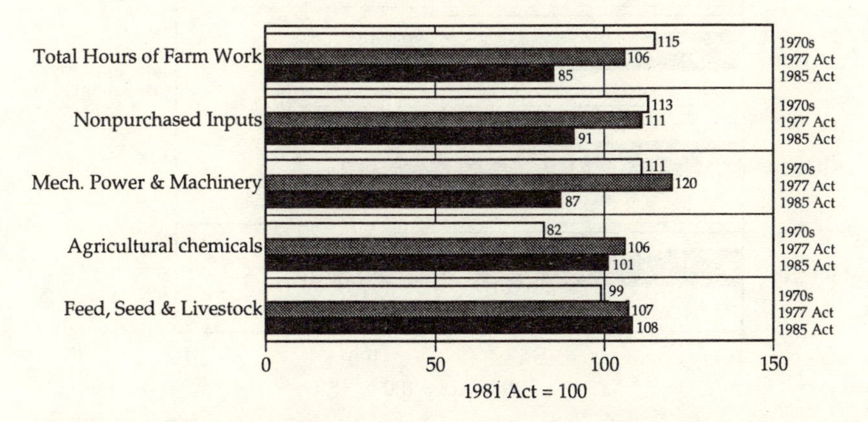

FIGURE 1.2 Indexed Changes in U.S. Agricultural Input Use from the 1970s to 1985 Act Period.
Sources: USDA (as compiled in Ray et al., 1993) and FAPRI 1993.

FAPRI wheat yield projections for the 1990 act period average the same as during the base (1981 act) period and increase 4 percent during the last 5 years of the decade (Figure 1.3). Corn yields are projected to continue their upward march during the 1990s, averaging 22 percent higher than the 1981 act average during 1996-2000. CBO projections for the 1990 act and 1996-98 are nearly identical.

Production of the major crops declined during the years covered by the 1985 act, compared with the 1981 act period. With both wheat average acreage and yield down during the 1985 act period, wheat production declined 15 percent between the two periods. Reduced acreage also caused corn production to decline an average of 1 percent between the years covered by the 1985 and 1981 acts.

FAPRI projections show average wheat production at 2.3 billion bushels during the 1990 act period and 2.6 billion bushels during the 1995 policy's tenure, compared with 2.2 billion bushels for the 1985 act period and 2.6 billion bushels during the 1981 act period. Hence, it takes until the latter half of the 1990s for wheat production to return to and probably exceed the average level observed during the 1981 act period.

Corn production is projected to increase substantially in comparison with averages for the 1981 and 1985 act years. FAPRI projects average corn production of 8.4 billion bushels and 9.1 billion bushels for the years

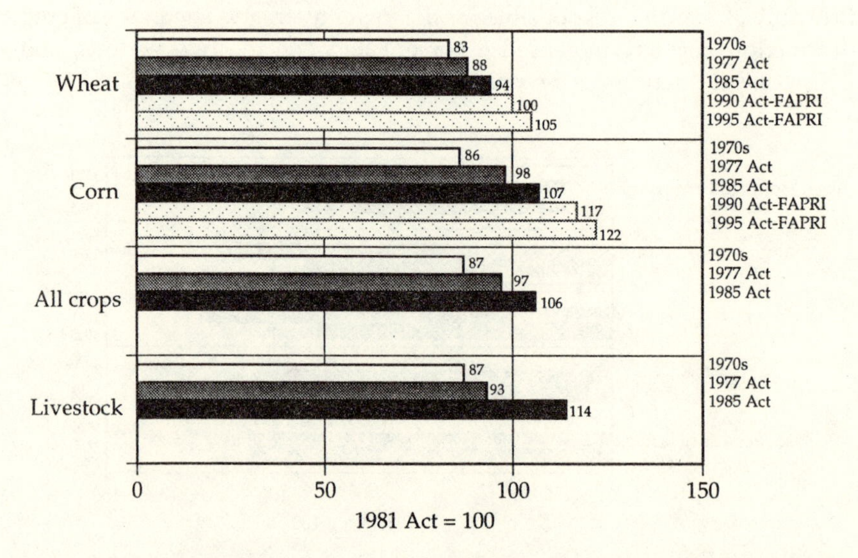

FIGURE 1.3 Indexed Changes in U.S. Crop Yields and Livestock Production Per Breeding Unit, 170-79, 1981 Act Period.
Sources: USDA (as compiled in Ray et al., 1993) and FAPRI 1993.

of the 1990 act and 1995 policy, respectively, compared with about 7.2 billion bushel and 7.1 billion bushel averages during the tenure of the 1981 and 1985 acts, respectively.

Total crop production declined 3 percent between the 1981 and 1985 act periods, but total farm output remained unchanged, with livestock production picking up the slack. Projections for total crop and livestock production are unavailable for the 1990s.

Output Mix

Major Crops

As total planted cropland declined considerably during the late 1980s, each of the major crops' share of the total of 15-crop planted acreage and idled acreage also declined. Between the periods of the 1981 and 1985 acts, corn's proportion of the 15-crop planted plus idled acreage decreased from 23.4 percent to 22.1 percent; wheat acreages declined from 24.3 percent to 22.1 percent of the 15-crop total; and soybeans fell from 20.4 percent to 18.3 percent.

FAPRI projections show relatively little change in the mix of major crops during the period of the 1990 act. Corn, wheat, and soybean planted acreages as a share of the 15-crop planted plus idled acreages increase between 0.5 percent and just over 4 percent during the period of the 1995 policy.

Major Livestock Enterprises

The mix of livestock enterprises also has remained relatively stable. Still, the data do confirm that red meats have lost ground to poultry, especially broilers (Figure 1.4). The proportions of livestock cash receipts from sales of cattle, broilers, and turkeys increased between the 1981 and 1985 act periods. Also between these periods, cattle rebounded only partially to the levels of the 1970s, while broiler and turkey shares increased significantly.

FAPRI projections for the 1990s show broilers and turkeys gaining additional share with slight reductions projected for hogs. The average share of cash receipts for cattle increases during the tenure of the 1990 act and the 1995 policy, but continues to remain below the level observed in the 1970s.

Demand for Major Crops

Domestic Demand

Domestic demand for feed grains is determined largely by livestock numbers and livestock production profitability. Cyclical lulls in livestock production and relatively expensive feed prices caused feed and crushed soybean demand to decline during the 1981 act years compared with the average of the 1977 act period (Figure 1.5). However, during the 1985 act period, when livestock prices and production rose and feed prices dropped sharply, demand for grains and soybean meal resumed their upward trends.

FAPRI projections for corn feed demand are 5.2 billion bushels and 5.4 billion bushels during the terms of the 1990 act and the 1995 policy, respectively. This shows an upward trend, when compared to 4.2 billion bushel and 4.5 billion bushel averages for the periods of the 1981 and 1985 acts. CBO averages for the 1990 act and 1995 policy periods are slightly higher than those of FAPRI (Figure 1.5).

Typically, wheat used for feed accounts for about 5 or 6 percent of total wheat utilization, but it is highly sensitive to the wheat-corn price ratio. During the term of the 1977 act—a time when the ratio of wheat to corn prices did not favor wheat feeding—utilization of wheat for animal feeds averaged about one-sixth that utilized for human foods. In contrast corn compared to wheat was expensive during the 1981 act period, and wheat feed demand averaged 314 million bushels, nearly one-half that (646 million bushels) for wheat food. During some years of this 1982-85 period, wheat feed demand approached two-thirds of wheat food demand. Of course, even during high-use years, wheat fed to livestock as a proportion of total grain fed is relatively small. During the 1981 act period, the wheat fed averaged 314 million bushels; while corn feeding averaged 4.2 billion bushels. FAPRI and CBO projections for the 1990s show wheat feed demand to be near the levels of the early 1980s.

Domestic demand for nonfeed corn has doubled since the late 1970s, increasing from an average of 600 million bushels during the 1977 act to

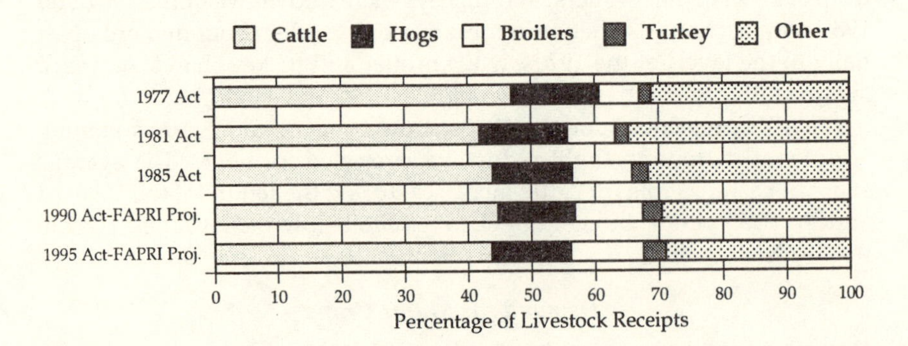

FIGURE 1.4 Changes in U.S. Livestock Production Mix, 1977 Act to 1995 Policy. *Sources:* USDA (as compiled in Ray et al., 1993) and FAPRI 1993. periods.

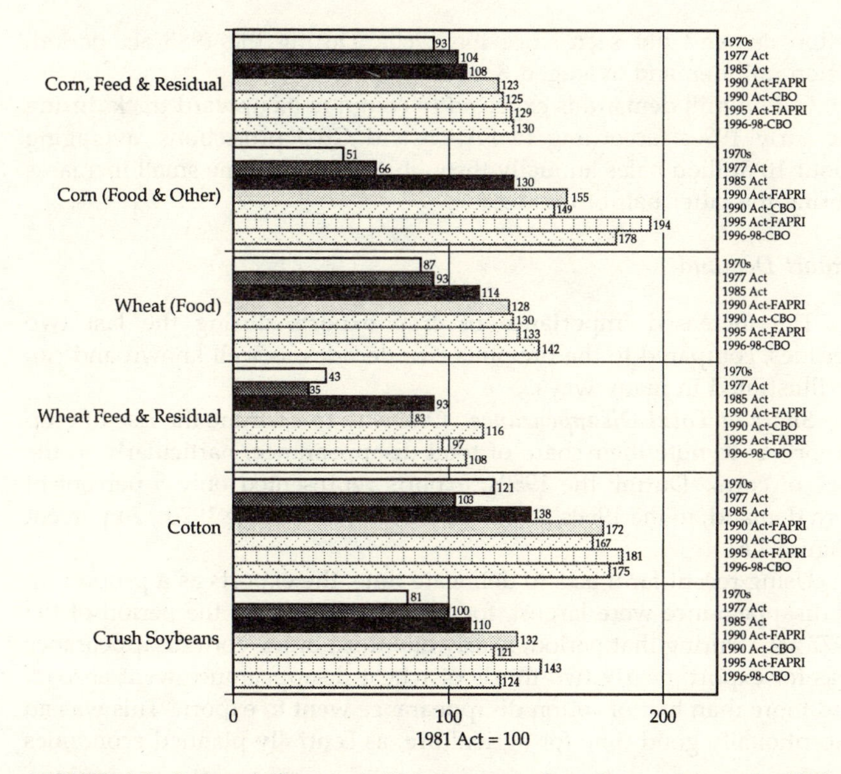

FIGURE 1.5 Indexed Changes in U.S. Domestic Crop Demand from the 1970s to 1995 Policy Period.
Sources: USDA (as compiled in Ray et al., 1993), FAPRI 1993 and CBO 1993.

1 billion bushels during the 1981 act period and 1.3 billion bushels during the 1985 act period. Increased production of sweeteners and ethanol largely are responsible for the sizable increase. These uses for corn increased demand at an average compound annual rate of 10.5 percent between 1982 and 1985. FAPRI projections suggest a similar growth rate for most of the 1990s, with nearly twice as much corn used for domestic non-feed purposes during the 1995 policy years, in comparison with the years of the 1981 act.

Cotton domestic demand has been on a roller coaster ride during the past two decades. During the 1970s, cotton mill demand averaged 7.1 million bales annually. During the first half of the 1980s, average annual mill demand fell off sharply to 5.8 million bales during the 1981 act period. The movement toward natural fibers in the late 1980s and the explosion in demand for denim jeans ignited a resurgence in domestic cotton demand not seen since the 1960s. During the 1985 act period, cotton mill demand averaged 8 million bales.

Cotton mill demand is expected to continue its upward track during the early 1990s, according to FAPRI and CBO projections, averaging about 10 million bales annually through 1995 with some small increases during the latter half of the decade.

Export Demand

The increased importance of crop exports during the last two decades, compared to the previous two decades, is well known and can be illustrated in many ways.

Share of Total Disappearance. One way to examine the rise of crop exports is to note their share of total disappearance, particularly in the case of corn. During the 1950s, exports represented only 5 percent of corn demand; in the 1960s, it was 12 percent; and in the 1970s, 24 percent (Figure 1.6).

Using recent farm acts to delineate time, the exports as a proportion of disappearance were largest, for most crops, during the period of the 1977 act. During that period, nearly one-third of the corn disappearance was for export, nearly two-thirds of wheat disappearance went abroad, and more than half of cotton disappearance went to export. This was an exceptionally good time for agriculture, as centrally planned economies

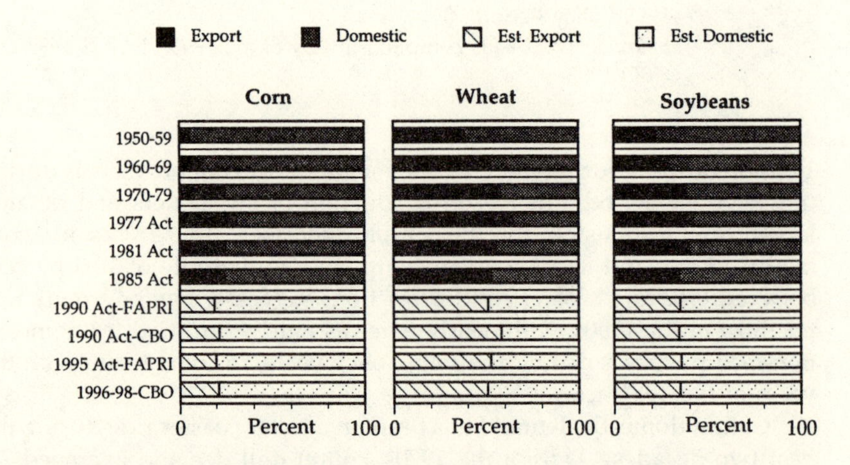

FIGURE 1.6 Exports and Domestic Demand as a Percent of Total Disappearance for Wheat, Corn, and Soybeans from the 1950s to 1995 Policy Period.
Sources: USDA (as compiled in Ray et al., 1993), FAPRI 1993, and CBO 1993.

were buying feed for expanding livestock herds, incomes in the developing countries were increasing, and loans from large U.S. and international banks to these countries were flowing freely. These fundamental changes, coupled with periodic yield fluctuations in major crop-producing countries, resulted in an explosion in world crop demand. The United States had the reserve capacity to fill a large portion of that demand.

But the increase in grain prices sowed the seeds of overexpansion. Price increases and low inventories during the 1970s caused countries around the world to expand agricultural production capacity. Governments offered price guarantees, expanded market-promotion programs, and provided other incentives which further encouraged their farmers to increase production. As we will see, increase production they did and, once started, there is an inertia which is not overcome easily. The net result was a dramatic decrease in demand for U.S. grain during the years of the 1981 act, and the export share of a crop's disappearance returned roughly to 1970s proportions. Even with the advent of the Food Security Act of 1985—promoted heavily as a catalyst for rejuvenating export markets—corn, wheat, and soybean exports as a share of total disappearance continued to decline.

Export Quantities. Quantities of exported wheat and soybeans were greatest during the years of the 1977 act. Exported volumes of major crops declined during the term of the 1981 act and averaged somewhat lower during the years of the 1985 act (Figure 1.7). However, the reduced exports during the period of the 1981 act cannot be attributed to reduced worldwide use of grain. World wheat disappearance, for example, increased from an annual average of 434 million metric tons during the period of the 1977 act to an average of 477 million metric tons during the period of the 1981 act. And in the case of wheat, worldwide wheat trade was larger. Hence, the size of the wheat export pie increased, but the United States received less business in terms of both worldwide share of demand and quantity (Figures 1.7 and 1.8). During the 1981 act period, the decline in U.S. wheat exports stemmed from increased grain production by other producers—like the EEC (24 percent) and Argentina (53 percent), which are wheat export competitors; and wheat export customers like China (40 percent) and India (25 percent). This increase in non-U.S. production largely was a reaction to higher market prices, low inventories, and fears of uncertain grain availability during the 1970s and early 1980s. Another factor was expansive in-country farm policies which set countries on a steady course of production expansion which continued largely unabated into the 1990s.

Of the major wheat producers, the United States, a large wheat producer, decreased production (from 69.4 million metric tons to 58.7 million metric tons) between the periods covered by the 1981 and 1985 acts. Argentina and Australia, relatively small wheat producers, showed a decline in production between the periods covered by the 1981 and 1985 acts (from 12.4 million metric tons to 9.4 million metric tons for Argentina, and from 16.4 million metric tons to 14.4 million metric tons

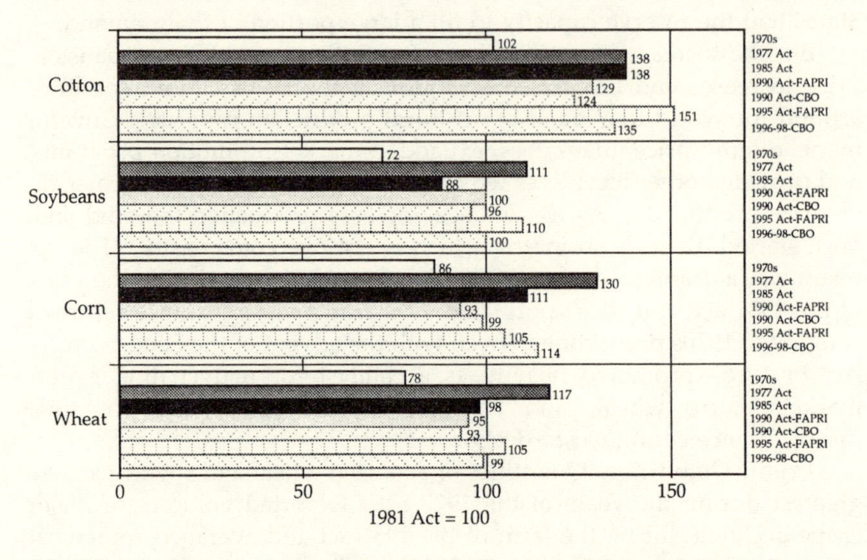

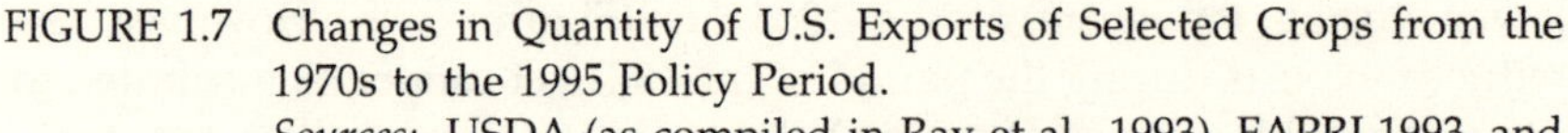

FIGURE 1.7 Changes in Quantity of U.S. Exports of Selected Crops from the 1970s to the 1995 Policy Period.
Sources: USDA (as compiled in Ray et al., 1993), FAPRI 1993, and CBO 1993.

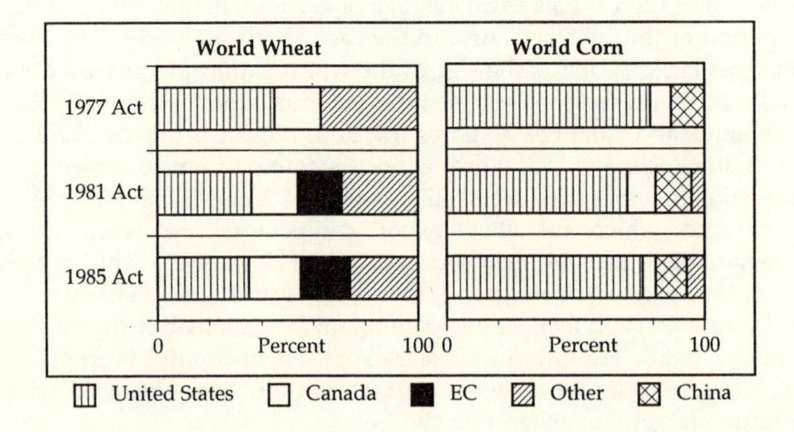

FIGURE 1.8 Changes in Selected Countries' Percent Market Share of World Wheat and Corn Exports, 1977 Act to 1985 Act Periods.
Sources: USDA (as compiled in Ray et al., 1993) and FAPRI 1993.

for Australia). The other major wheat-producing countries continued large production increases during this period. The EEC produced 74.3 million metric tons during the 1981 act period and 79.4 million metric tons during the 1985 act period. Other comparisons of wheat production between the years of the 1981 and 1985 acts, all in million metric tons, are:

- Former USSR, 72.0 and 86.3.
- China, 80.9 and 90.1.
- India, 42.4 and 48.3.
- Canada, 24.7 and 26.1.
- All other wheat producing countries, 64.3 and 75.8.

Furthermore, wheat production is expected to continue to increase in China, the EEC, and India during the 1990 act period (Figure 1.9). FAPRI projects a decline in EEC wheat production during the last half of the decade, which in turn largely is responsible for projected increases in U.S. wheat export demand during the period (Figure 1.9).

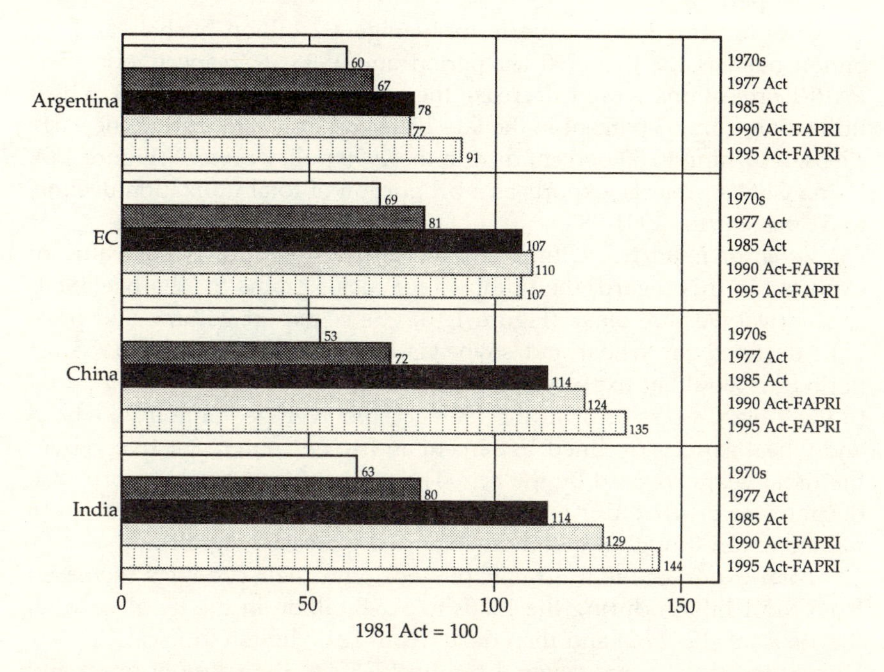

FIGURE 1.9 Indexed Changes in Wheat Production for Selected Producers from the 1970s to 1995 Policy Period.
Sources: USDA (as compiled in Ray et al., 1993) and FAPRI 1993.

U.S. average corn export volume declined 22 percent between the periods covered by the 1977 and 1981 acts. But in contrast to wheat and soybeans, average corn export volume increased during the fiscal years covered by the 1985 act. These gains were realized largely because the United States is the dominant corn export market player, with more than a two-thirds share of world trade.

Also, major corn export competitors tend to have relatively open markets. Hence, the price-reducing effects of lower support prices, Commodity Credit Corporation (CCC) and Farmer-Owned Reserve (FOR) inventory releases, export subsidies, and other export-expansion programs of the period of the 1985 act at best slowed the rate of decline in U.S. wheat and soybean exports. But in the case of corn, export volume increased.

There is little difference between the FAPRI and CBO projections of wheat exports for the remainder of the 1990s. FAPRI sees wheat exports averaging 1.3 billion bushels during the 1990 act period and 1.4 billion bushels during the 1995 policy tenure. This compares with an average of 1.263 billion bushels exported during the 1985 act period, 1.316 billion bushels during the 1981 act period, and 1.464 billion bushels during the 1977 act period.

CBO projects wheat exports to average 1.2 billion bushels and 1.3 billion bushels for the 1990 act period and 1996-98, respectively. The FAPRI projections would decrease the export share of total U.S. wheat utilization from 53 percent in the late 1980s to 51 percent during the early 1990s, returning to 53 percent during the 1995 policy period (Figure 1.6). Using CBO's projects, exports as a proportion of total utilization decline, to 51 percent for 1991-98.

Value of Exports. Ultimately, the critical measure is the value of exports. In this regard, the numbers for major grains for the late 1980s and early 1990s are bleak (Figure 1.10). Since export volume and price both declined for wheat and soybeans between the 1981 and 1985 act periods, it would be expected that their value would be down. Compared to the 4 fiscal years prior to the 1985 act, the average value for soybean and wheat exports declined 18 percent and 16 percent, respectively, over the fiscal years covered by the act. The value of feed grain exports also declined somewhat during the two periods—from $6.3 billion to $6 billion—even though the volume exported increased significantly.

Total export value of grains, oilseeds, and their products increased from $13.1 billion during the 1970s to $26.8 billion in the fiscal years of the 1977 act (1979-82) and then declined to $21.2 billion in fiscal 1983-86. During the fiscal years covered by the 1985 act, the value of grain and oilseed exports declined further to $19.7 billion. Of course, using averages does not reflect systematic and stochastic nonprice effects. But

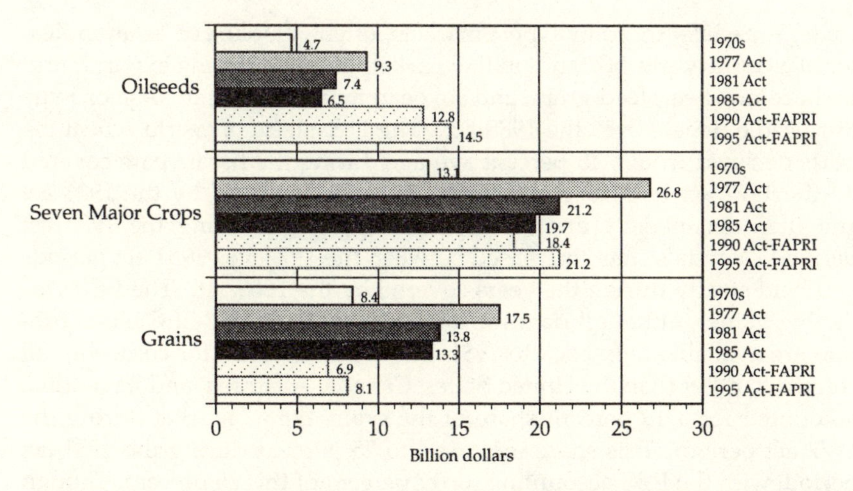

FIGURE 1.10 Seven Major Crops, Oilseed, and Grain Export Value, in the United States from the 1970s Through 1995 Policy Period.
Sources: USDA (as compiled in Ray et al., 1993) and FAPRI 1993.

it is reasonable to assume that the lower effective export prices flowing from the 1985 act period did not reduce the grain collectively supplied by competing countries. Nor did it sufficiently increase the collective grain demand of the grain importing countries to expand U.S. grain exports and offset the price declines. The working hypothesis would be that the short-run export demand for grains was inelastic during this period.

Projections by FAPRI suggest somewhat increased export values for grains and soybeans during the 1990 act period compared to that of the 1985 act period, but these values are projected to remain below 1981 act levels. FAPRI projections for the term covered by the 1995 policy show a 17 percent increase in grains and soybean export values compared to 1985 act levels. CBO projections suggest that total grain and soybean export values during the 1990 act period and 1996-98 will average well below the levels of the 1980s—23 percent lower during the 1990 act period and 16 percent lower during the term of the 1995 policy. Obviously, during the 15 years between the 1985 act and the century's end, significant changes will occur in the determinants of U.S. grain export demand. But even with a 15-year time horizon, it is unclear if U.S. export demand will be elastic in the long run.

World Market Share. Market share, aside from classifying an economic sector as perfectly competitive, oligopolistic, or monopolistic is not a concept or measure that is part of the profit- and utility-maximization models of traditional economic theory. Maximizing market share as a policy goal is akin to an entrepreneur maximizing production. Yet, market share, especially in grain export markets, often is looked to as an indicator of a country's importance in the marketplace. Regaining export market share in wheat, feed grain, and soybean markets became a major issue during the debate over the 1985 act. The U.S. share of world wheat exports declined from a 45 percent average during the fiscal years covered by the 1977 act to 36 percent during the years covered by the 1981 act and then declined by one more percentage point during the 1985 act period. Canada's share increased between the 1977 and 1981 act periods and held steady during the years covered by the 1985 act. The EEC was the big gainer, although data for the current 12-country EEC configuration are available only back to 1981 for wheat and 1985 for corn. But all countries, other than the United States, Canada, Australia, and Argentina, accounted for a 19 percent share of the grain export market during the 1977 act period. This share increased to 25 percent during the 1981 act period, with the EEC accounting for 17 percent of this 25 percent. Though much of the U.S. export subsidy outlays during the 1985 act period were designed to take business from the EEC, the EEC's share of world wheat exports actually increased 2 percent.

Share projections are unavailable for the periods of the 1990 act and the 1995 policy, but the export numbers suggest the U.S. share would increase somewhat by the end of the decade under FAPRI assumptions and decline in the CBO baseline.

100 Years of U.S. Agricultural Exports. Viewed over the last century, agricultural exports as a percent of all U.S. exports have been declining steadily, falling from about 45 percent in 1889 to less than 10 percent in 1991 (Figure 1.11). The rate of decline accelerated during the late 1980s. Figure 1.12 shows the 1970s and early 1980s were not the only periods in which exports were an important demand component of the U.S. agricultural sector. During the late 1800s and early 1900s, the value of agricultural exports as a percentage of farm cash receipts was similar to corresponding percentages computed for the last decade and a half.

Prices and Stocks

Figures 1.13 and 1.14 show trends in wheat and corn market prices and how those prices have compared with corresponding announced loan rates and target prices. Grain prices generally followed loan rate declines in the late 1980s, have remained within the loan rate and target price band, and are projected to remain within the loan rate and target price band during the 1990 act and 1995 policy periods. Commercial stock levels, as a percentage of total disappearance, have remained relatively steady through time. In the case of corn, commercial stocks (including grain) pledged for CCC loans generally ranged from 17 percent to 18 percent of disappearance and for wheat, from 20 percent to 24 percent.

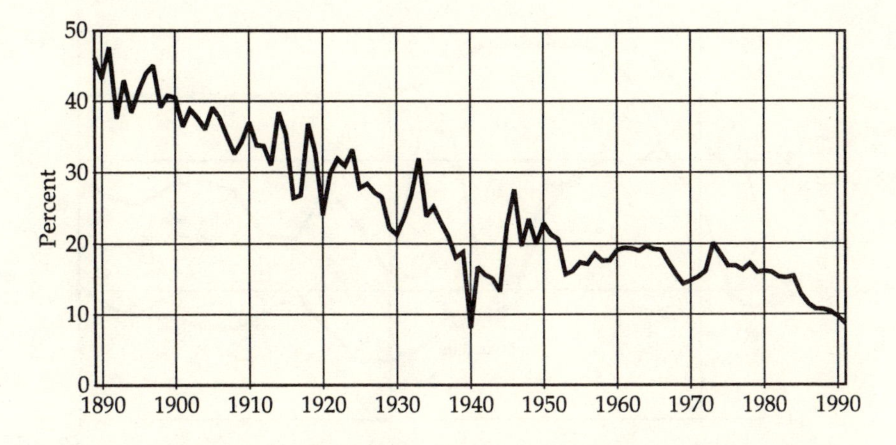

FIGURE 1.11 Agricultural Exports as a Percentage of All U.S. Exports, 1989-1991.
Source: Hallberg 1991.

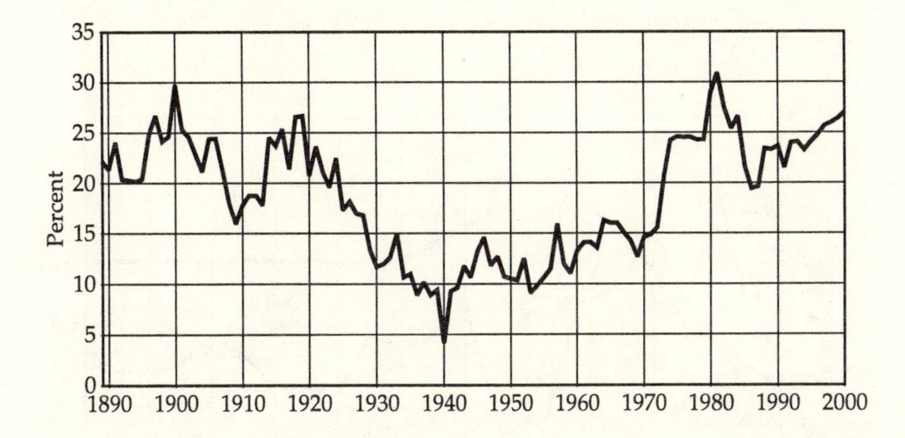

FIGURE 1.12 U.S. Agricultural Export Value as a Percentage of Agricultural Cash Receipts, 1889-2000.
Sources: Hallberg 1991, FAPRI 1993.

CCC and FOR corn stocks amounted to about 20 percent of disappearance during the 1980s and about half that during the 1970s. Wheat CCC and FOR stocks averaged 44 percent during the 1981 act period and dropped to 26 percent during the 1985 act period. FAPRI and CBO expect corn and wheat commercial stocks to be in the range of 20 percent to 24 percent—near historical levels for corn, though somewhat below historical wheat levels. Projected buffer stocks in the form of CCC and FOR inventories are inconsequential, particularly during the term of the 1995 policy.

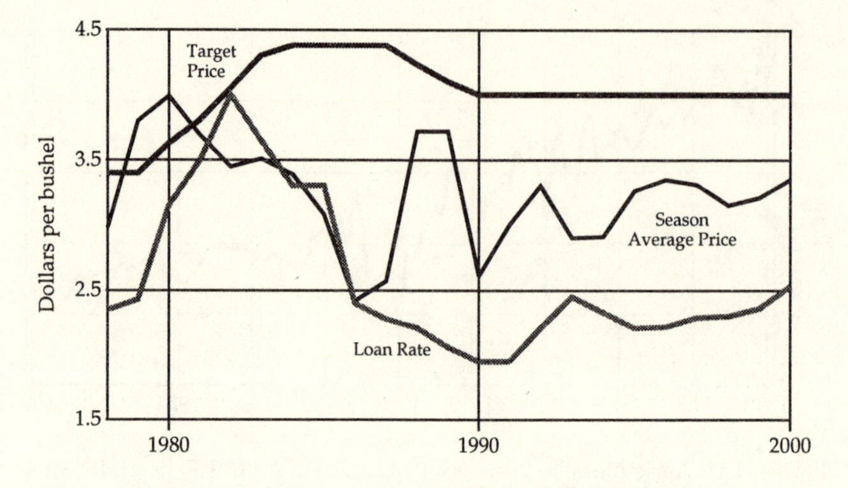

FIGURE 1.13 U.S. Wheat Target Price, Season Average Price, and Loan Rate, 1978-91.
Sources: USDA (as compiled in Ray et al., 1993) and FAPRI 1993.

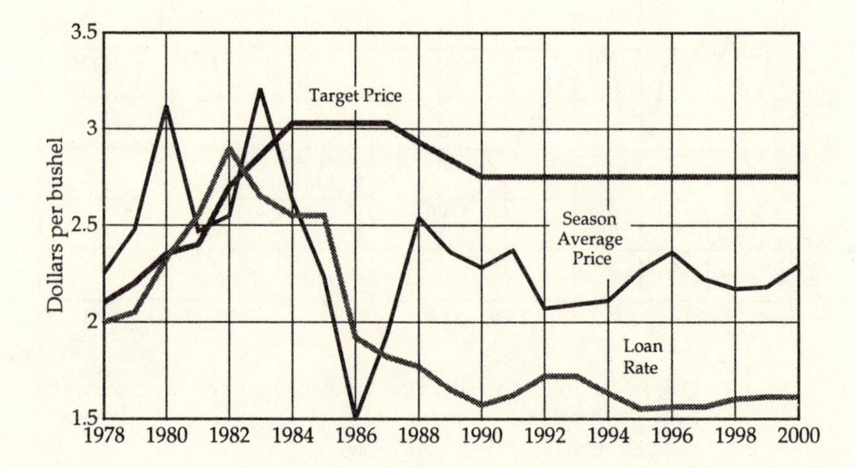

FIGURE 1.14 U.S. Corn Target Price, Season Average Price, and Loan Rate, 1978-91.
Sources: USDA (as compiled in Ray et al., 1993) and FAPRI 1993.

Farm Income

Measured in 1987 dollars, net farm income averaged $49 billion during the 1970s, dipped substantially during the latter years of the 1970s

($35 billion during the 1977 act period) and early 1980s ($26 billion during the 1981 act period) before climbing again later in that decade ($40 billion during the 1985 act period). In nominal dollars, net income during the 1981 act period averaged $23 billion and increased to a $43 billion average during the 1985 act period. Of the $20 billion dollar increase in income, $10 billion was due to increased livestock receipts, $6 billion to increased receipts for nonprogram crops, and the remaining $4 billion to decreased production expenses. In fact, the sum of cash receipts and government payments for the seven major crops (see footnote 1) averaged over the years of the 1981 and 1985 acts is identical at $47 billion. The increase in government payments following the 1985 act just offsets reductions in cash receipts from lower loan rates and market prices (Figures 1.13 to 1.15).

Livestock prices generally increased during the late 1980s. Beef prices, for example, increased from an average of $55 per hundredweight during the 1981 act period to $65 per hundredweight during the tenure of the 1985 act. FAPRI expects livestock prices to increase slowly throughout the period. Beef prices are expected to weaken during the first 4 to 5 years, then strengthen toward the end of the decade.

Based on CBO's projected price, production, and deficiency payment levels, the sum of program crop cash receipts and government payments is projected to change very little during the 1990s. The sum of government payments and cash receipts for the seven major crops during the 1990 act period averages 4 percent less than during the period of the 1981 act; for 1996-98, CBO projects this sum to increase 2 percent above the 1981 act average. FAPRI projections show increases in the sum of cash receipts for the seven major crops plus government payments of 6 percent from the 1981 act period to the 1990 act period. This sum averages 5 percent higher for 1996-98.

FAPRI net income projections averaged over the years of the 1990 act are $3.8 billion higher than for the 1985 act period, but net income is projected to be only $2.8 billion higher during the last half of the decade. Production expenses are projected by FAPRI to increase more slowly than livestock and crop cash receipts. By the latter half of the decade, government payments are expected to drop to four-fifths of the direct payments paid during the 1981 act period.

Summary and Conclusions

In general, the resource base, input use, and productivity growth will allow U.S. agricultural production capacity to continue to expand. If currently idled lands are brought on line, considerable productive capacity

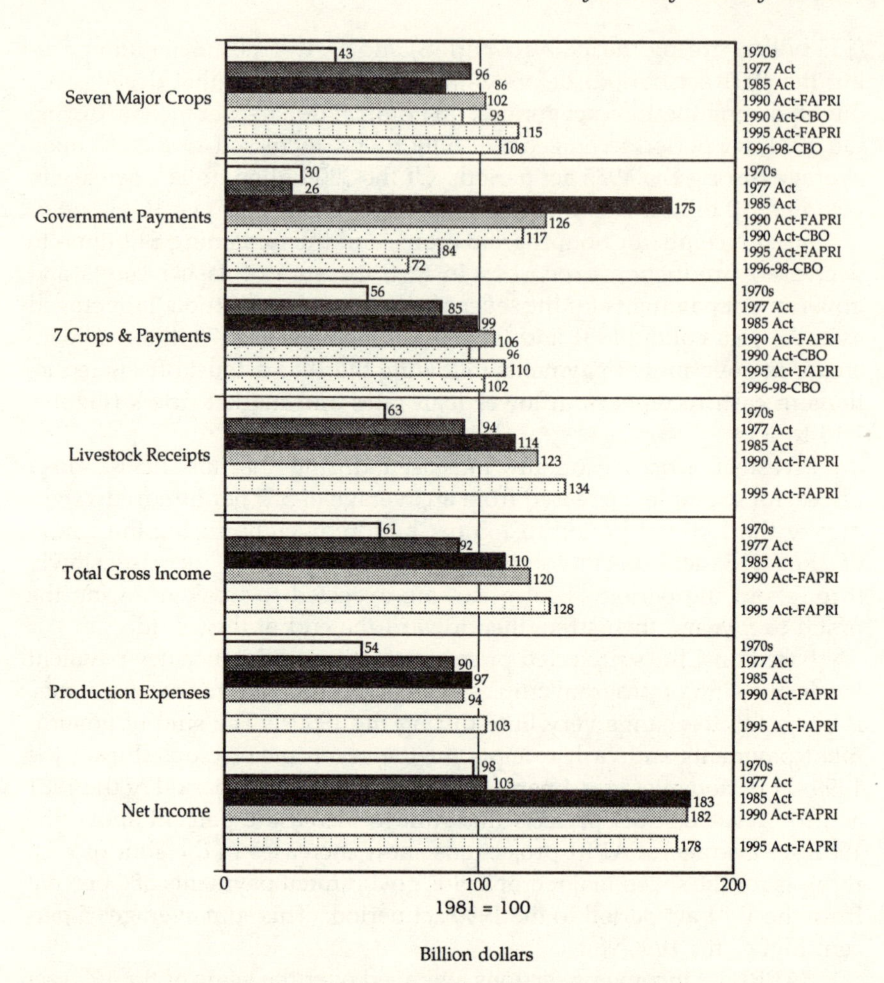

FIGURE 1.15 U.S. Farm Cash Receipts, Gross Income, Production Expenses, and Net Income from the 1970s to 1995 Policy Period.
Sources: USDA (as compiled in Ray et al., 1993) and FAPRI 1993.

will be available. On the demand side, both domestic and export demand are expected to grow steadily but not dramatically. Although the supply-demand balance may tighten somewhat, the agricultural sector is likely to face continued excess capacity and declining real prices and incomes.

In fact, nominal farm price and income levels for producers of farm program crops are likely to stay flat at best—much as gross income for producers of the seven major crops has been holding steady since the period of the 1981 act. Projections for the rest of the decade suggest flat nominal aggregate net income and decreasing net returns for the crops examined in the FAPRI and CBO projections. Actions to reduce government expenditures on farm programs may further reduce net incomes.

In addition to declining incomes in this decade, the stage is set for increased commodity price instability. Prices for major crops now can drop lower or soar higher than they could during the 1980s. Prices can drop because, since 1985, support prices have been reduced by a third, and prices can increase because government and FOR stocks have intentionally been reduced. A few years of weather-related low yields could cause prices to explode much as they did in the 1970s. This is unlikely but if it happens, consumers and public officials could demand an explanation. Reserve stock programs were designed to prevent a repeat of the abnormally high prices of the 1970s that increased consumer food prices in the short-run and created farm output overexpansion problems during the next decade. Price instability also interferes with efficient production. For example, if planning prices are not realized, bankers may become cautious in granting loans, and farmers may hesitate to risk investment in new technologies.

In retrospect, the prospect of an export-led rejuvenation of agriculture to free market prosperity through the lower loan rates and export subsidies of the 1985 act was embarrassingly oversold. So far, lower loan rates and crop prices have added billions of dollars to the federal budget deficit by widening the spread between target prices and loan rates; while in the export market lower prices failed to garner a proportional increase in export volume. This failure reduced export value and increased the nation's trade deficit.

With the emphasis on budget deficit reduction in the early 1990s, there is perhaps a parallel between how the commodity program portions of the 1985 and 1990 acts will play out and efforts to persuade countries to replace relatively inconspicuous import quotas with highly-visible tariffs and then use public pressure to reduce or eliminate tariffs. Of course, there is nothing wrong with this, but it is a different argument than that used to promote the legislation. And this argument may be quite different from the understanding of the 1985 and 1990 farm legislation by many of those who voted for it.

In light of the experience of the 1985 and 1990 acts with regard to export value and the outcome of the General Agreement on Tariffs and Trade (GATT) negotiations, the nature of the export market may be reexamined during the debate for the 1995 food legislation. It may also be time to reevaluate the objectives of farm programs. Economists traditionally have used structural and treadmill arguments as frameworks to justify programs. Under these frameworks, programs to moderate farm price and income instability compensate farmers for their inability to capture technology-driven lower costs. The benefits of agricultural technology advances are wholly passed to consumers as reduced real food prices.

Several questions may be pertinent:

- Can grain markets now accommodate large variations in supply without steep price changes?
- Do grain farmers quickly move resources out of and into agricultural production as output prices change?
- Will demand grow as fast or faster than productivity growth?

If the above questions can be answered "yes," the traditional justifications for farm programs may no longer be valid.

References

Committee for Economic Development. 1974. *A New Farm Policy for Changing World Food Needs*. A statement by the Research and Policy Committee of the Committee for Economic Development, Washington, D.C.

Congressional Budget Office. 1993. *Baseline for Programs of the Commodity Credit Corporation*. Washington, D.C.

Emerson, Peter M. 1978. *Public Policy and the Changing Structure of American Agriculture.* Washington, D.C.: Congressional Budget Office background paper.

Food and Agricultural Policy Research Institute. 1993. *FAPRI 1993 U.S. Agricultural Outlook.* Iowa State University and University of Missouri-Columbia, Staff Report No. 1-93.

Hallberg, M.C. 1991. "Agricultural Trade and Trade Agreements." In proceedings of *Pennsylvania's Agricultural Economy: Trends, Issues, and Prospects.* University Park, PA: Penn State University, College of Agriculture, pp. 31-40.

Martinez, Doug. 1992. "Fewer Owners Hold More U.S. Farmland." *Farmline* 8(7): 14-17. Washington, D.C.: USDA, Economic Research Service.

O'Brien, Patrick M. 1992. "Longer Term Prospects for the U.S. Agricultural Sector." In *Agriculture Outlook Conference 93*. Washington, D.C.: USDA.

Ray, Daryll E., Robert M. Pendergrass, Melissa R. Harding, and Richard L. White. 1993. *An Analytical Database of U.S. Agriculture.* The University of Tennessee, forthcoming.

Wallace, L. Tim. 1986. "Agriculture's Future: America's Food System." *Agricultural Research* 1(4) [Special Issue].

2

World Food and Agricultural Economy

Bruce L. Gardner and Abner W. Womack

The policy options that will appear most sensible and most needed in the debate leading up to the 1995 food and agricultural legislation will depend crucially on the world agricultural situation and outlook at that time. This chapter reviews the prospects for the 1995 situation and the longer-term outlook toward the year 2000. Despite being only 2 years away, forecasting the economic situation in 1995 is fraught with difficulties. Projections for the following 5 years are even more uncertain.

Our discussion deals in part with the overall world agricultural economy, and in part with commodity-specific issues for key agricultural products, particularly the grains. We consider first the outlook for global food demand, second the supply situation, and third policy considerations—all three of which are necessary for an overall picture of the agricultural economy in 1995-2000. The demand assessment turns on macroeconomic conditions—real income growth, inflation, and exchange rates. The supply side is more narrowly described, depending primarily on productivity growth, resource availability, and environmental constraints in farming. Policy issues involve both international issues, notably agriculture in the GATT, and national policies.

World Demand for Agricultural Commodities

The main factor in demand growth is income growth because income constrains expenditures on food. The income elasticity of spending on food is higher for low- and middle-income countries than for the high-income industrial economies. For the world, over the 1965-90 period, real income (gross domestic product or GDP) grew at an average annual rate of 3.5 percent, while purchases of agricultural products grew at an annual rate of 2.4 percent. This suggests a worldwide income elasticity of demand for farm-based products of about 0.7. This is a crude estimate, of course, since it assumes the relative price of farm products has remained constant and that no structural or income distributional changes have biased the estimate.

The overall growth in income can be decomposed in two ways: nominal versus real income, and real income as the product of population times per capita real income. The extent to which nominal income growth exceeds real income growth is a measure of inflation. While inflationary shocks can have large short-term effects, our projections for 1995-2000 do not incorporate a burst of inflation. We project steady, moderate inflation in dollar prices of about 4 percent per year in 1995-2000, and assume the real effects of inflation at this rate will be negligible.

Table 2.1 shows current projections of real gross domestic product (GDP) growth for foreign economies generated by Project Link, a collation of country projections published by the United Nations, used by the Food and Agricultural Policy Research Institute (FAPRI). The industrial countries are projected to emerge from recession in 1993, and GDP growth rates are assumed to be maintained through the 1990s at about the same rate as through the 1980s. Developing countries collectively are projected to have quite vigorous growth in 1993-2000.

One of the major disappointments of the 1990s to date has been worldwide real income growing at rates slower than had been expected. As of December 1990, when the 1990/91 recession was already established in forecasters' expectations, USDA projected real GDP growth outside the United States at 2.3 percent for 1991 (U.S. Department of Agriculture 1990 p 6). In early 1991, the forecast was cut to 1.4 percent, and as of January 1993, USDA estimates a 0.9 percent rate of growth for 1991, increasing to 1.3 percent for 1992 and 2.2 percent for 1993. The key problems have been the weakness of the European and Japanese economies in 1991/92, and GDP declines in Eastern Europe of 10 to 20 percent annually compared to 5 to 10 percent formerly anticipated.

The real income outlook for 1995-2000 envisages these economic problems being largely overcome. However, it cannot be said that these projections are wildly optimistic about 1995-2000 growth prospects. The 1995-2000 projections for rates of growth are actually a little below the historical rates of the 1980s (which are already below 1965-1980 rates)[1] for all developed market economies except Canada and South Africa. For developing countries, the projections of GDP growth shown in Table 2.1

---

[1]Proponents of the idea of American economic decline in the 1980s may wish to note that the United States is the only large industrial market economy to grow faster in 1980-90 than in 1965-80. (One country within the EEC, i.e., the United Kingdom, did see accelerated growth in the 1980s, compared to 1965-80.)

TABLE 2.1 Real GDP: Historical Data (1965-91) and Projections (1991-2000).

	1965-1980	*1980-1990*	*1991*	*1992*	*1993*	*1994*	*1995*	*1996-2000*
Developed								
Australia	4.0	3.4	-1.1	0.9	3.5	4.3	2.7	2.7
Canada	4.8	3.4	-1.7	0.9	3.9	4.8	4.3	3.4
EC-12	4.0	2.4	0.8	1.3	2.1	2.2	2.4	2.4
Japan	6.4	4.1	4.4	2.3	3.7	3.2	3.4	3.5
South Africa	3.7	1.3	-0.9	-0.7	1.4	2.2	2.3	2.2
United States	2.7	3.4	-1.2	2.0	3.0	3.3	2.8	2.9
Eastern Europe		2.2	-9.6	-2.9	1.2	2.5	3.5	3.7
Former USSR		1.7	-15.5	-17.6	-4.5	0.5	1.2	3.9
Developing								
Asia								
China	6.8	9.5	7.0	12.1	11.8	9.6	8.7	8.1
India	3.6	5.3	2.7	3.6	4.6	5.7	6.1	6.0
Indonesia	7.0	5.5	6.6	6.1	6.3	6.3	7.1	7.5
Pakistan	5.2	6.3	6.0	6.1	6.2	6.5	6.0	6.1
South Korea	9.9	9.7	7.6	7.1	7.8	8.0	8.0	8.3
Thailand	7.3	7.6	7.7	6.7	7.9	8.2	8.0	7.5
Latin America								
Argentina	3.4	-0.4	5.0	5.8	3.4	4.0	3.3	3.0
Brazil	9.0	2.7	1.5	0.9	2.4	4.2	5.5	4.2
Mexico	6.5	1.0	3.6	3.1	4.5	5.4	5.4	6.1
Paraguay	7.0	2.5	3.5	3.5	3.4	3.4	3.3	2.5
Africa								
Egypt	7.3	5.0	2.3	2.7	3.6	3.3	3.8	3.7
Morocco	5.7	4.0	4.5	1.0	3.8	3.6	3.5	4.2
Nigeria	6.0	1.4	4.4	2.3	3.0	3.2	3.1	3.5
Saudi Arabia	10.6	-1.8	1.2	2.3	4.7	1.1	1.6	2.7
Tunisia	6.5	3.6	3.5	-1.1	6.1	5.8	5.1	5.2

Source: World Bank, pre-1990 dates; Wharton Econometric Forecasting Associates, October 1992, for U.S. data; Project LINK, October 19, 1992, for all other countries, 1991-2000.

are slightly more optimistic than those of the Economic Research Service of USDA. O'Brien (1992) places annual GDP growth in 1990-2000 at 5.6 percent for Mexico, 6.1 percent for South Korea, 6.8 percent for Thailand, and 7.3 percent for China.

If world economies recover as generally expected in the latter half of this decade, it is very likely that the most significant growth in trade will be from the higher income developing economies. Although China is on the export side of the market, the substantial GDP growth rates projected indicate that internal demand could absorb all Chinese domestic production of feed grains and soybeans by the end of the decade.

Grains and Oilseeds. With respect to key commodities, soybean and meal exports tend to reflect the stability of developed economies. The largest portion of soybeans and meal flows into these markets. However, by the end of the decade, it is very likely that the gap between developed and developing country imports will be closing with about the same level of meal moving into developed and developing markets. Soybeans are also likely to benefit from this growth, but at a lesser rate than meal.

The significant deterioration in demand from the Former Soviet Union is partially offset by the Common Agricultural Policy (CAP) reforms of the European Economic Community (EEC). The FAPRI baseline assumes that a 15 percent land set-aside and a 30 percent reduction in support prices will prevail in the EEC in the latter half of the 1990s. Because of lower-yield land being in the program and planting by farmers who do not participate in the program, the effective set-aside rate is anticipated to be about one-half of this level. Lower support prices are also anticipated to influence yield levels—i.e., lower support prices leading to reduced levels of inputs with a correspondingly lower yield growth path. This combination implies declining excess capacity and declining exportable supplies from the EEC.

Overall, worldwide imports are expected to increase during the 1990s after being relatively stagnant in the 1980s. Figures 2.1-2.3 show projected trends for wheat, feed grains, and soybeans, respectively. In each case, developing country demand provides the impetus for this growth. The developed market economies show little prospective growth as net importers or as net exporters. The countries of the former USSR are anticipated to import decreasing quantities of wheat, about a constant amount of feed grains, and increasing quantities of soybeans and soybean meal.

Rice demand is expected to continue a steady growth path reflecting the second most consumed grain in the world. China and India are the largest producers at about 60 percent of the world production. Their policies have been self-sufficiency with very little trade coming from these regions. The largest world exporter continues to be Thailand, even though it produces only 4 percent of the world's rice. Attempts have been made to expand the area for rice production in Thailand in the last few years and this trend is expected to continue.

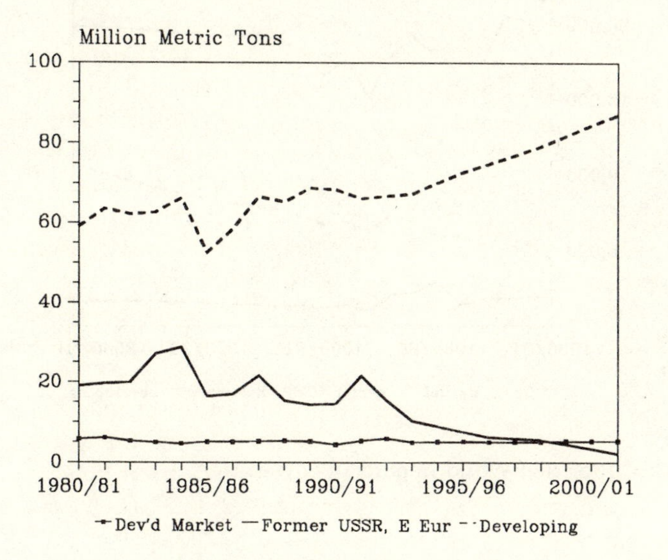

FIGURE 2.1 Projected World Imports of Wheat.

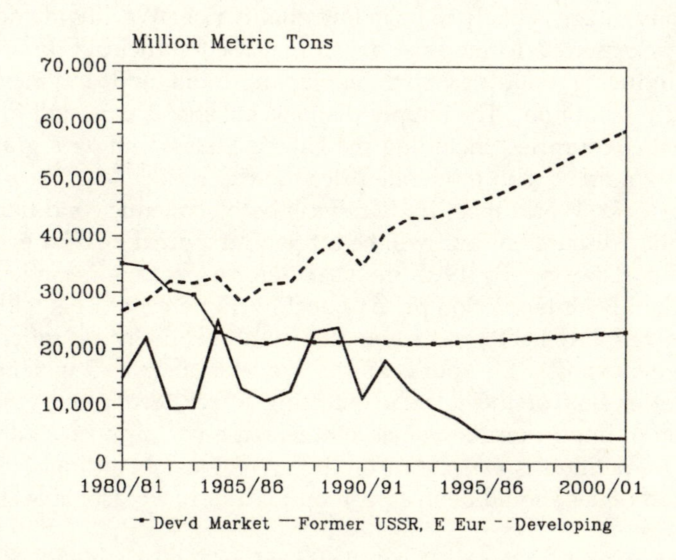

FIGURE 2.2 Projected World Imports of Feed Grains.

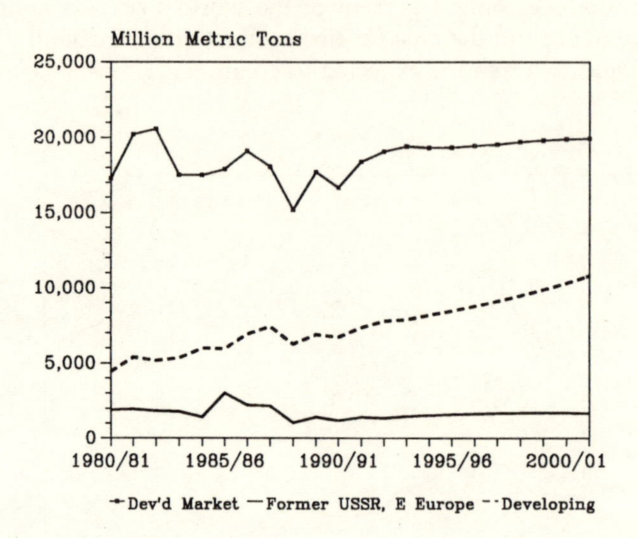

FIGURE 2.3 Projected World Imports of Soybeans.

A recent major contender for world trade in rice has been Vietnam. Its growth has contributed to a recent erosion in the U.S. rice export base. Vietnam is expected to increase its market share over time; however, its market advantage is likely to be in low quality rice. Worldwide demand for rice is expected to increase over time. Self-sufficiency policies by large producing countries such as Japan, India and Indonesia are expected to continue. The supply-demand balance is expected to result in exporting countries, including the United States, finding a gradually expanding market with moderate price increases.

Livestock. World meat trade reflects both economic conditions and trade policy discussions and will be influenced a great deal by events in the EEC and Japan. By 1995, the CAP reform package is projected to reduce the EEC intervention price of beef by 15 percent. This will likely result in reduced beef production that will eventually be reflected in reduced net exports. EEC pork exports, however, may very well increase in the latter half of the decade reflecting lower feed prices. Japan's reduction of import tariffs on beef will increase beef imports. Although strong competition is expected from other regions, including Australia, the United States should be in a position to capture a reasonable share of this market.

The dairy side of the market will also reflect CAP reform by the latter part of the 1990s. Reduced milk quotas will cause the dairy herd to decrease, slowing or stopping the expansion of milk production in the EEC. Some cutback is likely in exportable supplies of butter and cheese.

Longer-term estimates of world food demand growth are somewhat more secure than supply projections because of the role of population growth on the demand side. Even if policy reforms falter and growth in income per capita stagnates as compared to what is now projected, world population until the year 2000 will continue to grow at about 1.6 percent annually (compared to rates of 1.7% in 1980-90 and 2.0% in 1965-80), and this will certainly add to food demand.

The main demand-side uncertainty that is relevant to 1995 U.S. food and agricultural legislation is the potential for demand weakness in the 1993/94 and 1994/95 marketing years. The relatively strong GDP recoveries that Table 2.1 shows for Europe, Australia, and Canada, may not come to pass so soon. This could not only weaken immediate demand but also reduce the likelihood of reaching a trade-expanding GATT agreement. Failure of the Russian economy to turn around in 1994 as forecast would create special problems because in that case the credit-assisted agricultural imports that have so far helped insulate U.S. agriculture from the economic fallout of Soviet collapse could probably not be maintained.

Other short-term uncertainties for 1993-1995 involve inflation and exchange rates. The value of the dollar in terms of other currencies is the most volatile factor. The Federal Reserve Board's index of the trade-weighted value of the dollar has recovered from its low point in the third quarter of 1992, but remains below the levels of recent years (Figure 2.4). Further strengthening could be a significant obstacle to U.S. farm exports.

World Supply of Agricultural Commodities

On the supply side there exists no variable whose rate of change is as predictable as is population on the demand side, but nonetheless there has been a quite steady increase in agricultural production. Aggregate world agricultural output grew at an estimated average annual rate of 2.3 percent over the 1965-1990 period, with an acceleration from a 1.7 percent growth rate in 1965-80 to 2.7 percent in 1980-90 (World Bank 1992 pp 220-221). Some question exists as to whether this rate of increase can be maintained. The World Bank projects total grain production in the developing countries to grow at a 2.3 percent annual rate between 1992 and 2000, compared to 2.8 percent between 1970 and 1985 (Mitchell 1993 p 103). The reasoning is that the developments referred to as the "Green Revolution" are now maturing and no comparable new technologies are on the immediate horizon. This scenario is limited, however, in that it

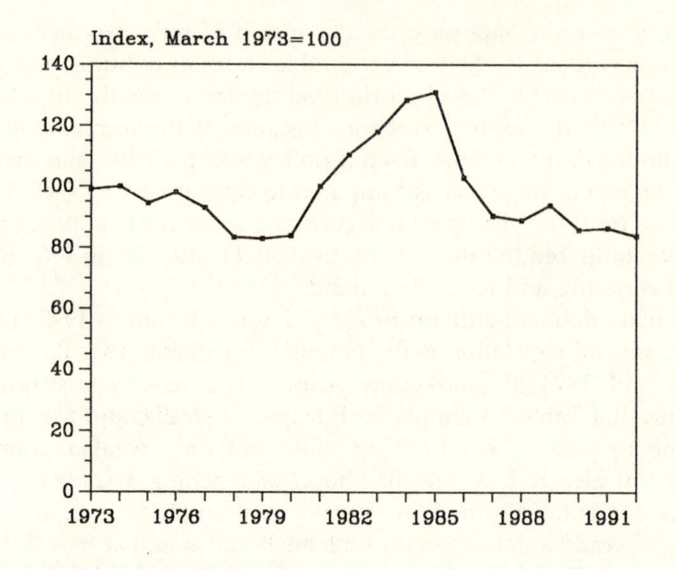

FIGURE 2.4 Real Trade-weighted Value of the Dollar.
Source: Board of Governors of the Federal Reserve System.

focuses on productivity—output per unit of inputs—while production can increase even with constant productivity if the commitment of inputs to agriculture increases.

Reliable data on aggregate input growth in the developing countries are not available. Nonetheless, given the qualitative depictions and some quantitative evidence of declining growth rates in the use of energy inputs, declining rural populations, and indications of land degradation and other environmental problems, it is likely that the resource base for agricultural production in the developing countries is not growing as rapidly as formerly. If so, the apparent acceleration of agricultural output growth during the 1980s indicates that agricultural productivity growth has so far continued at a nondeclining rate and indeed is probably accelerating.

More precise quantitative evidence on productivity growth is available for cereal yields (Table 2.2). As was the case with the aggregate agricultural output indicators, no evidence exists of a slowdown in productivity growth, and there may even be an acceleration. The more detailed country data on cereal yield trends along with fertilizer use data again indicate surprising resilience of output growth while the growth of fertilizer use slows up after 1980 (U.S. Department of Agriculture 1991).

TABLE 2.2 Annual Growth Rates in Cereal Yields Per Hectare, 1970-92.

	Worldwide	United States
1970-80	2.0%	1.8%
1980-92	2.1%	2.6%

Source: FAO (preliminary estimates for 1992).

We do not anticipate any other supply-side constraints likely to come into play in 1994-2000. Environmental factors will undoubtedly restrict the creation of new arable lands and even cause currently cropped land to be converted to noncropland uses. Heavy metal contamination and salinization in the former Soviet Union, problems of deforestation in tropical areas, endangered species protection in industrial countries, and water quality preservation in many locations will restrain cropland development. But world agricultural production growth has not relied upon cropland expansion. FAO data indicate the world's arable and permanent cropland have grown only from 1.41 billion hectares in 1970 to 1.47 billion hectares in 1987 (latest data available), an annual rate of increase of 0.3 percent.

Policy Variables

The U.S. policy baseline assumed in our projections extends the provisions of the Food, Agriculture, Conservation, and Trade Act of 1990. Since 1985, U.S. food and agricultural policy has been on a path that is less government-oriented with a greater possibility of market signals influencing supplies. Budget pressures have accelerated this pace. Current Administration proposals suggest a continuation of this theme. It is likely that U.S. agriculture will in the future be more in tune with domestic and global economic signals than other major competitive regions around the world.

Global economic conditions and foreign policies become much more significant under these circumstances. Fairly optimistic economic growth, with declining government supports in the rest of the world, could place the United States in a quite favorable trade position. U.S. agriculture has maintained the productive capacity and program flexibility for quick response to trade expansion signals, generally at the expense of major competitors with more rigid controls on the input and supply side of the market.

Four international policy issues that could be important are:

- The evolution of the General Agreement on Tariffs and Trade (GATT) negotiations on agriculture,
- European Economic Commodity (EEC) policy reforms,
- Broader governmental reforms in the former Soviet Union and Eastern Europe, and
- The North American Free Trade Agreement (NAFTA).

The GATT appears to be an easy policy forecast in that even if successfully concluded and ratified before 1995, it would not cause significant changes in agricultural trade before 2000. A ratified GATT agreement could impose some disciplines, particularly on the EEC, but the effects of these are unlikely to go beyond the effects of EEC policy reforms currently scheduled to occur anyway. For our analysis, we follow FAPRI's projection that the EEC's internal support price for wheat will fall from $216 per metric ton in 1992 to $118 per metric ton in 2000 (and a similar decline for feed grains), with a set-aside area of about 3 million hectares (7.5 million acres) each year in 1994-2000.

The result, again following FAPRI's baseline analysis, is that EEC net wheat exports fall from 20 million metric tons in 1992/93 to 10 million metric tons in 2000/01. Feed grain net exports by the EEC fall from 6 million metric tons in 1992/93 to net imports of 2 million metric tons in 2000/01. The net swing in the EEC's net grain export position of 18 million metric tons over this period is as much as the United States could have reasonably expected from a GATT agreement, even if it is generated by EEC unilateral reforms rather than by GATT. Another important assumption associated with this baseline is that other government policies are frozen at current levels. For the United States this implies frozen target prices, an extension of the Export Enhancement Program and maintenance of loan rates near current levels.

Price Implications

The demand and supply projections taken together indicate a trend of declining real commodity prices through the year 2000. FAPRI projects that the rest of the world will buy more corn and wheat from the United States as we move toward the year 2000, but only at lower prices (see Table 2.3 and Figure 2.5). Nominal prices rise but at less than the rate of inflation. The projections imply U.S. corn and wheat prices falling 1 to 2 percent annually in real terms. This is about the long-term trend rate of decline for the 20th century. With no significant policy barriers between U.S. internal and border prices, the world price trends are the same as internal U.S. price trends.

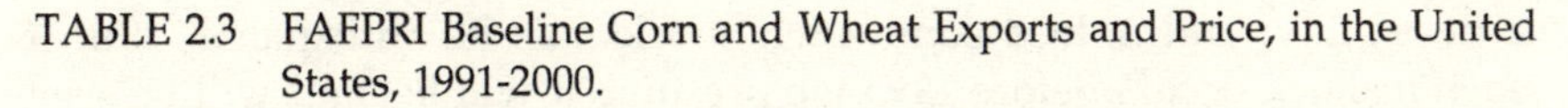

TABLE 2.3 FAFPRI Baseline Corn and Wheat Exports and Price, in the United States, 1991-2000.

Crop Year	*U.S. Exports*			*U.S. Farm Price*		
	Corn	*Wheat* (*million bushels*)	*Soybeans*	*Corn*	*Wheat* (*$/bu.*)	*Soybeans*
1991	1584	1281	685	2.37	3.00	5.60
1992	1657	1341	745	2.07	3.30	5.40
1993	1508	1133	752	2.09	2.90	5.55
1994	1565	1231	764	2.11	2.91	6.00
1995	1562	1283	773	2.26	3.26	6.05
1996	1615	1293	787	2.36	3.35	6.03
1997	1705	1341	802	2.22	3.31	5.69
1998	1787	1389	822	2.17	3.15	5.68
1999	1871	1428	838	2.18	3.21	5.71
2000	1952	1462	853	2.29	3.35	5.86

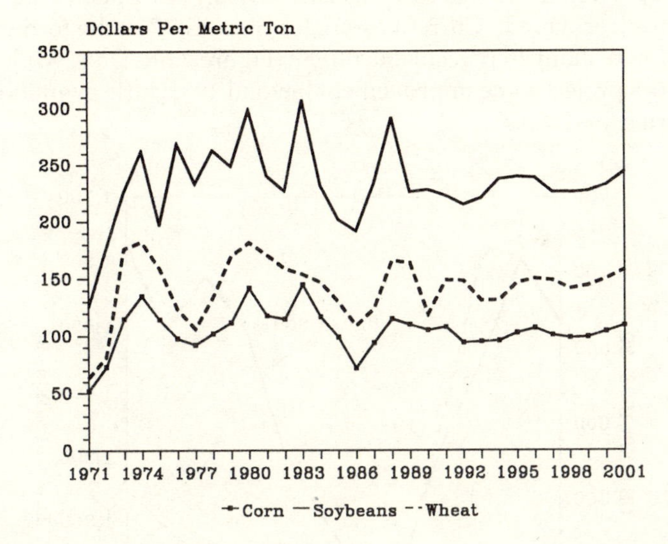

FIGURE 2.5 Gulf Port Prices, 1971-2001.

If U.S. farm costs continue to decline at the current total factor productivity growth rate of 1.5 to 2.0 percent annually, U.S. farm income can be maintained. Nonetheless, the preceding may be seen as a rather gloomy world economic environment in which to make U.S. food and agricultural policy in 1994/95 for the period through 2000. What contingencies could change the situation? Weather or other random shocks are always a possibility, and indeed some such shocks can be anticipated. But if such a shock occurs before 1995 we presume it will be, and will be seen as, a transitory phenomenon that would not radically alter provisions of 1995 legislation.

Longer-lasting deviations from our 1995-2000 world scenario could arise from: (1) Eastern Europe and the former Soviet Union, (2) the EEC, and (3) the developing countries.

The projections for the former Soviet Union and Eastern Europe assume their economies experience a turnaround to real income growth in 1994. While this is most plausible for Hungary and Poland, there is little of substance with which to support this assumption for the former Soviet Union. Best guesses about the future of these economies remain highly uncertain. The Project LINK projections we have used places the turnaround capability quite low. Indeed, the baseline shown in Table 2.1 places the former Soviet Union on a real income path comparable with that of the United States during the Great Depression. U.S. GDP declined about 30 percent from 1929 to 1933. By 1939, ten years later, GDP growth had almost recovered. Current estimates place GDP in the former Soviet Union down about 40 percent since 1989 (Figure 2.6). Thus, while Project LINK does project some improvement beyond 1993, little overall recovery is projected by 2000.

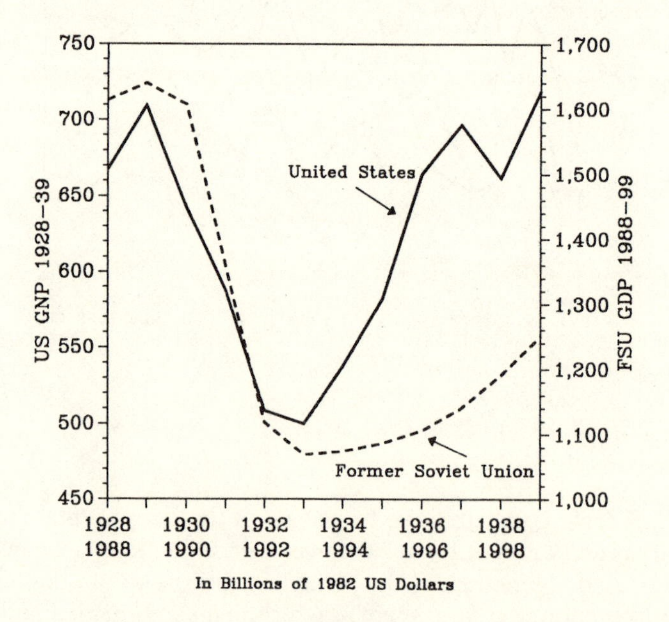

FIGURE 2.6 The U.S. Great Depression Versus the Current Economy of the Former Soviet Union.

Still, even the Great Depression scenario may be too optimistic. Any real growth requires privatization of state-owned enterprises, demonopolization (the introduction of competition in place of the existing monopoly structures), and price liberalization. None of these will be in place by 1994. Indeed, a prior question is whether the basic law, order, and property rights necessary for markets to function efficiently can be maintained. In short, there is considerable downside risk in the projected real income growth of these countries. This places in jeopardy the maintenance of current levels of U.S. export sales to the area. These sales have so far been financed with government-generated credit, and it is unclear how these countries will repay.

The risk in the EEC projections also seems largely on the unfavorable side for U.S. agriculture. While the estimates given earlier are the most likely scenario, we believe it is possible that even with lower intervention prices the EEC will continue to generate exportable surpluses of grains not so much less than current levels.

Import demand for agricultural products by developing countries is the quantitatively dominant source of uncertainty. But here deviations from current projections could equally well go in either direction. These countries could upgrade their diets sufficiently under stronger economic growth to place substantially larger demands on world agricultural production capabilities. But they also could develop their own agricultural sectors more rapidly, especially if they follow the well-worn pattern of Taiwan and Korea in converting former economic discrimination against agriculture to the industrial-country norms of subsidizing farmers and moving in the direction of producing commodity surpluses.

Conclusions

The former Soviet Union and Eastern Europe will likely remain a major source of concern in the international market. But with CAP reform played out according to current policy expectations, moderate growth in developed economies, and stronger growth in developing economies, we see a favorable market for U.S. trade in the latter part of this decade. While export growth is likely to be moderate, we expect to see expanding world trade in agricultural commodities, and an expanding U.S. share of this trade. Figures 2.7 and 2.8 show the FAPRI projections for world wheat and feed grain trade. Both grow quite substantially in 1995-2000 as compared to the stagnation of 1980-93. Figures 2.9 through 2.11 show projected market shares for leading exporters. The United States recovers some of its wheat export share, mainly at the expense of Canada and the EEC. The United States maintains its position in feed grains and soybeans, and even increases its dominance of the world feed grain market.

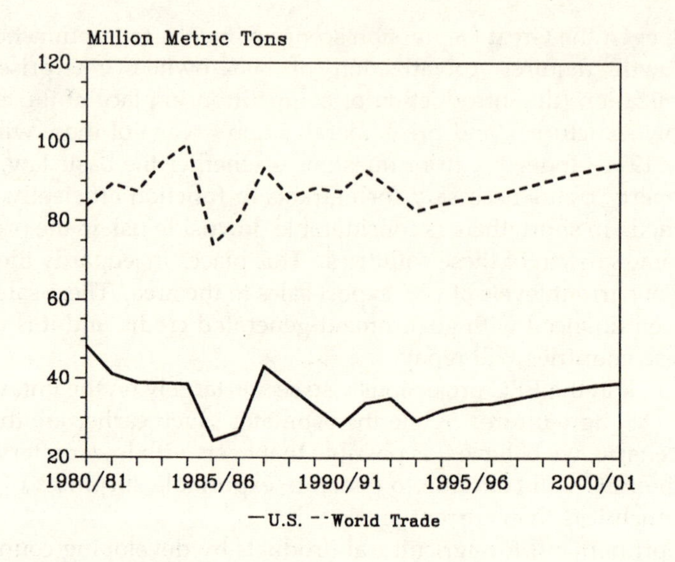

FIGURE 2.7 U.S. and World Wheat Trade, 1980-2001.

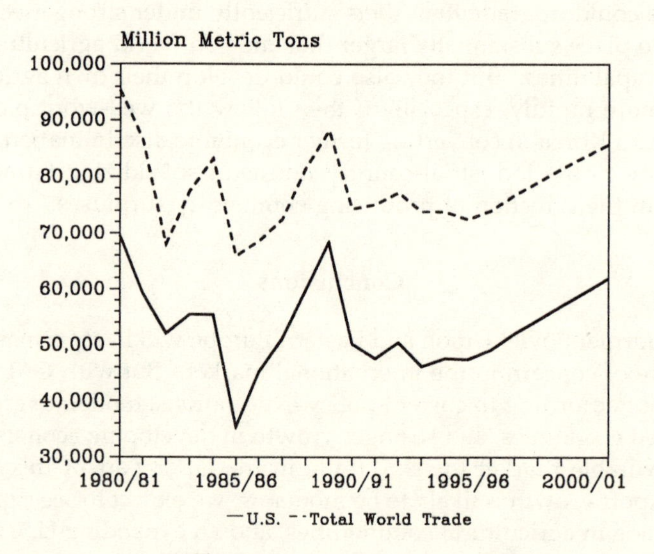

FIGURE 2.8 U.S. and World Feed Grains Trade, 1980-2001.

Nonetheless, demand is not strong enough to prevent real commodity prices from falling at a rate of 1 to 2 percent annually through the 1990s. Many uncertainties exist which could cause departures from the projected trends or even change their long-term direction. Because of the longer-term uncertainties, and the ever-present uncertainties of short-term shocks, a key feature of 1995 legislation, as was the case in the 1985 and 1990 legislation, is how flexibly its provisions will allow farmers and markets to operate.

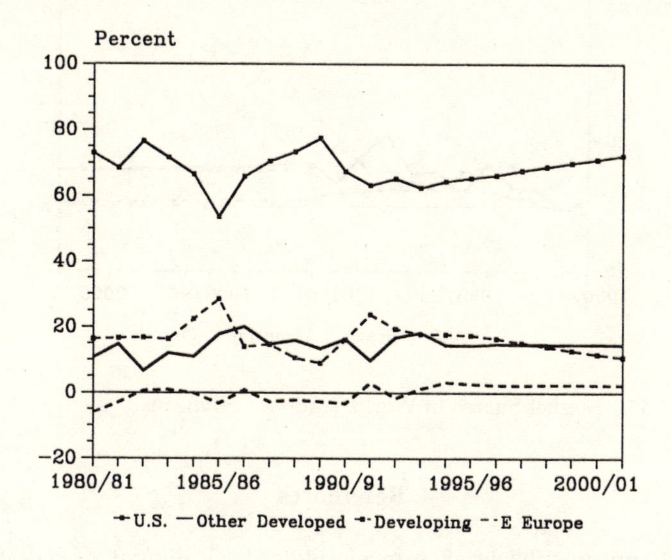

FIGURE 2.9 Market Shares in World Trade for Feed Grains.

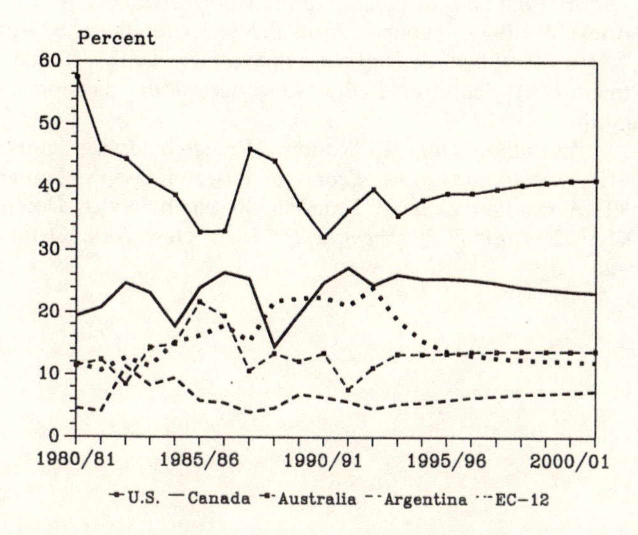

FIGURE 2.10 Market Shares in World Trade for Wheat.

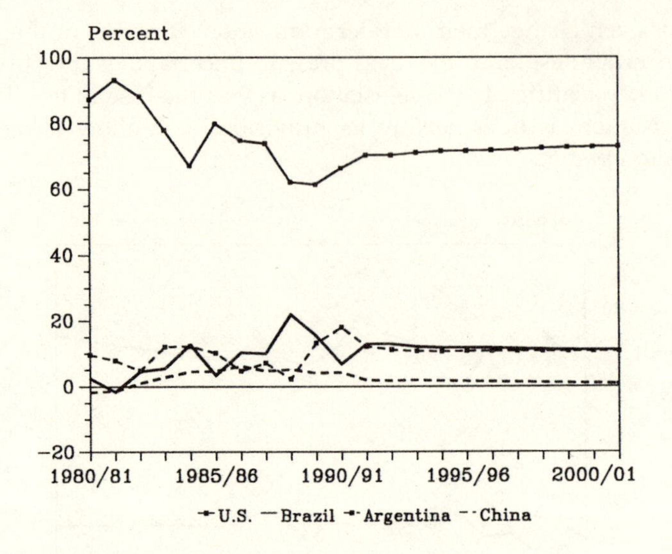

FIGURE 2.11 Market Shares in World Trade for Soybeans.

References

Food and Agricultural Policy Research Institute. 1993. *Agricultural Outlook*. Iowa State University and University of Missouri-Columbia. Staff Report #1-93

Mitchell, Donald. 1993. "World Macroeconomic Performance and Agricultural Trade," *Agricultural Outlook Conference Proceedings*. USDA, March.

O'Brien, Patrick M. 1953. "Longer Term Prospects for the U.S. Agricultural Sector," *Agricultural Outlook Conference Proceedings*. USDA, March.

U.S. Department of Agriculture. 1991. *World Agriculture*. Economic Research Service, June.

______. 1993. *Agricultural Outlook*. Economic Research Service, January.

______. 1991. *Agricultural Outlook*. Economic Research Service, January.

______. 1990. *Agricultural Outlook*. Economic Research Service, December.

World Bank. 1992. *World Development Report, 1992*. New York: Oxford University Press.

3

Evolution of U.S. Food and Agricultural Policy: 1970s to 1990s

Robert G. F. Spitze and Barry L. Flinchbaugh

Public food and agricultural policies have been developed in response to economic, social, and political problems throughout the history of the United States. These policies have, over the centuries, evolved from the early concerns as to how to distribute the vast public domain to the recent choices embodied in the Food, Agriculture, Conservation, and Trade Act of 1990.

There are five general determinants of these policies: (1) past and existing policies; (2) current economic and social conditions; (3) the knowledge base of the individuals and interest groups participating in the policy-making process; (4) the values and beliefs of these participants; and (5) the influence levels of these participants (Halcrow, Spitze, Allen 1993). Our purpose here is to identify and analyze, within an historical setting, the primary forces underlying U.S. public agricultural and food policies in the 1970s to the 1990s.

Nature and Purpose of Public Policy

Fundamental to this inquiry is the recognition that, in the United States, public agricultural and food policy is developed in the policy-making processes of a participatory political system. These processes, in turn, establish the institutional structure within which generally a market economic system functions. These public policies (hereafter referred to as policies) emerge as compromises among numerous private interests of individuals and groups and their governmental representatives. The resulting decisions and actions embody simultaneously economic, social, and political considerations of participants. These policies are not focused on predetermined goals, although the private participants do, indeed, have goals. Furthermore, the policies are not determined by theories of scientists, although reliable relationships from systematic analysis are usually part of the policy-making process.

This nation's agricultural and food policies have exhibited three distinctive characteristics: (1) evolving content, changing but gradual; (2) increasing comprehensiveness, arising from the complexity of a developing, commercialized, interdependent society; and (3) an expanding array of participant groups associated with an urbanized society and its communication technology. Thus, we are likely to have these policies in the future as long as diversity and problems exist in the private arena, and further, they are likely to be indeterminate as long as they arise from a participatory system.

Foundations of Policies Leading up to 1970

The U.S. policies of the past two centuries are of two general types: (1) developmental or input market, and (2) price and income or output market policies.

Developmental Policies

Developmental policies dominated the period from the founding of the nation in 1776 through the early decades of the 1900s (Benedict 1953). These policies represented the public response to problems arising out of the settlement of the frontier as the nation established its basic institutions that were to become the foundation for its highly productive, efficient, and adaptable agricultural sector. This historic process, unique in the history of world civilizations, occurred for the most part when an invading culture conquered and largely obliterated an indigenous culture.

This era of policies displayed a unique partnership between private property and public institutions that continues to the present, shaped by private problems and public concerns, and private decisions and public actions. These developmental policies emerged around three groups of problems and governmental responses: (1) land use and conservation, (2) education and research, and (3) finance and credit.

The earliest policies represented the struggle over whether the vast public domain was to be sold to those with money, thus encouraging a landed aristocracy, or to be granted to land users, thus fostering a family-farming structure with broad participation in the economic system. That controversy culminated in the Homestead Act of 1862. Deliberate farmland conservation policies, implemented, for example, by the Soil Conservation Service of 1935, did not appear until decades later when there was essentially no more land to be distributed free to farmers, and a highly commercial, technologically-driven farming system was in place and serious water and wind erosion of farmland became generally known.

An unprecedented concern about free public education for all citizens and about research knowledge for a family structure of farming led to a series of historic policies establishing the common schools (1787), the Land Grant College and University system (1862), and vocational education in the public schools (1917 and 1963). Finally, among these developmental policies, growing financial needs of an emerging commercial agriculture resulted in varied credit policies for farmers through the cooperative Farm Credit Administration (1916, 1923, 1933), Farmers Home Administration (initially in 1935), and Rural Electrification Administration (1936).

Price and Income Policies

In retrospect, it almost seems as though the U.S. policy-making public deliberately set about establishing its productive agricultural system through developmental policies before shifting its attention to problems of product markets, farmer incomes, and food distribution. The sequence is probably more random than it appears, but the U.S. experience is indeed at variance with that of most other countries that often launch major trade protection and farmer subsidization schemes long before an efficient production system is in place. Crises precipitated by market and trade disturbances from the earliest years spawned farmer protests, such as the historic Whiskey Rebellion, political populism, and repeated intervention proposals. However, concerted governmental action did not emerge until the post-World War I agricultural depression of the 1920s. Then, with the Agricultural Marketing Act of 1929 and its Federal Farm Board, a different type of policies was launched, often called price and income policies. The 1929 Act was followed by some twenty distinct Acts, the latest being the 1990 Food, Agriculture, Conservation, and Trade Act of 1990. (See Table 3.1 for the evolution.)

These price and income policies have focused on the major crops of grains and cotton, and some others, such as tobacco, sugar, and peanuts. Dairy products and many fruits have been affected by somewhat different policies known as marketing orders and agreements, often under state laws. Even though price and income policies have not directly focused on livestock and poultry, these industries have been indirectly impacted the altered grain prices.

TABLE 3.1 Continuity and Change in U.S. Agricultural and Food Policies 1970-1990.

1970 Act	*1973 Act*	*1977 Act*	*1981 Act*	*1985 Act*	*1990 Act*
Dairy	→ Dairy	→ Dairy	→ Dairy	→ Dairy	→ Dairy
Wheat	→ Wheat	→ Wheat	→ Wheat	→ Wheat	→ Wheat
Feed grains	→ Feed grains	→ Feed grains	→ Feed grains	→ Feed Grains	→ Feed Grains
Cotton	→ Cotton	→ Cotton	→ Cotton	→ Cotton	→ Cotton
Wool/mohair	→ Wool/mohair	→ Wool/mohair	→ Wool/mohair	→ Wool/mohair	→ Wool/mohair
Payment limits	→ Payment limits	→ Payment limits			
		Rice	→ Rice	→ Rice	→ Rice
		Peanuts	→ Peanuts	→ Peanuts	→ Peanuts
		Soybeans and Sugar	→ Soybeans	→ Soybeans	→ Oilseeds
			Sugar	→ Sugar	→ Sugar
		Grains reserve	→ Grains reserve		
	Food programs	→ Food programs	→ Food programs	→ Food programs	→ Food programs
	Educ./Res./Ext.	→ Educ./Res./Ext.	→ Educ./Res./Ext.	→ Educ./Res./Ext.	→ Educ./Res./Ext.
	Grain Inspection				→ Grain Quality
				↱ Credit	→ Credit
			Rur.Dev./credit		→ Rur. Dev.
Rural development	→ Rural Dev./cons.	→ Rur.Dev./cons.	→ Conservation	→ Conservation	→ Conservation
P.L. 480	→ P.L. 480	→ P.L. 480	→ P.L. 480	→ Trade	→ Trade
				Gen. commod.	→ Gen. commod.
				Marketing	→ Fruit & veg. mkt.
					Honey
					Forestry
					Ag. promotion
					Organic certif.
					Crop insurance
					Global warming
Miscellaneous	→ Miscellaneous	→ Miscellaneous	→ Miscellaneous	→ Miscellaneous	→ Miscellaneous

NOTE: Primary titles, often abbreviated, of major price and income policies, identified by year of the major acts and connected by lines specifying similarity in names, although not necessarily in content.

During this foundation period of price and income policies, several key provisions, some continuing and some significantly changed, should be noted. First, price supporting of major commodities, largely through nonrecourse loans, has continued to the present. Second, direct income-support and cost-sharing payments have continued, but with varied purposes and means. Third, food distribution and purchase assistance programs have remained, but have substantially broadened in scope.

Fourth, "production balancing" or "supply management" policies have changed significantly from a variety of attempts at compulsory controls to a system of voluntarily-induced controls. Fifth, the "measures of economic well-being" for price or income support have shifted from the concept of parity to cost-of-production, and then to compromised ad hoc levels of support.

Sixth, the level of direct price support provided via nonrecourse loans has changed from relatively fixed levels to more variable levels. Finally, the focus of price and income policies has shifted from only commodities, to now include a broadening array of other domestic issues, such as food assistance, resource conservation, and rural development. They also encompass international issues, including food aid, concessional sales, export promotion, and technical agricultural assistance.

Forces Shaping Policies Leading up to 1970

A study of the foundation period of policy prior to 1970, reveals several distinct, over-riding forces: (1) the need to secure control of the land that was acquired through purchase, conquest, armistice, and merger, beyond the original colonies; (2) a desire to establish an efficient, productive agricultural sector to feed a rapidly growing population, to release labor resources to an industrializing and urbanized economy, and to support an expansion of exports to obtain needed foreign exchange and profitable markets; (3) the recognition of the interdependence among the aspirations for a democratic society, a proprietor-based farming system, and an educated rural population; (4) the importance of the values of a unique and fundamental agricultural sector to the viability of the social, economic, and political structure of the nation; (5) a commitment to provide greater stability and economic returns to the agricultural sector than otherwise would flow from the market, with its exposure to unpredictable natural conditions, its variability, inelastic demand, and competitive structure; and (6) a disproportionate representation of the farm and nonfarm rural populations in the structure of state and federal governments, deliberately built into the U.S. Senate and permitted in all other legislative bodies by failure to reapportion prior to the mandated reapportionment decision of the Supreme Court in 1962. This build-up of disproportionate representation continues because of the lag in reapportionment between decennial censuses.

Policy of the Early 1970s

In order to understand the decades central to this chapter (1970 to 1990), the format for the remaining sections will be to first identify the critical policy forces and problems of distinct periods, and then to follow with the policies that were adopted.

Policy Problems and Forces

Forces affecting policy development in the early 1970s included: (1) the rapid introduction of new resource-saving and output-increasing technology of the post-war era; (2) the associated increasing commercialization and specialization of farming; and (3) the persistence of periodic imbalances resulting from more rapid growth of farm product supply than of product demand (Hallberg 1992). It was also a relatively prosperous period for the U.S. economy, not fully evident in its rural component, and a period with an emerging concern about the magnitude of Treasury payments flowing to a minority of the farmers participating in the various price and income programs (Cochrane and Ryan 1976). Furthermore, a new tension began to emerge between a Democratically-controlled Congress desiring more public involvement and a Republican President supporting more market-oriented policies that except for the 4 years of the Carter Administration, persisted for 24 years.

Agricultural Act of 1970

The first major set of agricultural and food policies of the 1970s was the Agriculture Act of 1970 that set the policy course for the next three years. In retaining the narrowly-focused commodity policies of the previous decades, this Act continued voluntary land retirement for "balancing production" with slowly growing demand, payments to participating producers, and a slight lowering of the price support level. The most important changes were: use of a set-aside of total farm acreage (whole farm base) with farmer discretion in using the remaining cropland instead of a required shift in acreage of a specific program crop; initial disengagement of price supports from the parity concept; and the introduction of a limitation on program payments of $55,000 per commodity per farm. At the outset, the latter affected a limited number of large farmers.

Unique Policy Period of 1973-74

Policy Problems and Forces

By the time the next major set of policies was to be debated and launched in 1973, a major disturbance later seen as an aberration, instead of a new trend, occurred. Simultaneously, major crop-producing regions of the world experienced short-falls due to adverse weather. Among other rapidly occurring economic and social events, the central purchasing agency of the former Soviet Union bought most of the grain stocks available on the open U.S. market, unbeknownst to the trade generally, and even at subsidized levels (Robinson 1989). The world commodity markets reacted with price jumps, and domestic food prices began to escalate. Farmers were encouraged by the Administration to plant "fencerow to fencerow," consumer food boycotts occurred, and the first export embargo was announced on certain grains and oilseeds by the Republican Administration to protect "domestic food supplies."

Agriculture and Consumer Protection Act of 1973

With the bullish economic conditions for agriculture during this brief period of prosperity, there was little call for substantive changes during the next policy cycle, which produced the Agriculture and Consumer Protection Act of 1973. Thus, there were only minor alterations in policy to continue for another four years. These included a launching of target prices and deficiency payments as minimal income protection for voluntary program participants; a short-lived factoring of some production costs into target price levels; and a reduction of the maximum direct program payment limit to $20,000.

Inflationary and Expansion Policy Period of the Late 1970s

Policy Problems and Forces

With the exuberance of the high prices and farmer incomes of the mid-1970s beginning to fade, the agricultural sector during the second half of the decade settled into another unique period of consolidation of gains and capital expansion (Tweeten 1989). Economic forces at play and problems beginning to emerge included: (1) the value of the dollar declined; (2) agricultural exports soared; (3) real interest rates were low; (4) oil prices skyrocketed under the influence of a new oil exporters' cartel and escalation of mid-East conflicts, and farm input prices began to rise; (5) inflation rose to record post-war levels; (6) agricultural credit expanded and land values jumped; and (7) agricultural production began out-pacing demand with the reappearance of a serious cost-price squeeze.

Food and Agriculture Act of 1977

Although the rapidly changing economic environment did not signal a major policy change when the next policy cycle began, the four-year Food and Agriculture Act of 1977 did chart a path toward increasing comprehensiveness of policy issues. Also, the very uncertainties and shocks of the decade led to some new policy thrusts. The unique provisions of this Act may be summarized (Spitze 1978) as follows: (1) acreage bases for crop production control purposes were shifted from a fixed historical base to a moving-year average; (2) payment limits were gradually raised to $50,000; (3) the first national grain reserve of specified minimal grain stocks was established (Meyers, ed. 1988); (4) the first mandated price support loan for soybeans appeared; (5) export embargoes of agricultural products with a domestic market rationale were seriously discouraged by trigger provisions in existing programs; (6) the previously separate food stamp programs were folded into the on-going price and income policy stream and all stamp purchase requirements were eliminated, permitting eligible recipients to obtain their allotment without monetary cost; and (7) national agricultural teaching, extension, and research provisions were loosely tied into this policy for the first time, with USDA as lead agency.

Federal Crop Insurance Act of 1980

A long standing effort to launch federally sponsored crop insurance was expanded nationally to cover most crops (Knutson et al. 1983).

Contraction and Farm Crisis in the Policy Period of the Early 1980s

Policy Problems and Forces

Few periods in U.S. agricultural history have been as tumultuous as the early 1980s, when the "farm crisis" was generally viewed as the worst since the Great Depression (Flinchbaugh 1986). Many of the problems and forces in play had their roots in the previous decade's monetary policies, spending excesses, and entrepreneurial mismanagement. This led to: (1) restricted money supply to stem inflation and rapidly rising interest rates; (2) increase in foreign exchange value of the dollar and plummeting agricultural exports; (3) increases in agricultural production abroad, often subsidized [as in the European Economic Community (EEC)] by protectionist policies and loss of U.S. markets, thereby precipitating various U.S. export subsidy programs and renewed international trade negotiations; (4) spread of the U.S. general recession throughout the trading world; (5) price support levels rising while the general price level began slowing; (6) decline in farm product prices, farmer incomes, and farmland values; (7) farm bankruptcies increasing; and (8) a second agricultural embargo on exports to the Soviet Union, this time by the Democratic Administration in response to the U.S.S.R.'s invasion of Afghanistan.

Agriculture and Food Act of 1981

The four-year Agriculture and Food Act of 1981 was shaped during an inopportune transition period, i.e., one following high inflation and record exports, but before both inflation and exports had faded (Langley and Baumes 1989). Thus, the following policy changes in fact lagged behind rapid economic shifts, some soon becoming candidates for revision: (1) target prices and price support levels set to escalate 4 to 6 percent per year; (2) domestic market prices held above world levels; (3) additional compulsory supply control provisions replaced with voluntary programs; and (4) export embargoes on agricultural products implemented for foreign affairs reasons but strongly discouraged by additional trigger mechanisms in existing programs.

Other Relevant Acts of the Period

Due to rapidly changing economic conditions, unexpected emerging policy problems, and a worsening of the national fiscal situation, the Omnibus Budget Reconciliation Act of 1982, Balanced Budget and Emergency Deficit Control Act of 1985, and others with an agricultural focus were enacted. All had agricultural impacts as follows: (1) price support and target price levels were frozen; (2) loan and acreage reductions were mandated for 1983 while an unprecedented payment-in-kind (PIK) program was administratively launched to reduce record governmental stocks; (3) paid land diversion was authorized; and (4) farm program outlays were reduced (Flinchbaugh 1985).

Policies of the Recovery Period of the Late 1980s

Policy Problems and Forces

The policy development period of the last half of the 1980s experienced no unique economic shocks such as those that characterized the previous decade and a half. Instead, it revealed a society and an agricultural sector grappling with the consequences of past turbulent economic conditions and the policies spawned by them. Specifically, these included: (1) the still lingering farm crisis with periodic burdensome supplies; (2) the escalating commodity support levels of the previous policy, now indefensible in the presence of the lingering recession; (3) the decline of market export shares; (4) budget costs; and (5) growing national concerns about conservation of natural resources. In addition, the policy development period of 1984-85 offered another of those eras when policymakers were confronted with strongly conflicting directions represented by the President's proposal to phase out government intervention, in contrast to advocacy by some members of Congress for a return to compulsory production controls. These forces at play can be summarized as (1) the recent farm economic crisis, the substantially reduced inflation rate, the sharp loss of exports, and the potential for agricultural supply out-pacing demand; (2) the need to evaluate the market-oriented and deregulation philosophies being vigorously advocated in national policy by the incumbent administration; (3) the threat to program funding by the worsening national budgetary crisis and restraints on entitlement programs; and (4) the necessity to respond to the widely expressed preferences for stronger conservation of natural resources, often referred to as the "green movement."

Food Security Act of 1985

When the President signed the five-year Food Security Act of 1985, it closed one of the most contentious, and yet important, policy debates in U.S. price and income policy history. This revealed a unique interplay of the forces of economic conditions, ideological commitments, and a global environment. Yet, when completed, the new Act was neither revolutionary nor did it uphold the status quo. However, it did signal a shift in emphasis away from reliance on price supports to greater dependence on deficiency payments. It bore further evidence of the evolutionary nature of policy, continuing essentially all major provisions of the previous policy but with notable changes and additions as follows: (1) duration of five instead four years; (2) reversal of price-support and target price level escalation embodied in the previous Act by mandating a gradual lowering of both; (3) one-time whole herd buy-out program to accompany the on-going dairy price support program; (4) major export subsidization and promotion initiatives involving donations, short and longer term credit sales, market development, and export payment-in-kind (PIK), with the policies being implemented by such programs as P.L. 480, GSM-2, GSM-3, EEP (Export Enhancement Program), and TEA (Targeted Export Assistance), now MPP (Market Promotion Program); (5) cropland conservation compliance on erodible farmland acreage if operators are to qualify for future commodity program benefits; (6) swampbuster and sodbuster restrictions on farmers to qualify for future program benefits; and (7) the new Conservation Reserve Program (CRP) offering participating farmers cost-sharing for establishing acceptable conservation cover along with annual leasing payments for ten years on up to 45 million acres of highly erodible land.

Other Acts of Late 1980s

Other important policy developments of this period include: successive farm disaster assistance acts for producers experiencing adverse weather difficulties; the Agricultural Credit Act of 1987, restructuring the Farm Credit Administration and changing the legal and credit terms for borrowers, launching a secondary farm mortgage market, and providing a "bail-out" loan to bolster the financial base of the Farm Credit System; various acts to strengthen domestic food assistance programs, particularly for women, infants, and children; and the United States—Canada Free Trade Agreement Act of 1988 to move toward substantial lowering of trade barriers between the two countries.

Recovery and Uncertainty Underlying the Policy Period of the 1990s

Policy Problems and Forces

Economic and political forces alike dominated the 1990 policy development cycle. The two economic factors were the farm sector recovery from the devastating financial crises of the past decade and the budget constraints arising from the national consensus to attempt to reduce the federal budget deficit. Economic conditions of the 1987-88 period preceding Food, Agriculture, Conservation, and Trade Act of 1990 were generally the opposite of those preceding the 1985 Act: (1) slower growth in farm production, (2) increased export demand, (3) strengthened farm product prices, (4) increased net farm income per farm family, (5) declining total farm debt and rising assets, and (6) lower government commodity stocks (Spitze, ed. 1990).

Four political forces were at work. Policy decisions had to be made, somewhat inopportunely, due to the approaching termination of the 1985 Act. Off-year Congressional elections were to be held in the year of the policy decision with 11 of 19 of the Senate agriculture committee members up for reelection. The four year protracted GATT negotiations were at a critical stage. Publicly-inspired environmental and food safety concerns were rising (see Environmental Coalition 1990).

In contrast to recent policy-making cycles, the usual proposals for a free market approach or for mandatory production controls (or even farmer welfare-oriented programs) were not vigorously pursued. In fact, unprecedented consensus prevailed among both Executive and Congressional leaders, as well as within both Houses of Congress and political parties: the 1985 Act was a satisfactory guide for policy decision-making. The mood was to improve, not to replace. Yet, the cloud of budget constraints hanging over the two years of debate finally descended with a mandated cut of 25% in outlays below proposed budgets across all commodity program entitlements and a rescinding of proposed increases in food programs. These adjustments resulted from the Omnibus Budget Reconciliation Act of 1990, reflecting the agreements in the late summer of 1990 among Congressional and Administration leaders.

Food, Agriculture, Conservation, and Trade Act of 1990

Even though each of the varied economic and political forces had impacts on the final version of the 1990 Act, it was clearly a continued evolution in content and directions of the 1985 Act and of the previous six decades of price and income policy (Spitze 1992). The 1990 Act is the most comprehensive case or example of price, income, food, conservation, and trade policy in our history. No substantial existing programs were eliminated, but there appeared among its 25 titles seven new ones for forestry, fruit and vegetable marketing, grain quality, organic certification, rural development, global climate change, and crop insurance and disaster assistance.

No abrupt change in U.S. agricultural and food policy for the 1990s was heralded. Commodity programs were continued, generally at the same per unit targeted return levels but with more flexibility in planting provisions and a reduction in Treasury costs through the elimination of 15 percent of the acreage base eligible for program payments. Food programs, both domestic and foreign, were continued at about the same levels. The extensive CRP was slightly strengthened. Aggressive export subsidization and promotion programs were set to continue and, in the absence of a GATT agreement, even at higher mandated trigger levels. The most substantive changes from previous policy were in three areas: (1) flexibility to permit, through triple-base and optional set-aside provisions, more producer discretion in using program acreage for profitable nonprogram crops, and also to restrict Treasury costs which, incidentally, reduced income support levels for farmers; (2) stronger environmental protection through the CRP and water quality improvement programs; and (3) authorization of new initiatives, yet to be funded and implemented, in grain quality, reforestation, pesticide use, rural development, organic food certification, sustainable agriculture, and global warming.

Summary and Conclusions

The public agricultural and food policies of the United States in the 1970s to the 1990s were generally a continuation of two centuries of governmental involvement in that sector, and specifically, a further evolution of six decades of price and income policy. (See Table 3.1.)

Major acts during these recent years appeared in 1970, 1973, 1977, 1981, 1985 and 1990, with other policies, such as the Agricultural Credit Act of 1987, focusing on specific problems. Each of these installments became more comprehensive and involved a wider array of interest groups that were concerned about a myriad of issues, such as food security, food aid, conservation, food safety, and the environment. Increasingly, budget constraints and issues of international trade hung over these policy developments (Flinchbaugh 1992).

The dominant forces that pervaded the policy process during these two decades can be summarized as (1) impacts of monetary policies upon trade, product markets, input costs, and asset values; (2) globalization of product markets, increased agricultural production and subsidization abroad (especially in the EC), and the GATT negotiations; (3) budgetary pressures on all entitlement programs associated with the escalating national budget deficit; (4) continued public commitment for a public "safety-net" for farm families, their institutions, and rural communities, but simultaneously giving expression to persistent values among many policy participants for a "market-oriented," flexible, and adaptable policy; (5) a stronger public resolve to improve food accessibility for disadvantaged citizens; and (6) broadened societal concerns about conservation of natural resources, environmental quality, and food safety.

Thus, these policy forces and the resulting policy evolution are the heritage, the foundation, and useful indicators for public agricultural and food policy development of the United States throughout the 1990s.

References

Benedict, M. R. 1953. *Farm Policies of the U.S., 1790-1950.* New York: The Twentieth Century Fund.

Cochrane, W. W. and M. E. Ryan. 1976. *American Farm Policy, 1948-1973.* Minneapolis: University of Minnesota Press.

Environmental Coalition. 1990. *Farm Bill Agenda for the Environment and Consumers.* Washington, D.C.: Island Press.

Flinchbaugh, B. L. 1985. "Impact of Budget Proposals on Agricultural and Farm Commodities Issues." 99th Congress, 1st Session, House of Representatives, Committee on the Budget, February 15.

______. 1986. "Crisis in the Heartland—The Real Story." Proceedings, Agricultural Law Symposium. Cambridge: Harvard University, April.

______. 1992. "Disorder in the New World Order." Address to National Institute of Cooperative Education, Denver, July 23, 1992.

Halcrow, H. G., R. G. F. Spitze, and J. E. Allen. 1993. *Agricultural and Food Policy: Economics and Politics.* New York: McGraw-Hill Book Co. Chapter 3.

Hallberg, M. 1992. *Policy for American Agriculture.* Ames, IA: Iowa State Univ. Press.

Knutson, R. D., J. B. Penn, and W. T. Boehm. 1983. *Agricultural and Food Policy.* Englewood Cliffs, N.J.: Prentice-Hall, Inc.

Langley, S. and H. Baumes. 1989. "Evolution of U.S. Agricultural Policies in the 1980s." Chapter 2, *Agricultural-Food Policy Review: U.S. Agricultural Policies in a Changing World.* USDA ERS Agri. Econ. Rpt. 620.

Meyers, W. H., ed. 1988. *The Farmer-Owned Reserve After Eight Years: A Summary of Research Findings and Implications.* North Central Regional Research Publication No. 315. Ames, IA: Iowa State University Agriculture and Home Economics Experiment Station.

Robinson, K. L. 1989. *Farm and Food Policies and Their Consequences.* Englewood Cliffs, N.J.: Prentice-Hall, Inc.

Spitze, R. G. F. 1978. "The Food and Agriculture Act of 1977: Issues and Decisions." *American Journal Agricultural Economics* 60:225-235.

_____, ed. 1990. *Agricultural and Food Policy: Issues and Alternatives for the 1990s.* Urbana: Department of Agricultural Economics, University of Illinois.

_____. 1992. "A Continuing Evolution in U.S. Agricultural and Food Policy—the 1990 Act." *Agricultural Economics* 7:125-139.

Tweeten, L. 1989. *Farm Policy Analysis.* Boulder: Westview Press.

4

Structure of Food and Agricultural Sector

Milton C. Hallberg and Dennis R. Henderson

The economic performance of an industry has several dimensions, including productive efficiency, progressiveness of producers in adapting to technical and economic change, full employment of resources (especially human resources), and equitable distribution of income. While causal relationships accounting for satisfactory or unsatisfactory economic performance of an industry are sometimes difficult to establish, the structural character of an industry is of vital importance in conditioning economic performance. Knowledge of an industry's structure is thus crucial to assessments of whether or not that industry is evolving in directions consistent with society's broader aims, and to judge if special policy action is needed to encourage it to perform more in line with society's expectations. The objective of this chapter is to identify the major structural characteristics that should be examined in assessing the economic performance of the U.S. food and agricultural sector.

The Issues

This nation has long pursued the ideal of a farm sector populated with privately operated "family farms," as opposed to corporate farms or large public units. Have today's farms become so large that they can no longer be considered "family farms"? Have they evolved into corporate-like business organizations which threaten to weaken productive efficiency as well as contribute to an unacceptable distribution of income? Farm people and farm policymakers alike have long been concerned about an imbalance in the market-power relationships in agriculture. Have the firms with which farmers must deal in the market place become so large and/or so powerful as to render obsolete the competitive structure most conducive to good economic performance? Is the resulting structural character of the sector such as to diminish its ability to pursue the most efficient levels of production, to provide consumers with the lowest-priced product, to ensure rapid adjustments of resources in response to external forces, to permit easy entry into and exit from the industry, to ensure resource owners a competitive return, and to provide farm families and hired workers an income comparable to that of individuals or families in other sectors?

The food and agricultural sector consists of many internal as well as many external linkages that condition both firm behavior and structure. The number, size distribution, and geographic dispersion of farms is certainly part of the structural picture, but so also is the number, size distribution, and geographic dispersion of buyers of farm produce and sellers of farm inputs. The distribution of farmers by age, ownership, and type of business organization is important as are the debt position and capital requirements of farmers. The existence of scale economies that force firms to larger and larger sizes, and the extent of vertical control in agriculture also help determine its structure. Agriculture is not "an island empire" (as Mayer and Mayer 1974 argue, for example), but rather part of a complex web of interconnected parts with domestic as well as international links. The sector must be viewed in this context rather than as an isolated industry immune to conditions in other domestic industries and independent of events and structural conditions in international markets.

Structure of Agricultural Production

Size Distribution of Farms

For statistical purposes, a "farm" is defined as any place from which $1,000 or more of agricultural products were or normally would have been produced and sold during one year. The Bureau of Census tabulates farms by size based on reported sales, and classifies them by type of farming activity using the standard industrial classification (SIC) system. A farm is classified in the commodity or enterprise group (called farm type in this chapter) which accounts for 50 percent or more of the total value of sales of the farm's agricultural products. When ambiguity exists, a farm is classified as a general crop farm or as a general livestock farm. The specific farm types into which farms were classified in the 1987 census are listed in Table 4.1 along with the distribution of farms by value of sales in each farm type category.

TABLE 4.1 Percentage Distribution of Farms by Value of Sales and by Farm Type, 1987.

	Value of Sales per Farm				
Farm Type	*Under $20,000*	*$20,000- $49,999*	*$50,000 $99,999*	*$100,000- $499,999*	*Over $500,000*
	-----------------percent-----------------				
Cash grain	44.7	22.6	16.6	15.6	0.5
Cotton	22.1	17.6	19.1	36.2	5.0
Tobacco	80.8	10.9	4.6	3.6	0.1
Other field crops	77.5	8.7	4.8	7.5	1.5
Vegetables & melons	57.9	15.4	8.7	12.4	5.6
Fruits & tree nuts	63.7	13.9	8.8	11.0	2.6
Horticulture specialties	47.8	16.8	10.3	17.7	7.4
General crop farms	71.0	11.3	6.9	9.3	1.5
Beef cattle (not feedlots)	81.4	10.2	4.2	3.7	0.5
Other livestock	53.9	14.7	11.5	17.3	2.6
Dairy	7.2	19.9	32.1	38.1	2.7
Poultry & egg	18.3	5.0	10.9	51.9	13.9
Animal specialties	91.1	4.4	2.0	1.9	0.6
General livestock	62.0	10.4	11.4	15.2	1.0
All farms	61.3	14.2	10.4	12.6	1.5

Source: (U.S. Department of Commerce 1987).

The data in Table 4.1 provide a vivid reminder that the majority of farms (and for some farm types, the vast majority of farms) in the United States are extremely small units. Those with sales of less than $20,000 annually, for example, are so small that they cannot generate, from farm sales alone, sufficient revenue to support a family of even two. The percentage of poultry and egg, dairy, and cotton farms in the smallest size group is much less than is the case for other farm types. Had rice farms been separated out by the census reports, they too would have shown a very small percentage of farms in the smallest size category, although for cash grain farms as a whole, nearly 45 percent of the farms had annual cash sales of less than $20,000.

Farms selling $50,000 or more of farm products, on the other hand, clearly accounted for the bulk of total sales of the majority of the farm types listed. For agriculture as a whole in 1987, 88 percent of the sales of farm output was accounted for by the 25 percent of the farms having sales of $50,000 or more. There is a great deal of variation in these percentages among the different farm types. For example, slightly more than 8 percent of the tobacco farms had annual sales of $50,000 or more and accounted for 52 percent of total sales by tobacco farms. Similarly, just over 8 percent of the beef cattle farms had annual sales of $50,000 or more; these farms accounted for 65 percent of total sales by beef cattle farms. In contrast, about 77 percent of the poultry and egg farms and 73 percent of the dairy farms had annual sales of $50,000 or more, and these farms accounted for 99 and 91 percent, respectively, of the total annual sales on these two types of farms.

We emphasize that the data in Table 4.1 do not include government payments to farmers. To the extent that such payments are made to farmers who voluntarily restrict production, government payments should be added to the value of sales for a more accurate reflection of farm size. This is most relevant in the case of cash grain and cotton farms, and thus the size distribution shown for these two farm types in Table 4.1 is somewhat inaccurate. Unfortunately, the data on government payments by size of farm and farm type were not available for this analysis.

With few exceptions, the majority of farm output is produced by a minority of the farms. Farms of all types have been getting fewer in number and larger in size. Nevertheless, the number of farms with annual sales of $50,000 or more (or even with $100,000 or more) in the major farm type categories is so large that no single farm is able to significantly influence the terms of trade in the market place. From this perspective, then, the issue of fewer and larger farms does not pose a significant problem for the foreseeable future.

Most of the farms with annual sales of farm produce of less than $50,000 fail to generate an income sufficient to support a farm family. Yet these farms not only persist, they make up the bulk of farms in the Nation. As we shall see in a later section, these farms are able to persist because the farm operator and/or operator's spouse are able to supplement their farm earnings with part-time or full-time nonfarm jobs. So long as rural communities providing nonfarm jobs to farm families remain viable, then, many of these smaller farms will also remain viable.

Type of Farm Organization

In 1987, 87 percent of the 2.1 million farms in the United States were operated by individuals or families, 10 percent by partnerships, 3 percent by family-held corporations, and less than one percent by nonfamily-held corporations and other business forms. Of the Nation's nearly one billion farm acres, 65 percent were operated by individuals or families, 16 percent by partnerships, and 8 percent by nonfamily held corporations and other business forms.

Of the total market value of agricultural products sold, individuals or families accounted for 56.3 percent, partnerships accounted for 17.1 percent, family-held corporations accounted for 19.5 percent, and nonfamily held corporations and other business forms accounted for 7.0 percent.

Data is not available with which to establish long-term trends; however, there has been no significant change in these percentages since 1978. Corporations or other nonfamily business forms are not taking over farming in the United States. Rather, the vast majority of farms are still family-oriented units.

Farm Ownership and Tenancy

About 59 percent of the farms in the United States are operated by full owners, but only about one-third of the acreage is operated by full owners. This suggests that the percentage of the small farms operated by full owners is much greater than is the case for medium-sized and large farms. In 1987, for example, 74 percent of the farms with annual cash sales of less than $20,000 were operated by full owners whereas only 30 percent of the farms with annual cash sales of $50,000 or more were operated by full owners. There are substantially more part-owners and slightly more tenants among the large farms than among the small farms. These percentages have remained fairly stable since 1950.

Apart from a general interest in who owns farm land or how much farm land is rented, all this is important in gaining a better perspective on the distribution of the benefits of farm programs. It is well known that since the supply of farm land is so inelastic, most of the benefits of farm programs accrue to the land (i.e., are capitalized into the value of land) rather than to hired or self-employed farm workers or other farm inputs (see, for example, Hallberg, 1992, pp. 104-115). Since a much higher percentage of the operators of the smaller farms are full owners than is the case on the larger farms, we can infer that small farms are not treated inherently unfairly even though they receive a relatively small portion of the direct government payments from farm programs.

Part owners now operate about 30 percent of the farms and about 54 percent of the farm acreage—increases from 15 percent and 40 percent, respectively, in 1950. Part owners own 44.5 percent of the land they farm. The relative importance of full tenants has diminished. Full tenants now operate about 12 percent of the farms and 13 percent of the farm acreage—decreased from 27 percent and 20 percent in 1950, respectively. Apparently, farmers to a significant extent, have enlarged their operations by renting additional acreage rather than by buying out neighbors' farms. Many full tenants have transferred out of agriculture.

All this suggests that land tenure adjustments in U.S. agriculture have been reasonably smooth, with no major accumulations of land in the hands of the few. Further, full tenants unable to compete in farming have not been trapped in the sector.

Age Structure in Farming

Farm operators nationwide are getting older. In 1950, the average age of farm operators was 48.3 years; in 1987, it was 52.0 years. In general there has been a steady increase in the average age of farm operator. Between 1974 and 1978, however, there was a significant drop in the average age of farm operator that correlated, no doubt, with the change in census definition of a farm following 1974. The new definition excluded many smaller operations that previously would have been considered farms. The average age of farm operator is lowest on dairy, poultry, animal specialty, and cotton farms, and highest on beef cattle and fruit and nut farms.

The capital required to start up a farming operation is much greater (relative to the sales that this capital will generate) than for almost any other type of business in the United States. Thus if the average age of farm operator increases significantly in the future, it may suggest that young farmers are not entering the industry to replace older farmers. Capital requirements could be one of the more significant limitations here. Per-farm in 1987, machinery and equipment values represented 63 percent of the value of cash farm sales, while total per-farm asset values represented more than five times the value of cash farm sales. A young man contemplating entry into farming via a purchase of land, buildings, and equipment must not only have a huge amount of venture capital available, he must be willing to wait several years before the farm generates enough sales to repay that capital investment.

It should be noted, however, that the change in average age of farm operators regionally is not uniform. In Pennsylvania, for example, there was virtually no change in the average age of farm operators between 1950 and 1978; while between 1978 and 1987, the average age of farm operators increased by 0.5 years. Here a sufficient number of young farmers have been encouraged to enter farming upon the retirement of older farmers. Clearly though, Pennsylvania is one of the exceptions rather than the rule.

The increasing age of farm operators nationwide prompts two concerns: (1) will new and younger entrants replace the older farmers when the latter retire, and (2) are the older farm operators motivated to make the adjustments to changing economic conditions required to stay competitive and efficient, or are they simply hanging on until retirement?

Specialization in Farming

One of the more prominent features of the agricultural sector since the 1930s has been the increasing degree of specialization brought about by (1) improved and expanded communication and transportation that has facilitated rapid and low-cost movement of commodities over long distances, (2) development of specialized and capital-intensive production technologies that have increased size advantages, (3) government farm programs leading to reduced risks associated with specializing on the production of a single or a few commodities, and (4) the shifting of many functions formerly performed on farms to nonfarm firms, as evidenced by the increasing importance of purchased inputs to farm firms.

This increasing specialization has undoubtedly contributed to increased production efficiency. At the same time, however, increasing specialization means that farmers, in the absence of government programs, become subject to greater price risks than if they were more diversified. Further, farmers are now more vulnerable to rising input prices and to interruptions of input supplies.

Capital Requirements and Debt

Assets per farm have increased by almost 17 times since 1950, while debt per farm has increased even more. In fact, the debt-asset ratio has increased from 10 in 1950 to 24 in 1985. Since 1985, the debt-asset ratio has declined, but in 1991 was still greater than 17. The increase in debt-asset ratios, the increase in market rates of interest in the early to mid-1980s, and the declining value of land (leading in turn to an erosion of loan security values) highlight why several farmers have had financial difficulties in recent years. This is more of a problem for the larger farms than it is for the smaller farms. Debt-asset ratios rise directly with sales volume, as tabulated below (U.S. Department of Agriculture 1991):

Sales class	*1991 average debt-asset ratio* (*percent*)
Under $20,000	11.3
$ 20,000-$ 39,999	12.3
$ 40,000-$ 99,999	15.5
$100,000-$249,999	18.6
$250,000-$499,999	19.9
$500,000 and over	24.2

As Hallberg (1992) demonstrated, farming requires greater amounts of capital relative to the gross revenue this capital will generate than most any other industry in the Nation. Prospective farmers are expected to generate their own capital—they do not have the option, as do large nonfarm corporations, of relying on stockholders for their equity capital. This makes entry into agriculture much more difficult than entry into most other industries.

Size Economies on U.S. Farms

If farms are too small to capture available size economies, production efficiency will be sacrificed. If size advantages exist, the larger farm can operate at lower per unit costs than the smaller farm by purchasing the inputs it needs at lower per unit costs, by exploiting existing technology more fully, by combining lumpy inputs more effectively, and/or by more fully utilizing its fixed resources, including management.

A rough idea of the nature of size economies in agriculture can be obtained from an examination of the ratio of inputs to outputs on farms in different sales classes as developed from USDA's 1988 Farm Cost and Returns Survey data (Hallberg 1992, p. 283):

Average gross sales	*Average gross farm income*	*Ratio of cost of farm inputs to gross value of farm sales*
Less than $10,000	$ 11,580	1.98
$ 10,000-$ 19,999	22,589	1.66
$ 20,000-$ 39,999	34,413	1.59
$ 40,000-$ 99,999	72,192	1.20
$100,000-$249,999	155,973	1.06
$250,000-$499,999	314,299	1.05
$500,000 and over	1,042,879	0.97

The measure of inputs here consists of the 1988 dollar cost of all farm inputs, including a charge for depreciation and an estimate of the opportunity cost of equity capital (assessed at 9 percent of total farm equity). The measure of output includes the 1988 dollar value of all farm sales, estimated nonmoney income, and government payments.

This tabulation suggests that per unit costs decline rapidly as farm size increases, and that the cost curve becomes quite flat for the larger sized farms. Farms with annual gross sales of less than $40,000 would have an input-output ratio near 1 only if their return on equity capital was in the vicinity of 2.5 percent. Clearly the operators of these farms are willing to accept a return on capital that is nowhere near a competitive level (the rate of interest on AAA-rated corporate bonds in 1988 was 9.71

percent). It is in this sense that many will argue that farms selling less than $40,000 of farm produce in 1988 were "inefficient."

More-detailed studies confirm this general shape for the long run average-cost curve for most farm types in the United States. From these studies, it appears that the average dairy farm in the Northeast and the Lake States is probably somewhat below optimal size. Farms in the West and Southwest, with typically larger dairy herds, have probably captured most of the technical economies achievable. Tobacco farms have undoubtedly been maintained at a size well below the most efficient-sized unit because of the allotment policies that have been adopted for this subsector over the years. Economies of size have also been found to exist on wheat, feed grain, and cotton farms. The estimated long-run average-cost curves for these farms decline significantly at first and then become quite flat over a wide range of farm sizes. The general conclusion is that the majority of the commercially viable wheat, feed grain, and cotton farms are large enough to have captured most of the size economies achievable.

Farm Income and Off-Farm Income

A comparison of the average money income of farm households with the average money income of all U.S. households reveals much about the relative well-being of farm families. Money income of farm households includes income from both farm and nonfarm sources, but excludes all nonmoney income, such as the value of food produced and consumed on the farm, changes in the value of inventories, and estimated rental value of farm dwellings. Recent studies (Hallberg 1988 and Reimund and Gale 1992) indicate that money income of farm households was, on average, relatively low in the 1950s and early 1960s. Through the 1970s, average money income of farm households exceeded or nearly equaled that of the general population (Table 4.2). During the early 1980s, average money income of farm households fell below that of the general population but recovered again by 1985. Farm households' average money income was well above that of the general population in 1973-75 and again in 1985-91.

We know that farm income is much more variable than is nonfarm income because of the riskiness of farming. We also know that the distribution of income is more skewed among farm families than in other occupations so that merely examining ratios of averages, as was done above, can overlook much interesting detail. For example, Hallberg (1992, p 295) shows that average farm and nonfarm income per farm as a percent of average money income of all U.S. households increases steeply with farm size. Indeed, in 1988 this percentage was 87 for farms

TABLE 4.2 Farm Family Income as a Percent of Incomes of All U.S. Households, and Importance of Off-Farm Income and Government Payments to Farm Families, 1970-90.

	1970	*1975*	*1980*	*1985*	*1990*
			- - - - - - percent - - - - -		
Money income of farm households relative to money income of all U.S. households[a]	102.5	118.9	87.5	121.9	136.6
Off-farm income of farm families relative to total farm family money income[a]	63.1	60.4	77.4	68.5	61.6
Direct government payments to farm families relative to total farm family money income[a]	13.3	2.0	2.9	9.6	8.5

[a] Money income of farm families equals net cash farm income plus off-farm income plus direct government payments to farmers. Money income of all U.S. households is estimated from U.S. Department of Commerce. *Current Population Reports.* Bureau of the Census (various annual issues).

Source: U.S. Department of Agriculture. *Economic Indicators of the Farm Sector: National Financial Summary.* Economic Research Service (various annual issues).

in the $40,000-$99,999 sales category but 2,000 for farms in the $500,000 and over sales category. Finally, it is well known that there is a greater incidence of poverty (as measured by annual money income) in agriculture than in other occupations. Nevertheless, the gap between average money incomes of farm and nonfarm households has narrowed considerably since the 1950s and 1960s.

Government payments to farmers have certainly assisted many farm families, but in general government payments have not been the principal factor in reducing the gap between farm and nonfarm incomes. Off-farm income of farm families *has* been a principal factor. Off-farm income has constituted well over 50 percent of farm family money income since 1970 (Table 4.2)! In some years, off-farm income has approached or exceeded 70 percent of total farm family income. Direct government payments, on the other hand, have generally accounted for less than 10 percent of farm family money income. The distribution of off-farm income by farms in different sales classes shows that most of the off-farm income is earned by families on small farms. It is interesting to note, however, that even on the farms that we generally consider of commercial size (i.e., those with annual cash sales of $100,000 to $449,999), off-farm income constituted more than 18 percent of aggregate net farm income in 1990:

	Percent of total farm family income attributable to:	
Sales class	*Off-farm income*	*Government payments*
Less than $20,000	100.5	0.5
$20,000-$39,999	74.0	4.7
$40,000-$99,999	44.5	10.1
$100,000-$499,999	18.3	17.3
$500,000 and over	3.9	4.8

The crucial importance of off-farm income to most farm families has significant implications for the rural community. If nonfarm jobs are not available, these farm families will not be able to remain in farming. In general if the farms do not survive, neither will the rural community since the latter depends not only on the needs of farmers for market outlets, farm inputs, and services, but also on farm families for surplus labor needed to staff local businesses.

Use of Coordination Devices

Production, processing, distribution, retailing, and consumption activities must in some way be coordinated so that proper signals are sent through the economic system to guide output and consumption decisions. In a perfectly competitive system this coordination is accomplished via prices established in open markets—by the "invisible hand." When operative, this system is highly effective, fosters innovativeness and progress, and prevents abuse of one market participant by another. It is operative, however, only when there are numerous buyers and sellers who have access to full and adequate information and who interact for the purpose of exchanging information and establishing a truly "competitive" price.

Other coordinating devices, though, are used extensively in agriculture (see Table 4.3). These include grower-processor contracts and vertical integration. Contractual agreements take on a variety of forms depending on the commodity. Basically they specify the nature of the buyer-seller relationship. Some specify the date of product delivery, some the price at which the transaction is to be made, and some the type of production practices to be employed. In vertically integrated operations, decision-making at more than one stage in the marketing channel (e.g., apple production and apple processing) is in the hands of a single firm. Highly perishable commodities that use land and capital intensively are prone to be produced under contract or vertical integration.

In highly perishable commodities, contracts that are entered into before the crops are planted can control quantity and timing of production and eliminate expensive storage and handling operations as well. Contracts reduce price risks and ensure a market. However, formal coordinating devices require farmers to relinquish some (or all) of their independent decision-making responsibility to larger and more economically powerful nonfarm firms. Furthermore, when these devices are used extensively, there remain too few buyers and sellers using open markets with which to establish a competitive price. It is thus difficult for the market to reflect all of the forces of supply and demand. The limited market information available under these circumstances may be insufficient to serve as a base for rational resource allocation decisions.

Formal coordination as opposed to open market coordination of production and processing ranges from 100 percent for sugar beets and cane to less than 1 percent for hay and forage (Table 4.3). Formal coordination is quite extensive in eggs, broilers, turkeys, and fruits and vegetables. For the agricultural sector as a whole, farm output produced under contractual arrangements or by vertically integrated firms increased from 13 percent in 1960 to 18 percent in 1990.

The use of formal coordination devices such as these arose out of a felt need by the private sector to minimize risk to both producer and processor. The proliferation of these devices, however, is accompanied by new concerns about the adequacy or fairness of contract terms, concerns about the adjudication of contract disputes, and consequences of the resulting "thinness" of any still remaining open markets.

Structure of Farm Input and Food Processing Industries[1]

Firms Are Getting Larger

Food manufacturing establishments have diminished in numbers and grown in size at phenomenal rates since the 1950s, due in large measure to the existence of technical economies as well as buying and marketing economies. Since 1954, real value added per establishment has increased 2.6 times in meat processing, 4.4 times in dairy processing, 5.8 times in preserved fruits and vegetables, 4.1 times in the bakery industry, 3.7

[1] Much of this section draws heavily from Manchester (1992).

TABLE 4.3 Proportion of Output Produced Under Contract or Vertical Integration, Selected Commodities, 1960-80.

	Production and Contracts[a]				*Verticle Integration*[b]			
Commodity	*1960*	*1970*	*1980*	*1990*	*1960*	*1970*	*1980*	*1990*
	----------------- Percent ----------------							
Crops								
Feed Grains	0.1	0.1	1.2	1.2	0.4	0.5	0.5	0.5
Hay and Forage	0.3	0.3	0.5	0.5	0.0	0.0	0.0	0.0
Food Grains	1.0	2.0	1.0	0.1	0.3	0.5	0.5	0.5
Soybeans	1.0	1.0	1.0	0.0	0.4	0.5	0.5	0.4
Seed Crops	80.0	80.0	80.0	80.0	0.3	0.5	10.0	10.0
Fresh Vegetables	20.0	21.0	18.0	25.0	25.0	30.0	35.0	40.0
Processing Vegetables	67.0	85.0	83.1	83.0	8.0	10.0	15.0	15.0
Potatoes	40.0	45.0	60.0	55.0	30.0	25.0	35.0	40.0
Citrus Fruits	0.0	0.0	0.0	0.0	20.0	30.0	35.0	35.0
Sugar Beets	99.0	99.0	99.0	99.0	1.0	1.0	1.0	1.0
Sugarcane	24.4	31.5	50.0	37.5	75.6	68.5	50.0	62.5
Livestock								
Fed Cattle[c]	--	--	--	--	6.7	6.7	3.6	4.2
Hogs	0.7	1.0	1.5	5.0	0.7	1.0	1.5	5.0
Fluid Milk	0.1	0.1	0.3	0.1	0.0	0.0	0.0	0.0
Mfgering Milk	0.0	0.0	0.0	0.0	2.0	1.0	1.0	1.0
Eggs	7.0	20.0	43.0	45.0	5.5	20.0	45.0	50.0
Broilers	90.0	90.0	91.0	92.0	5.4	7.0	8.0	8.0
Turkeys	30.0	42.0	52.0	60.0	4.0	12.0	28.0	28.0
Total Farm Output	8.3	9.3	10.3	10.7	4.6	5.6	6.5	7.4

[a] Agreements between farmers and processors, dealers, or others including cooperatives.
[b] Production and marketing activities performed by the same firm. Products produced on farms and utilized by the livestock on these same farms (e.g., feed grains and hay) and products marketed directly by farmers to consumers are not included in these tabulations.
[c] Cattle fed by meatpackers some of which are under contract.

Source: (Manchester 1992a).

times in the sugar industry, and 5.8 times in the beverage industry (Table 4.4). The same phenomenon has occurred in most of the farm input industries, although the farm machinery and equipment industry represents a glaring exception. Here growth was strong through the 1970s, but since has declined quite rapidly in response to reduced demand by farmers and increased competition from foreign manufacturers.

Much of the growth in food processing and farm input industries has occurred through mergers, acquisitions, leveraged buyouts, and divestitures. Large manufacturing firms account for a much greater share of the market than they did in the 1950s. They are more diversified into non-food and nonagricultural products than was the case in the 1950s. Increasingly, the large food firms are specializing in a single market segment such as grocery store products, food-service products, or ingredients for other food manufacturers.

TABLE 4.4 Number of Establishments and Real Value Added Per Establishment in the Food Manufacturing Industries, 1954 and 1987.

Industry	*1954*	*1987*
Food and Kindred Products (SIC 20)		
Number of Establishments	36,785	20,624
Real Value Added ($1,000)[a]	$1,354	$5,210
Meat Industry (SIC 201)		
Number of Establishments (1,000)	4,992	3,267
Real Value Added ($1,000)[a]	$1,438	$3,702
Dairy Industry (SIC 202)		
Number of Establishments (1,000)	11,503	2,366
Real Value Added ($1,000)[a]	$1,004	$4,431
Preserved Fruits and Vegetables (SIC 203)		
Number of Establishments (1,000)	3,513	1,912
Real Value Added ($1,000)[a]	$1,377	$7,952
Bakery Industry (SIC 204)		
Number of Establishments (1,000)	6,414	2,850
Real Value Added ($1,000)[a]	$1,146	$4,711
Sugar Industry (SIC 206)		
Number of Establishments (1,000)	1,627	1,094
Real Value Added ($1,000)[a]	$1,935	$7,183
Beverage Industry (SIC 208)		
Number of Establishments (1,000)	5,384	2,214
Real Value Added ($1,000)[a]	$1,545	$9,023
Fats and Oils Industry (SIC 207)		
Number of Establishments (1,000)	931	595
Real Value Added ($1,000)[a]	$2,207	$4,802

[a] Value added by manufacture deflated by the CPI (1982-84=100).

Source: U.S. Department of Commerce (1954, 1987).

International Operations

It is widely recognized that U.S. agriculture has become internationalized and that U.S. farmers compete as much with foreign producers as with one another. Indeed, a sense of national pride emerges from observations that as much as one-fifth or more of the cash receipts from the sales of U.S. farm products is generated in foreign markets.

The industries downstream from the farm—those upon which farmers depend for their market—are even more globally oriented so that foreign markets have provided opportunities for expansion of many U.S. food processing firms. The U.S. food processing industry is, to be sure, still quite domestic-market oriented with exports accounting for only about 4 percent of U.S. production of processed foods. Large foreign companies have increasingly moved into the U.S. food business. There were only about 5 in the 1950s, but 42 in 1989. Their share of the U.S. food market, however, remains fairly modest—about 9 percent in 1989. Nevertheless, the value of international trade in processed foods is three times the total value of global trade in farm commodities (Handy and Henderson 1992). In 1992, U.S. exports of processed foods exceeded $22 billion, and about $21 billion worth of processed foods were imported into the United States from foreign manufacturers.

Value of exports and imports of processed food, however, does not reflect the total presence of U.S. firms in foreign markets or foreign firms in the U.S. market. Six of the world's 10 largest food processors in 1989 were U.S. firms. Of the world's largest 20, 12 were U.S. firms, and of the world's largest 50, 21 were U.S. firms. For the 64 largest U.S. food processing firms, 78 percent of their food sales were to other U.S. firms and 22 percent to foreign subsidiaries owned by these U.S. firms. Thirty-eight of these 64 firms owned food processing plants in foreign countries. Indeed, 27 percent of their processing plants were operated abroad. Finally, export sales plus sales from foreign subsidiaries understates the foreign perspective in that international licensing, joint ventures, and franchising operations are not reflected in the sales figures.

Most firms enter the export market by using *foreign agents or brokers*. As their export sales increase, these firms take the next step of setting up a separate United States-based *export office or division*. U.S. processors can also decide to pack *under contract* for a foreign firm. For example, several Japanese manufacturers of soda and fruit drinks contract out production of their Japanese brands to U.S. bottlers. Firms may also choose to have their branded products produced and marketed in foreign countries under a *licensing agreement* with a foreign firm. While this strategy requires no direct investment in foreign production facilities, considerable investment is required to identify appropriate licensees, develop production and marketing procedures for the product to be sold in a foreign market, and establish quality controls. *Joint ventures* allow a U.S. firm to tap into the production, marketing, and regulatory know-how of a host-country firm without the expense of acquiring a wholly-owned subsidiary. Finally, a U.S. processor can acquire or build foreign manufacturing facilities to operate as a *wholly-owned subsidiary*.

While few comprehensive data are available with which to document the extent to which firms engage in overseas licensing, contract production, or joint ventures, casual observation suggests that such activities are extensive. Henderson and Sheldon (1992), for example, have estimated that the value of foods produced under a foreign license arrangement exceeds the value of international processed food trade.

By contrast, foreign direct investment is extensively documented. The value of production by foreign affiliates of U.S. food manufacturing firms exceeded $69 billion in 1989, four times the total value of U.S. processed food exports. For the 34 largest U.S. food processing firms with foreign affiliates, the value of international sales originating in foreign plants was nearly 9 times larger than their exports from the United States. At the same time, more than 11 percent of U.S. production of processed foods originated in foreign-owned plants.

Food processing firms seek to produce finished products in foreign plants rather than export from domestic plants for several reasons. First, this is probably the easiest way to avoid trade barriers set up by foreign governments. Secondly, transportation costs are reduced. This is especially important for products for which packaging materials add considerable weight. Third, dealing with local governments and regulatory agencies is easier when the product is produced in the host country. Fourth, keeping abreast of local tastes and opportunities for new product development is easier when the processing facility is located locally. Fifth, locating the operation in a foreign country can improve access to local distribution and promotion facilities. Finally, firms may acquire established brands in a foreign country and use the facility as a base for further expansion.

In short, competition between U.S. and foreign food operations is extensive. Not only do U.S. and foreign food manufacturing firms ship products into each other's markets, they meet head-to-head with production facilities and distribution networks in the same countries. Not only do U.S. farmers have to compete among themselves for sales to domestic food processors, they must compete to supply foreign-owned processors at home and U.S.-owned food manufacturers in overseas markets.

Considerable research has been done to determine the reasons why food firms expand into foreign markets and to distinguish those that emphasize international trade from those that rely more heavily on foreign subsidiaries (see Conner 1983, Grubaugh 1987, Henderson and Frank 1990, and Voros 1992). The findings suggest that trade appears to be heavily driven by procompetitive factors such as lack of dominant firms, relatively low seller concentration, little product differentiation, and specialization by firms in food.

At the farm level, relatively open trade in basic farm commodities is generally consistent with the procurement needs of such firms, thus, the trade side of the global food marketplace seems to be compatible with a fairly independent system of farm production. In contrast, foreign direct investment is driven by factors that distinguish one firm from others: large size, substantial market shares, heavy investment in brand names and other intangible assets, and diversification across a wide band of food products. At the farm level, these characteristics are often associated with or call for specification buying and to contracts and other types of agreements with farmers regarding the type of products that farmers grow and how and when these products are to be shipped to the processor. Food firms with strong brand names, for example, cannot bear the risk of an insufficient supply of the specific varieties of farm products needed to meet the consumer expectations that go hand-in-hand with such brand images. Thus, at the farm level, this form of international commerce is significant in terms of restructuring toward the more interdependent, industrial-type system.

Summary

Farms in the United States are becoming fewer in number and larger in size as farm operators take advantage of new technologies to achieve additional size economies and become more specialized. This trend will likely continue into the foreseeable future as technological developments continue and as management levels improve. Most of the larger farm units have income levels well above that of the average American family. Most of the larger farms are still family units. Corporate takeover in farming is not an immediate nor long term threat. Farm numbers in all size categories are sufficiently large so that no one farm can significantly influence the marketplace.

The vast majority of farms are in the smaller size categories. There are over twice as many farms with annual farm sales of less than $50,000 than of farms with annual farm sales of $50,000 or more. The 75 percent of U.S. farms in the small size categories, however, produce only 22 percent of total farm output. Farm families on these smaller farms have a net income from farming that is in general much below that necessary to sustain the farm family. The vast majority of the individuals on these smaller farms, though, are able to supplement their farm income with income from nonfarm jobs so that in total they are well above the poverty income level. Government payments to these farms contribute relatively little to total farm family income.

Increasing specialization in agriculture leads to increased production efficiency, but also to increased vulnerability to price risks. A partial response to this increased risk has been increased use of coordination devices such as contractual arrangements with food processors and feed dealers. Capital requirements in agriculture are high putting added pressure on the potential entrant and disadvantaging farmers when interest rates rise. The average age of farm operator nationwide is increasing suggesting that new and younger entrants are discouraged from replacing older farmer operators who retire.

Food processing and retailing firms and farm input manufacturers are still fairly numerous, but they too are becoming fewer in number and larger at the expense of small, local firms no longer able to compete. Mergers have been a major force of change. Large companies increasingly handle a broader line of products combining food production as well as nonfood production facilities.

Larger firms are national or regional multiproduct firms, some operating internationally, that do not depend solely on any one production area for raw materials. On the contrary, they obtain their supplies anywhere they can get the volume and quality necessary to support a nationwide or regionwide marketing program. With ready access to markets thus reduced, small-scale producers for local markets are at a serious competitive disadvantage. Also, due to the decline in transportation services (particularly rail) some rural areas do not have ready access to production inputs or product markets. Farmers are increasingly paid on the basis of their ability to provide commodities that meet buyers specifications—e.g., leaner pork, pesticide-free products, organic foods, etc.

Of perhaps greater significance is the fact that for a production activity to be viable at all in a particular area, it must be undertaken on a large enough size that support services and processing capacity can be provided at an economically justifiable scale. Given the size economies in fluid milk processing, for example, a plant processing less than 50 million pounds of milk per year is not cost competitive. Thus, a minimum of 4,500 good producing cows must be in the region supplying such a plant.

Given these type of constraints, it is easy to see why in some areas with low production densities, support services have diminished or vanished. In such circumstances, farmers' adjustment possibilities are severely limited irrespective of their managerial ability or competitiveness.

References

Connor, John M. 1983. "Foreign Investment in the U.S. Food Marketing System." *American Journal of Agricultural Economics* 65:395-404.

Grubaugh, Stephen G. 1987. "Determinants of Direct Foreign Investment." *Review of Economics and Statistics* 89:149-152.

Hallberg, M. C. 1992. *Policy for American Agriculture: Choices and Consequences.* Iowa State University Press. Ames.

______. 1988. "The U.S. Agricultural and Food System: A Postwar Historical Perspective." Northeast Regional Center for Rural Development. Publication No. 55. University Park, PA.

Handy, Charles R. and Dennis R. Henderson. 1992. "Foreign Investment in Food Manufacturing." Ohio State University, North Central Regional Research Project NC-194, Report No. OP-41.

Henderson, Dennis R. and Ian M. Sheldon. 1992. "International Licensing of Branded Food Products." *Agribusiness, An International Journal* 8:399-412.

Henderson, Dennis R. and Stuart D. Frank. 1990. "Industrial Organization and Export Competitiveness of U.S. Food Manufacturers." Ohio State University, North Central Regional Research Project NC-194, Report No. OP-4.

Manchester, Alden C. 1985. "The Farm and Food System: Major Characteristics and Trends." In *The Farm and Food System in Transition: Emerging Policy Issues*, #1. Cooperative Extension Service, Michigan State University. East Lansing, MI.

______. 1992. "Rearranging the Economic Landscape: The Food Marketing Revolution, 1950-91." Commodity Economics Division, Economic Research Service, U.S. Department of Agriculture. Agricultural Economic Report No. 660.

______. 1992a. "Transition in the Farm and Food System." U.S. Department of Agriculture, Economic Research Service. Unpublished paper prepared for the National Planning Association, Committee on Agriculture.

Mayer, Andre', and Jean Mayer. 1974. "Agriculture, The Island Empire." *Daedalus* 103:3:83-96.

Reimund, Donn A., and Fred Gale. 1992. "Structural Change in the U.S. Farm Sector, 1974-87." Economic Research Service, U.S. Department of Agriculture. Agricultural Information Bulletin No. 647.

U.S. Department of Agriculture. 1991. *Economic Indicators of the Farm Sector: National Financial Summary, 1991.* Economic Research Service. ECIFS 11-1.

U.S. Department of Commerce. 1987. *1987 Census of Agriculture.* Vol. 1, Part 51. Bureau of the Census. Washington, D.C.

______. 1954. *1954 Census of Manufactures.* Bureau of the Census. Washington, D.C.

______. 1987. *1987 Census of Manufactures.* Bureau of the Census. Washington, D.C.

Voros, Peter. 1992. Industrial Determinants of International Trade and Foreign Investment by Food and Beverage Manufacturing Firms. Unpublished MS thesis. Ohio State University.

PART TWO

The Issues

Environment

5

Water, Agriculture, and Public Policy

Norman K. Whittlesey and Roy R. Carriker

The relationship of national agricultural policy and water resource issues is important to the nation. Agriculture in the United States is both a very large consumer and a major provider of water. In 1985, agriculture accounted for 42 percent of all freshwater withdrawals in the United States (U.S. Department of Interior 1990). Agriculture is also the largest single user of the land base that is the source of surface and groundwater available to all other uses. Approximately one billion acres in the United States are used for agriculture, with about 400 million acres in cropland (U.S. Department of Commerce 1982).

More than 50 million acres of cropland are irrigated, with about 90 percent of the irrigated acreage located in 17 western states. In these states, agriculture can account for as much as 80 percent of all water consumption. Nearly half the harvested crop acreage in Florida is irrigated and twelve other eastern states irrigate some cropland. Private investment in irrigation during the past several decades has been encouraged by improved irrigation technology and by production inputs, such as pesticides, fertilizers, and machinery, that are most profitable when combined with irrigation. Federal farm programs that restricted crop acreage but rewarded farmers for increased yields also contributed to this development.

Irrigation developments have increased agricultural production, created wetlands, recharged groundwater aquifers, and moved economic development to the west. But there were costs associated with these developments. Instream uses of water were diminished, streamflow patterns were changed to accommodate agriculture's needs, and water quality problems occurred. In view of today's environmental goals and patterns of water demand, the previous system of allocations may no longer be suitable.

Trends in public attitudes have ushered in a new era of natural resource concerns. In her 1988 presidential address to the Southern Agricultural Economics Association, Batie (1988) noted an increasing public perception that agricultural science and policy are having severe negative impacts on farm labor, the environment, the structure of agriculture, and rural communities. Public sentiment for agriculture is eroding as the farm population loses political strength, as environmental interests gain in legitimacy, and as commercial agriculture is increasingly perceived as a few "factory-like" farms which neither need nor deserve special socially-funded benefits and exemptions. There is growing sentiment for directing incentive payments to achieve environmental, resource, and social goals.

These changing relationships and perceptions have important political implications. The agricultural tradition emphasizes voluntary compliance and incentives as the preferred modes of policy implementation; those outside agriculture are more accustomed to a tradition of strict regulation. In policy discussions, agriculturalists seem most concerned about food supply, productivity, and farm income; others may emphasize environmental quality and food safety. Both groups may agree that sustainability and resource conservation are important public issues, but their perception of how agriculture should be controlled or affected may differ. The federal government's approach to these concerns will likely depend on which committee in Congress assumes jurisdiction. The institutional question, "Who decides?" is important.

Water Problems

Water policy issues relating to agriculture involve several different water problem areas. Farming's impact on water quality constitutes a major problem set that has become the focus of policy debate. Other problem areas include wetlands conversion or alteration, groundwater mining for irrigation, the efficiency and equity of federal water subsidies to agriculture, and growing competition for water that, until recently, has been reserved for agriculture.

Water Quality

Some by-products of agricultural operations can contaminate groundwater and surface water. These include soil sediments, nutrients, pesticides, mineral salts, heavy metals, and disease organisms for which water is a primary transport medium (Libby and Boggess 1990). When water containing mineral salts is used for irrigation, the salts are carried into the soil root zone and then left behind as the water evaporates or is taken up by the plants. Livestock production processes can be a source of nitrogen and phosphorous and of heavy metals (e.g., lead, zinc, mercury, and arsenic) and of disease organisms (e.g., coliform bacteria and viruses).

Groundwater contamination is similar to offstream contamination by nutrients, pesticides, and mineral salts, in surface water. Environmental Protection Agency (EPA) studies in 1988 and 1989 (U.S. Environmental Protection Agency 1990, 1993) estimated that 10 percent of the nation's community drinking-water wells and about 4 percent of the rural domestic wells have detectable residues of at least one pesticide, but less than 1 percent of the wells had pesticide residues at levels considered detrimental to human health. EPA also estimated that more than one-half of the nation's wells contain nitrate, with about 1.2 percent of the community wells and 2.4 percent of the rural wells showing detections above the established health standard of 10 mg/l nitrate.

Clark, et al. (1985), report a comprehensive attempt to quantify the annual damages caused by nonpoint source (NPS) pollution in the United States. In physical terms, they estimated that nonpoint sources dominate point sources, contributing about 70 percent of the total biochemical-oxygen-demand (BOD) loads, 99 percent of the suspended solids, 83 percent of the dissolved solids, 82 percent of the nitrogen, 84 percent of the phosphorus, and 98 percent of the bacterial loads in surface waterways of the United States.

The relationship between agricultural production and water quality were reflected in the Conservation Titles of the Food Security Act of 1985 and the Food, Agriculture, Conservation, and Trade Act of 1990. New policy initiatives were directed not at maintaining agricultural productivity, controlling surpluses, or maintaining farm income, but at a wider array of concerns, such as human health, environmental preservation, and resource amenity values related to off-farm impacts of agricultural practices. Another theme in public discussions of agriculture and water quality issues is the contention of environmentalists that the USDA lacks either the will or the ability to reduce effectively agricultural nonpoint sources of water pollution.

Water pollution from agriculture has some unique and problematic characteristics (Oh 1991). Being generally nonpoint source in nature, the source of contamination cannot be readily traced to individual sites. The area vulnerable to pollution is extensive and agricultural contaminants are influenced by stochastic and uncontrollable weather conditions (especially rainfall), making it difficult to control pollution even with good management. Although a regulatory agency can observe the quality of a stream or containment, it may be unable to determine if a reduction in quality is due to inappropriate actions, failure of appropriate actions, or to undesirable effects of natural events.

Wetlands Preservation

Wetlands perform many ecological services such as providing habitat for aquatic, terrestrial, waterfowl, and fur-bearing species; storing flood waters; improving water quality; groundwater recharge; and controlling erosion. Heimlich (1991b) states that of the 215 million acres of wetlands present during colonial times, less than 100 million acres remained in the 1970s. Today, annual losses of wetlands are estimated at between 100,000 and 300,000 acres, less the 450,000 acres per year during the 1970s. Agriculture accounted for 87 percent of the nearly 14 million acres of wetlands lost between the 1950s and mid 1970s (Heimlich 1991c), and still plays a major role in the continuing losses.

The public benefits of wetlands cannot be appropriated by landowners in the same sense that the monetary returns to crop production can be appropriated. This leaves landowners with incentives to convert wetlands, unless protected, to other uses. Economic factors affecting agricultural wetland conversion include technology, crop prices, farm programs, tax effects, and land values. Policies such as federal cost sharing for drainage have abetted conversion. The Internal Revenue Code formerly favored wetland conversion investments by allowing costs to be deducted from farm income, applied to investment tax credits, or added to the basis for land cost. Farm price and income support programs have contributed to wetland conversion economics by setting commodity prices higher than prevailing market prices and by requiring compliance through maintenance of base acreage and set-aside provisions.

The desire to regulate conversion or alteration of natural wetlands gives rise to the need for criteria that clearly specify the attributes of wetlands that are important for regulatory purposes. Preservation of existing wetlands (however finally defined) will require specific actions to prohibit their conversion, or provide incentives for their protection. In agriculture, such measures could include cross-compliance with income features of the food and agricultural legislation, strict cost-recovery rules, and cost sharing for restoration or protection of wetlands. Some agricultural interests and landowner groups argue that regulatory restrictions on uses of wetlands acreages may constitute an unconstitutional "taking" of private property for public use without just compensation.

Groundwater Mining

More than 97 percent of the U.S rural population's drinking water, 55 percent of the nation's livestock water, and 40 percent of the nation's irrigation water comes from groundwater sources (Nielsen and Lee 1987, p 1). Irrigation is heavily dependent on groundwater in the desert southwest and the Plains States. In many of these and other western groundwater pumping areas, overdrafting or "mining" the groundwater aquifers is lowering water levels. Increasing pumping costs and declining water supplies threaten irrigation (the major consumer of water) and domestic water supplies, as well. Groundwater mining produces other adverse effects, such as land subsidence, salt water intrusion, depleted stream flows, and loss of wetlands. Overriding the widespread dependence on groundwater is the fact that much of the groundwater law and the public institutional framework for managing groundwater use is archaic and, in many cases, based on faulty assumptions about aquifer characteristics and models of groundwater dynamics.

Support for imposed limitations on groundwater mining is growing, a result of increased realization of the damages to public interests. But public concerns regarding groundwater mining have not yet been addressed within the context of federal omnibus agricultural legislation. This is largely because virtually all groundwater use is controlled by the states, and little direct federal legislation can be invoked in pursuit of societal goals. If groundwater mining and related conservation issues were to be addressed by Congress, it is conceivable that eligibility for commodity (and other) program benefits could be tied to compliance with specified management practices designed to minimize groundwater mining.

Federal Water Subsidies

More than one-fifth of all the irrigated agricultural acreage in the United States is supplied with subsidized water provided by U.S. Bureau of Reclamation (USBR) projects. A wide variety of crops are grown with the federally subsidized water, including several supported by federal price and income support programs. Demand for forages and grains produced with federally subsidized water from USBR projects stems from dairy operations supported by the federal dairy program.

Federal water subsidies have long been controversial among persons concerned about the efficiency of government agricultural commodity programs. Discussions about allocative efficiency and equity inevitably turn to the role of USBR-served agriculture. Can water delivered through USBR projects be reallocated to alternative uses? Who owns the water and who should profit from water transfers? If transfers occur, should original cost-sharing subsidies be withdrawn? These and other important policy questions are raised and discussed in the debates over the appropriate allocations of water from federal projects.

Moore and McGuckin (1988) argue that federal policies coalesce to create three concerns: *consistency*, as farm programs offer incentives to limit acreage and production of program crops while the federal water program encourages expansion of irrigated acreage and increased production; *fairness*, as agricultural producers without USBR water must compete with recipients of subsidized water; and a *federal budget* issue, as expenditures on the two programs are additive. Farmers who now rely on reclamation water point out that most of the subsidies from federal water projects were capitalized into land values and captured by original land owners. Subsequent generations of farmers have not benefitted from the subsidies to the same degree.

In any case, there are agricultural irrigation development programs and commodity programs that continue to work at cross purposes (Cotner 1985). Farmers develop irrigation to increase yields, and then greater expenditures are required, through other programs, to reduce the surplus production and support commodity prices. In turn, the supported commodity prices have been used as justification for the profitability of irrigation development, overstating the benefits of such development. Such program circularity has been the subject of criticism in policy debates over patterns of water allocation.

Water Scarcity

In the western states water scarcity is a major policy issue. Irrigated agriculture frequently controls major portions of existing water supplies, creating potential conflicts when other highly valued economic, environmental, or social needs arise for the water.

Issues relating to competing demands for freshwater resources are generally the domain of state and local governments. States grant and enforce water rights, and the law differs from state to state. Growing competition for water used in agriculture has created pressures to increase water-use efficiency in agriculture, change water laws to foster reallocation of water to higher-valued uses, and increase federal cooperation with state and local governments in managing water resources. Reallocation issues become an item for federal government consideration when the water supplies in question are a part of a federal reclamation program, used to grow crops involved in federal commodity programs, or otherwise interact with other specific federal objectives, such as protection of endangered species.

In Congress, management of the federally financed Central Valley Project of California is under review (U.S. Department of Agriculture 1992, p 19). At issue is the amount of water to be diverted from present agricultural uses to maintain stream flows for fish, waterfowl, and wildlife. In the Columbia-Snake River Basin of the Pacific Northwest, the recent listing of several salmon subspecies as threatened or endangered has fueled an intense debate over environmental impacts of traditional river allocations to irrigation and hydropower. This has prompted federal, state, and public-interest groups to press for a reallocation of river flows an action that will involve a wide variety of federal, state, and local laws and institutions.

Present Policy

Conservation Provisions: 1985 and 1990 Food and Agricultural Legislation

In response to concerns about the environmental effectiveness of traditional conservation programs, Congress enacted four new conservation programs in the Food Security Act of 1985 and enhanced some in the Food, Agriculture, Conservation, and Trade Act of 1990. These were the conservation reserve, sodbuster, swampbuster, and conservation compliance programs. Each has potential impact on water use or water quality, though their main focus is on soil conservation.

Conservation Reserve. Under the Conservation Reserve Program (CRP), agricultural producers could bid to enroll land susceptible to erosion into the reserve for at least 10 years. The 1990 act added water quality improvement to the goals of the CRP (Ribaudo 1992, p 42). Farmers receive annual rental payments plus cost-sharing and technical assistance to establish vegetation that will effectively protect the enrolled acreages from further erosion. Prior to 1993, 36.5 million acres were placed in the program, but Congress has withheld new funding for additional enrollment.

Sodbuster and Conservation Compliance. Sodbuster provisions were designed to discourage farmers from bringing highly erodible land not farmed between 1981 and 1985 back into production. Producers who did so, with some exceptions, became ineligible for most major farm program benefits. Beginning in 1990, the sodbuster provisions were modified to apply to all highly erodible land, even if cultivated between 1981 and 1985. To become eligible, farmers were required to develop an approved conservation plan and have it fully implemented by 1995.

Swampbuster. Under the swampbuster provisions, a farmer who drains and then cultivates a wetland likewise becomes ineligible for farm program benefits. The Section 404 Dredge and Fill permit program, our only federal wetland regulation, is part of the Clean Water Act of 1977 administered jointly by EPA and the Corps of Engineers (Heimlich 1992a). However, numerous other informal programs and policies at state and federal levels attempt to influence farming practices and protect existing wetlands. The Water Bank program, authorized in 1970, makes annual per acre payments to landowners to protect wetlands from conversion (Heimlich 1991c).

Water Quality

Much of federal water pollution policy is based on the Water Pollution Control Act Amendments of 1972 (WPCA), and amended by the Clean Water Act of 1977 (CWA) and the Water Quality Act (WQA) of 1987. The WPCA directed states to identify NPS problems and encouraged them to implement appropriate controls. The CWA further emphasized the role of NPS controls and gave the USDA a role in providing technical and financial assistance for water quality management practices. These acts were largely focused on point source pollution, however, and not very successful in reducing NPS pollution (Malik, Larson, and Ribaudo 1992).

The CWA as amended in 1987 requires states to submit assessments of water pollution from nonpoint sources, including agricultural. Accountability and responsibility for these programs has not been fully registered at the state level, however, due to lack of an enforcement authority by EPA (Malik, Larson, and Ribaudo 1992); thus most state programs rely on voluntary programs of education, technical assistance, and cost sharing that minimize economic impacts on farmers. Failure of existing programs to control NPS pollution has prompted Congress and EPA to reassess the approach to NPS pollution problems in the current discussions concerning reauthorization of the CWA.

There are additional programs and legislative attempts largely directed at water quality issues. The Environmental Easement Program, created by the Food, Agriculture, Conservation, and Trade Act of 1990 and administered by the USDA, is designed to protect environmentally sensitive lands and reduce degradation of water quality by inducing producers to implement a natural resource conservation management plan in return for cost-sharing and compensation for any imposed loss in market value of the land.

The USDA established an interagency working group on water quality in 1989 to coordinate a "water quality initiative." To support these efforts, USDA has initiated Demonstration Projects and Hydrologic Unit Area Projects, as well as the Water Quality Incentives Program (Swader 1992). Landowners in Hydrologic Unit Areas receive technical, educational, and financial assistance to solve water quality problems connected with farming practices.

The Coastal Zone Act Reauthorization Amendments of 1990 required the 24 states and five territories with approved Coastal Zone Management Plans to develop strategies for protecting coastal water quality from NPS pollution, specifically including measures relating to agricultural sources (U.S. Environmental Protection Agency 1993). The coastal NPS program is administered jointly by the EPA and the National Oceanographic and Atmospheric Administration's Office of Ocean and Coastal Resources Management.

Policy Approaches and Tradeoffs

But as most water use is controlled by state laws and institutions, federal farm legislation cannot by itself be successful in solving problems of water conservation and optimal allocation. All agricultural production activity uses water and influences the quantity and quality of water available to others.

General Policy

If water allocation and conservation are issues of national concern, federal agricultural policy could constitute appropriate water policy. The programs authorized in the Food, Agriculture, Conservation, and Trade Act of 1990 affecting water quality and water conservation are currently small in scale, primarily due to limited appropriations. Expansion of these programs could help solve some of the more important problems of water resource management. Ribaudo (1992) identifies three categories of agricultural policy mechanisms to address water management objectives:

1. Make qualifications for all farm program benefits contingent upon taking certain steps to protect the environment. Existing compliance provisions include conservation compliance, sodbuster, and swampbuster provisions applied to USDA program participants.
2. Influence farmer behavior by altering price or income support programs to reduce incentives for production of specific crops. Crops and rotations that reduce erosion and chemical use should be encouraged.
3. Continue education, technical assistance, and financial incentives that encourage buffer strips, integrated pest management, soil nutrient testing, manure management, irrigation efficiency, conservation tillage, etc. These programs are not tied to commodity programs, but they provide additional measures with considerable flexibility in addressing water management problems.

Cotner (1985) described some issues of water policy linking long term agricultural capacity needs and short term agricultural stability and income objectives. His concerns centered on the importance of having a coordinated federal/state policy on irrigation development and groundwater use. Possible elements include:

1. Encourage states to establish water markets to facilitate allocation of water to the highest economic use.
2. Prohibit the use of newly developed water for those crops and products for which national programs exist to restrict their production.
3. Encourage states to establish water management programs to effectively utilize the water resource over time and in a manner consistent with regional and national needs.
4. Foster federal/state water programs and regulations that preclude or minimize an over-commitment of water use in agriculture. An approach similar to that of conservation compliance might impose standards of irrigation water use by farmers if they are to maintain eligibility for agricultural program benefits.

In this context, current-versus-future water use is an important national policy issue. Regions now dependent on a deteriorating groundwater resource will eventually face difficult adjustments to declining and inadequate water supplies. Government programs that contribute to groundwater mining, therefore, may encounter opposition from people who feel that long-term prospects for economic viability of a region require a more conservative approach to the withdrawal and use of groundwater supplies.

Water Use and Quality

Four basic instruments of public policy have been used to implement water quality objectives. These are: (1) regulation, (2) incentive payments and penalties, (3) acquisition, and (4) education and technical assistance. The essential purpose in any policy change is to alter those human actions that seem to cause problems. Any change has gainers and losers, costs and benefits.

Critics of the voluntary, incentive-based approaches of the 1985 and 1990 food and agricultural legislation to water quality issues argue that the programs have not been effective because of inadequate funding for the Water Quality Incentives Program (WQIP). A related criticism is that USDA has used the CRP instead of the WQIP as its primary instrument for pursuing water quality goals. Critics argue that it is neither possible or prudent to reduce agricultural sources of water pollution by idling all land associated with water quality problems, pointing out that farmers can substantially reduce pollution by more efficient management of water, nutrients, pesticides, and reducing sedimentation. A concern is that government price and income support or productivity enhancement programs may be incentives that intensify rather than reduce the detrimental water quality impacts of agriculture. For example, deficiency payments based on yield may encourage use of more chemicals, and acreage set-aside provisions can cause land to be left fallow and exposed to erosion.

As the Clean Water Act (CWA) is reauthorized and new legislation for food and agriculture is developed, possibilities for program coordination are suggested by Malik, Larson, and Ribaudo (1992):

1. Define appropriate agricultural sources, such as irrigation return flows and small feedlots, as point sources henceforth subject to the permitting requirements of the National Pollutant Discharge Elimination System, established in 1972 to combat point sources of pollution.
2. Strengthen those provisions of the CWA devoted to state planning for control of NPS pollution by linking state compliance with federal funding for the program. States will have a greater incentive to vigorously implement their plans if federal funding is available to support the program.
3. Mandate minimum enforceable best management practices (BMPs) rather than rely on voluntary measures of education and technical assistance. Failure to adopt minimum BMPs could be subject to civil or criminal sanctions, or as cause for disqualification from participation in federal farm income support programs.
4. Implement a deposit/refund program to assure safe disposal of pesticide containers.
5. Utilize market trading policies involving transferable emissions permits, similar to programs implemented through the CWA.
6. Impose taxes on fertilizer, pesticides, irrigation water, or other proxies for NPS loadings to create incentives for management of inputs that are at risk of becoming pollutants.

On the latter point, advocates of effluent fees or taxes on agricultural chemicals and fertilizers argue that such measures will create an incentive for farmers to use these products more efficiently. The tax revenues could be earmarked to help pay for research, education, technical assistance, and incentive programs to aide the adoption of pollution-reducing practices. Critics complain that taxes on fertilizers and pesticides would add unnecessarily to production costs of all farmers, including those whose operations do not now pollute the water.

In any case, fees on polluting inputs such as fertilizer, chemicals, or irrigation water can influence the level of their use. Fees on polluting outputs can cause producers to recognize pollution as a cost of doing business. The effectiveness of using such proxies depends upon how closely they are linked to the pollutant loadings.

The regulatory approach to pollution abatement requires controlling the right to use water in ways that diminish its utility to others. Design standards designate specific technology and management to assure that water discharges do not carry unacceptable levels of pollutants. Land use zoning and controls are traditionally used by local units of government to protect sensitive watersheds, shorelands, and groundwater recharge areas by regulating the kinds of activities that may occur on the overlying land.

Some proponents of regulation for pollution control argue that the nature of NPS pollution calls for a grassroots approach to land use management for pollution control. However, there are problems with this approach. First, a locality has little incentive to control NPS pollution if the pollution problems are borne primarily by residents outside the community. Second, a community may tolerate a pollution-causing activity to attract development or to avoid politically unpopular actions. Third, local governments may not have the financial, technical, or legal ability to investigate environmental damages and to design effective solutions. The case for a federal role rests primarily on a need for coordination, uniformity across state and regional boundaries, and the capacity to address trans-state pollution problems.

Acquisition of land ownership or easement rights that secure protection of water quality can help to assure performance in particularly sensitive areas. Education and technical assistance can also influence the behavior of farmers in a positive way. Sometimes farmers can even improve net farm income while moving to more environmentally sensitive practices.

Supporters of current voluntary, incentives-based approaches to agricultural soil and water conservation argue that current efforts will be effective in the long run and must be given a fair chance to succeed (Swader 1992). Farm technologies that are environmentally benign and economically feasible are still evolving. Further, it takes years of education, demonstration, and technical assistance to change the management-decision processes of thousands of independent farmers. To require adoption of specified management practices before those practices are technically proven and understood by farmers may impose additional costs without producing long-term water quality benefits. Supporters of a voluntary approach concede that recent assessments of water quality impacts demonstrate some reason for concern, but argue that the evidence does not justify rapid movement to more stringent and potentially less efficient approaches to the control of agricultural nonpoint sources of water pollution.

Conclusions

Future demands for water will reflect a wide range of factors such as population, technological development, economic growth, environmental standards, consumer preferences, and governmental policies (Anderson 1983). Issues of water supply, demand and quality always present alternatives and tradeoffs in the solution process. Selection of water management options should consider the economic factors, but decisions in the arena of public choice most ultimately be directed by ethics, conscience, and personal preferences that do not easily lend themselves to economic measurement. Such questions as "how clean is clean enough, who decides, who benefits/pays, and ultimately who owns the water," are the elements that will drive the public policy debates relating to water management and allocation.

The agriculture committees of Congress must face several fundamental questions concerning the nexus of agricultural policy and water policy. Should agricultural policy be proactive in pursuit of water policy objectives or should it simply remove the contraventions to water policy goals? Deciding the appropriate government agency for implementing water management programs involves fundamental issues of public policy, and must be addressed at both the federal and the state levels. For example, the Departments of Agriculture (Soil Conservation Service), Interior (Bureau of Reclamation), Defense (Army Corps of Engineers) and the Environmental Protection Agency have responsibilities for water resources programs. Each is characterized by its historical mission, staffing patterns, and relationships to political constituencies and committees of Congress. Each is uniquely designed to achieve specific types of policy objectives and cannot easily adapt to changes in societal goals.

Agricultural commodity programs to be considered in the new food and agricultural legislation will have relevance for the issues of national water policy. These programs, directly or indirectly, affect water quality, water use in irrigation, water available for alternative uses, and opportunities for reallocation. Policymakers will be challenged to insure that commodity programs compliment water policy goals. Commodity programs that control or influence price, production, and farm income are not, in themselves, good tools for achieving water policy goals. Where the two meet or interact, however, there will be calls for consistency of federal policy and for a minimum of conflict with state and regional goals.

References

Anderson, Terry L. 1983. *Water Crisis: Ending the Policy Drought*. Baltimore, MD: Johns Hopkins University Press.

Batie, Sandra S. 1988. "Agriculture as the Problem: New Agendas and New Opportunities." *Southern Journal of Agricultural Economics* 20:1:1-11.

Clark, E. H., II; J. A. Haverkamp, and W. Chapman. 1985. *Eroding Soils: the Off-Farm Impacts*. Washington, D. C.: The Conservation Foundation.

Cotner, M. L. 1985. Irrigation Development and Farm Programs—A Catch 22. Working Paper, ERS, USDA.

Heimlich, R. E. 1991a. "Wetlands and Agriculture: New Relationships." *Forum for Applied Research and Public Policy* (Spring).

______. (ed.) 1991b. A National Policy of "No Net Loss" of Wetlands: What do Agricultural Economists Have to Contribute? Staff Report No. AGES 9149, Resources and Technology Division, Economic Research Service, USDA.

______. 1991c. Wetlands and Agriculture: Recent Federal Policy and Emerging Issues. Paper presented to The Southern Extension Public Affairs Committee, Clearwater, Florida.

______. 1992. Wetlands Policies in the Clean Water Act. Paper presented to National Symposium on New Directions in Clean Water Policy, University Council on Water Resources, Charlottesville, Virginia.

Libby, Lawrence W., and William G. Boggess. 1990. "Agriculture and Water Quality: Where Are We and Why?" Pp 9-37 in John V. Braden and Stephen B. Lovejoy (eds). *Agriculture and Water Quality: International Perspectives*. Boulder, CO: Lynne Rienner Publishers.

Malik, A. S., B. A. Larson, and M. Ribaudo. 1992. Agricultural Nonpoint Source Pollution and Economic Incentive Policies in the Reauthorization of the Clean Water Act. Staff Report No. AGES 9229, Resources and Technology Division, Economic Research Service, USDA.

Moore, M. R., and C. A. McGuckin. 1988. "Program Crop Production and Federal Irrigation Water." *Agricultural Resources: Cropland, Water, and Conservation Situation and Outlook Report*. AR-12. Washington, DC: Economic Research Service, USDA.

Nielsen, E. G., and L. K. Lee. 1987. *The Magnitude and Costs of Groundwater Contamination From Agricultural Chemicals: A National Perspective*. AER Report No. 576, Washington, D.C.: Economic Research Service, USDA.

Oh, S. 1991. "Managing Nitrate Groundwater Pollution From Irrigated Agriculture: An Economic Analysis." Ph.D. Dissertation, Department of Agricultural Economics. Pullman, WA: Washington State University.

Ribaudo, M. O. 1992. "Options for Agricultural Nonpoint-Source Pollution Control". *Journal of Soil and Water Conservation* 47(1).

Swader, Fred. 1992. "The USDA Water Quality Initiative and Public Policy." Paper presented at the 47th Annual Meeting of the Soil and Water Conservation Society, Baltimore, MD.

U.S. Department of Agriculture. 1992. *Agricultural Resources: Cropland, Water, and Conservation: Situation and Outlook Report*. AR-27. Washington, D.C.: Economic Research Service.

U.S. Department of Commerce. 1982. *1978 Census of Agriculture (Vol. 4: Irrigation)*. Washington, D.C.: Bureau of the Census. U.S. Govt. Printing Office.

U.S. Department of Interior. 1990. *National Water Summary 1987—Hydrologic Events and Water Supply and Use.* Water Supply Paper 2350. Washington, D.C.: U.S. Geological Survey.

U.S. Environmental Protection Agency. 1990. *National Survey of Pesticides in Drinking Water Wells*. EPA570/9-9-015. Washington, D. C.

______. 1993. *Guidance Specifying Management Measures For Sources of Nonpoint Pollution In Coastal Waters*. 840-B-92-002. Washington, D. C.: Office of Water.

6

Land Use Issues

Michael R. Dicks and C. Tim Osborn

Influencing land use has been a principle focus of agricultural policy from the time of the first settlers in America. During the first 150 years following independence, public policy promoted agriculture as the primary engine for economic growth and development. Growth was accomplished by encouraging the privatization of newly acquired public lands by anyone willing to farm them. Following the closing of the American frontier in the late 1800s, agricultural land use policy shifted to incentives that encouraged farmers to improve their land and increase the amount of productive cropland through such means as irrigation, wetland drainage, and land clearing. By the 1920s, due to the success of these efforts, gains in the productive capacity of U.S. agriculture had outpaced increases in demand, leading to periods of large commodity surpluses and low prices. Against this backdrop, the Agricultural Adjustment Act of 1933 contained new policy to control output, primarily through the idling of cropland. The 1933 Act, with the demand enhancement and federal crop storage additions of the 1950s, are the foundation for our Nation's current agricultural policy.

The 1980s was a decade of transition for agricultural land use policy. Empowered by a growing awareness of environmental problems associated with agricultural production, environmental organizations became an active voice in the agricultural policy debate. At their insistence, reducing the adverse effects of agriculture on natural resources and the environment achieved priority status in agricultural policy, along with retention of price stabilization and farm income support. Land use provisions in prior agricultural acts were used mainly as a preventative measure to maintain a portion of total cropland with the "potential to produce" and prevent surplus production. Today the focus, with respect to land use, is not only on adjusting the number of acres devoted to crop production, but also on how the acres devoted to production are managed and the resulting effects on the environment, i.e., the focus has shifted from production control to production management.

Scrutinization of the performance of the new multiple-objective federal land use policies has continued. In fact, federal agricultural policy has become an increasingly favored target in recent years, accused of causing the farm crisis of the 1980s, crop surpluses, low commodity prices, loss of global competitiveness, environmental degradation, and disappearance of the family farm. That such policy can come under such varied attack is indicative of its complexity.

The purpose of this chapter is to examine current land use issues and their association with federal agricultural policy objectives. A brief historical overview is provided to illustrate the importance of land use policies within the agricultural policy setting, followed by discussion of current land use concerns. The final section discusses the effectiveness, efficiency, and economic consequences of attempting to solve current agricultural problems using various agricultural policies that attempt to influence land use.

Historical Overview

The expansion of agricultural output through increased cropland area was the major focus of federal agricultural policy during America's first 150 years. Numerous policies were put in place to provide incentives to develop croplands from forestlands, woodlands, marshes, and prairies. The Land Ordinances of 1785 and 1787 were the first national land use policies; they were developed to encourage expansion of the agricultural sector. While vestiges of the farmland expansion period survived into the 1980s (e.g., tax incentives for wetland draining and reclamation of arid western lands through water subsidies), the Homestead Act of 1862, and the land runs of the 1890s were the last major federal policies providing incentives to increase farmland.

By 1890, the Census announced the end of the American frontier (Nash 1968). This termination of farmland expansion, coupled with a continued need for growth in agricultural output, gave birth to the 20th century emphasis on increasing the amount of cropland on farms (shifting from range, pasture, and forest to cropland) and increasing agricultural productivity. Land reclamation legislation subsidized irrigated agriculture in 1902, making conversion of marginal lands such as the open plains, deserts, and some woodlands to cropland economical. The federal government also moved to provide additional sources of credit (Federal

Farm Loan Act of 1916) and skilled labor (Smith-Hughes Vocational Education Act of 1917) to assist a booming agricultural sector (1914-20).

The encroachment on pristine areas by urban sprawl, crop farming, heavy grazing, mining, and manufacturing during the cropland expansion era gave rise to a conservation movement for the protection of resources and primitive areas in the United States (Petulla 1977, Hays 1959). The Forest Reserve Act of 1891 provided the first federal protection for specific areas. However, the first tangible efforts to control land use were initiated in 1905 with the transfer of the "forest reserves" to USDA, and their subsequent renaming as "national forests." Over the next three decades, Congress considered re-establishing forest and grasslands for both conservation and supply control purposes.

The great depression and severe drought in the 1930s brought forth a significant shift in agricultural policy, from output expansion to stabilization. Again land use policies were used as one of the main instruments in stabilization. In 1933, $25 million was transferred from the Public Works Administration to the Federal Emergency Relief Administration for the purchase of submarginal lands. These submarginal lands were actually removed from production in 1937 under the Bankhead-Jones Farm Tenancy Act because they could not be economically farmed. The land purchased under this Act later formed the National Grasslands in the Great Plains.

During the 1930s several agricultural acts, including the first comprehensive and first permanent agricultural legislation, were written. The Agricultural Adjustment Act of 1933 established the generic price support and acreage reduction mechanisms used today. The Soil Conservation and Domestic Allotment Act in 1936 was legislated to pay farmers for land use improvements and adoption of conservation practices (Agricultural Conservation Program). The Agricultural Adjustment Act of 1938 (the first permanent agricultural law) was to "provide for the conservation of natural resources and to provide an adequate and balanced flow of agricultural commodities in interstate and foreign commerce and for other purposes" (Cochran and Ryan 1976).

The depressed farm economy of the 1920s and 1930s was followed by the agricultural boom of the 1940s. World War II triggered a second American agricultural revolution, bringing major changes in land use, farm policies, agricultural production, farm management, and farm life. The rapid changes in technology during this period were driven by tremendous growth in export demand. Surplus production all but disappeared. Even before the United States became directly involved in the war, high support prices were provided and guaranteed for a period of 2 years after the end of hostilities (Rasmussen 1985).

The Agricultural Act of 1949 came as the export demand generated by WWII began to evaporate, and set the stage for agricultural policy into the 1980s. This 1949 Act made permanent several stabilization tools, including government stocks, crop loans, demand enhancement (food aid), and support of basic commodities.

The Agricultural Act of 1956 introduced set-aside and land diversion programs (Soil Bank) to "protect and increase farm income, to protect the national soil, water and forest and wildlife resources from waste and depletion, . . . and to provide for the conservation of such resources and an adequate, balanced, and orderly flow of such agricultural commodities in interstate and foreign commerce" (Cochran and Ryan 1976). The main components of the 1956 Act were the Acreage Reserve (ARP) and the Conservation Reserve (SBR) programs. The ARP was used in 1957 and 1958 and set aside 21 million acres while the SBR occurred from 1957 to 1972 and idled nearly 29 million acres at the peak enrollment in the early 1960s. The Great Plains Conservation Program (GPCP), also created by the 1956 Act, was designed to stimulate a permanent shift in land use for the ten Great Plains states. Farmers and ranchers were to put land poorly suited for cultivation back into permanent grass. Nearly 60,000 contracts, covering more than 110 million acres, have been involved in the GPCP.

By the 1960s, crop surpluses and storage had reached a level such that the cost of storing an additional unit of grain would have been greater than the value of the commodity itself (Paarlberg 1980). Trade-offs between supply management tools, stocks, and land retirement demonstrated the cost effectiveness of land use policies. An emergency set-aside program (Emergency Feed Grain Act of 1961) provided farmers a small income supplement to set aside not less than 20 nor more than 50 percent of their cropland acreage.

The Cropland Adjustment Program (CAP) of the Food and Agricultural Act of 1965 again attempted to cut increasing stock levels by providing 5- to 10- year contracts to producers willing to convert cropland into conserving uses. Some 14 million acres were idled by 1967.

By 1973, the export demand seen during the war years for U.S. farm products returned, due to world crop shortages. Once again commodity surpluses appeared to have vanished and great concern for our ability to produce enough food to meet a growing world demand led to the call to plant "fence row to fence row." The Agriculture and Consumer Protection Act of 1973 emphasized production to respond to growing world wide demand for food and fiber and permitted substantial changes in the ways programs were implemented (Rasmussen 1985). Cropland use in the 1970s was encouraged and land use policies provided farmers with considerable planting flexibility.

By 1982, stocks of the basic commodities had again reached excessive levels. A Payment-In-Kind (PIK) program implemented by USDA in 1983, offered government-owned surplus commodities in exchange for agreements to reduce production by cutting crop acreage. Government outlays on commodity programs jumped from an annual average $1 to 2 billion to more than $12 billion, farm prices continued to decline, surplus production continued to accumulate, and land use policies increasingly constrained farmers' ability to plant the crops of their choice and maintain access to commodity program benefits.

The Food Security Act of 1985 was written during a period of increasing environmental concerns, excessive stocks, farm financial stress, and enormous and growing government outlays. In response, target prices (deficiency payments) were lowered, program yields were frozen, numerous land retirement programs were legislated [e.g., Acreage Reduction Program, Paid Land Diversion, 50/92 (later changed to 0/92 for wheat and feed grains), multiyear set-asides, and the Conservation Reserve Program] and several new land use policies were legislated (conservation compliance, swampbuster, and sodbuster).

The Food, Agriculture, Conservation and Trade Act of 1990 continued the change in focus begun in the Food Security Act of 1985, with new reductions to income support and with increased restrictions on land use to obtain environmental benefits.

Agricultural Land Use Issues for 1995

The litany of agricultural acts, programs, and provisions presented above represents the major components of agricultural land use policy, but is by no means comprehensive. However, it does provide a clear picture of how important land use policies (policies that influence, regulate, or restrict land use options) have been in attempting to manage agricultural production capacity in order to meet agricultural policy objectives. The future use and importance of land use policies will be conditioned by an ever-evolving policy environment. That environment likely will be characterized by the following points:

1. *The agricultural land use/conservation agenda has broadened.* For 50 years conservation land use policies focused mainly on minimizing soil loss as a conservation goal. The conservation title in the 1985 act expanded the conservation goals to include the preservation of wetlands. The conservation title of the 1990 act was expanded to include the minimization of the adverse impacts of agricultural land use on water (ground and surface) quality. This trend will likely continue in the next

agricultural legislation and expand to "total resource management" to include concerns about the effects of agricultural land use on water quality, wildlife habitat and biodiversity, and sustainability.

2. *A continuing trend toward greater free market orientation in agriculture.* Most other industries have enjoyed more liberalized trade since WWII. Up to the present time, agriculture is one of the few industries where a General Agreement on Tariffs and Trade (GATT) has not been negotiated. Continued budget and political pressures will increase the certainty of a GATT agreement in agriculture in the future. The movement toward greater free market orientation implies, for farmers, a decreasing attractiveness of commodity programs and a diminished leverage for compliance provisions such as sodbuster, swampbuster, and conservation compliance. If a true free trade environment is established, so-called "green payments"—payments for environmental performance—may be among the few forms of agricultural income support allowable.

3. *Few new policy tools will be available for influencing agricultural land use.* Traditional tools include education, technical assistance, cost-sharing (ACP), short-term voluntary land retirement (annual set-aside, paid land diversion, 0-50/92), and medium-term voluntary land retirement (CRP). Newer tools include direct payments not linked to cost sharing (e.g., Water Quality Incentives Program), compliance mechanisms (conservation compliance, sodbuster, and swampbuster), and permanent easements (e.g., Wetlands Reserve Program). Although many farmers view the current compliance policies as regulatory, no true regulatory approaches have been implemented to date. This may not hold true in the future, however.

4. *Concern regarding use of CRP acreage after expiration of present contracts.* This concern arises from farm and from environmental groups. Because of the potential production increase from land released from the CRP, consideration of supply control objectives (maintaining farm incomes and minimizing federal outlays) will once again become more explicit and may once again outweigh conservation goals.

5. *Dissatisfaction with compliance provisions established by the 1985 act.* Conservation groups claim that USDA has been lax in enforcing conservation compliance, sodbuster, and swampbuster, thus reducing their effectiveness. Farmers feel strapped by inflexible and often impractical restrictions on land use. Both groups are piqued by the lack of uniformity, between counties, in conservation compliance plans.

6. *Budget reductions continue to be felt first in conservation programs.* Several 1990 act programs (e.g., Water Quality Incentive Program, Environmental Easement Program) have received either partial or zero appropriations. Fiscal year 1993 appropriations for new enrollment under

the CRP and the Wetlands Reserve Program were cut by the House-Senate conference.

7. *Increasing emphasis on a new approach to agricultural conservation and land use problems.* Budget pressures and perceived ineffectiveness of some of the current conservation programs are forcing consideration of regulatory policies rather than incentives. The possibility of obtaining the goals of the conservation title by adding new provisions to the Clean Water Act is a current example.

Development of the next omnibus food and agricultural legislation will include considerable debate over a number of issues relating to use of agricultural lands, including not only what land should be used but how that land should be used. For purposes of discussion, these issues can be categorized, somewhat arbitrarily, into two areas: those dealing with the future of existing programs and policies, and those dealing with broader concerns about how agricultural land use will affect other resources. The most obvious example of the former is concern over what, if anything, the government should do in response to the expiration of Conservation Reserve Program contracts. These contracts will terminate beginning in 1996. An example of the latter is the approaches to be considered for reducing contamination of ground and surface waters by agricultural chemicals.

Existing Program Issues

Conservation Reserve Program (CRP)

Under the voluntary CRP established by Food Security Act of 1985, USDA has established contracts with farmers to retire highly erodible or environmentally sensitive cropland from agricultural production for 10 to 15 years. Through the 12 sign-up periods, approximately 36 million acres were enrolled in the program. Upon retirement from crop production, CRP land is to be converted to grass, trees, wildlife cover, or other conservation uses. This shift in land use has provided society with numerous environmental benefits including improvement of surface water quality, creation of wildlife habitat, preservation of soil productivity, protection of groundwater quality, and reduction of offsite wind erosion damages. The program has also assisted farmers by providing them with a dependable source of income from their retired land and has reduced commodity surpluses.

At the end of the CRP contract period, annual rental payments will be no longer made by USDA and CRP farmers will no longer be under obligation to keep CRP acres in grass or trees. Depending upon the level of commodity prices and commodity program provisions, a significant amount of CRP land will return to crop production. Even though most of the land will be subject to conservation compliance if returned to crop production, erosion will increase and many of the environmental benefits (e.g., wildlife habitat) provided by the CRP program will probably end.

Options for dealing with expiring CRP contracts range from doing nothing to extending contracts for varying lengths of time to placing certain CRP lands under permanent conservation easements. Complicating the decision about what to do is the fact that Congress is under increasing pressure to reduce the federal budget deficit and the CRP is a costly program. Currently, the government spends nearly $1.8 billion annually in CRP rental payments. Although up to 50 percent of this cost may be offset by reduced commodity program deficiency payments, the potential cost of extending contracts or establishing permanent easements on all CRP land makes such options unlikely. For budgetary reasons, a more feasible approach may be to target programs to those CRP acres that provide the greatest benefits (environmental, supply control, etc) at the lowest possible costs.

Compliance Provisions (Conservation Compliance, Sodbuster, Swampbuster)

According to the Soil Conservation Service (SCS), as of mid-1993, conservation compliance plans had been fully implemented on approximately 61 percent of the 141 million acres with plans. As the January 1, 1995 deadline for full implementation of conservation compliance plans approaches, the response of the remaining farmers to either implement plans or forego commodity program benefits will become more apparent.

In addition to meeting the 1995 deadline, conservation compliance also requires that farmers of highly erodible land "actively apply" their plans, making progress in the interim. To check for noncompliance, SCS annually conducts status reviews on a random 5 percent of plans. Producers not meeting their plan on schedule risk loosing commodity program benefits. In 1991, SCS conducted status reviews on 71,000 conservation plans. Because of differing time frames and procedures, state SCS offices identified more instances of unsatisfactory or noncompliance than did the local SCS field offices. The State reviews estimated that the rate of farmers not actively applying their conservation plans ranged from 3.6 to 4.6 percent.

While SCS status reviews show low rates of noncompliance, estimates by other agencies appear to challenge these results. Studies by USDA's Office of Inspector General (OIG) and the Soil and Water Conservation

Society (SWCS) question USDA's enforcement of Conservation Compliance and Sodbuster programs. In 1990, OIG concluded that in four of ten SCS status reviews conducted in an Iowa county, the approved Conservation Compliance plans had not been applied, but the producers had not been found to be out of compliance. In 1990 and 1991, the SWCS reviewed conservation compliance plans and progress on a 123-farm sample and concluded that only 54 percent of the farms were in compliance. Only one violation had been reported by SCS.

Beyond the question of USDA enforcement of conservation compliance is the larger issue of whether compliance mechanisms are viable long-term policy tools. Continuing pressures to reduce federal budget deficits together with the continued movement towards free trade will lead to reduced federal commodity program benefits. This reduction in commodity program benefits will reduce the bite in the current compliance mechanisms. Further, the compliance mechanism is only cyclically effective. Commodity program participation varies with the health of the agricultural economy, rising when prices are depressed and falling when prices rise. Commodity program participation is greatest when market incentives to farm marginal lands or convert new lands are weakest (returns are lowest), providing the strongest incentive for compliance. On the other hand, farmers are more likely to expand acreage and farm marginal lands when prices are high, at which point program participation is less attractive and the incentive to comply is weak.

Environmental Assistance Programs

The 1990 act contains numerous programs and provisions designed to target financial and technical resources towards specific environmental objectives. These environmentally targeted programs were designed to directly influence the use of farm resources (including land). The Water Quality Incentives Program, Wetlands Reserve Program, Integrated Farm Management Program, Forest Stewardship Program, and extension of the Agricultural Conservation Program and Great Plains Conservation Program are the best known of the new land use programs targeted to environmental objectives. The basic objective of these programs is to provide farmers with the resources required to change to production practices thought to be more environmentally benign. Funding for many of these programs is relatively small and often uncertain. Targeted are areas most likely to have environmental problems associated with agricultural production.

The effectiveness of these programs in reducing agriculture's contribution to environmental degradation is rarely known or understood. However, a major benefit of the programs may simply be that they allow farmers to try new production practices without the increased risk normally associated with the adoption of new technologies. The debate leading up to legislation for 1995 will very likely include vigorous arguments for keeping and for eliminating these programs.

Broader Issues

Several additional issues, including global warming, water quality, wildlife and biodiversity, and private property rights, may also affect land use and be affected by land use. These issues may well be included as objectives and/or constraints in the production management policies of the next food and agricultural legislation.

Global Warming

The potential causes and impacts of global warming will continue to be debated in the future. Already, various policies and programs, such as those designed to promote tree planting (to increase carbon sequestering capacity) or partial conversion of petroleum to ethanol, are supported for their ability to reduce the trend towards global warming. However, U.S. programs to promote carbon sequestration such as tree planting on croplands may reduce commodity output and increase prices, providing our foreign competitors with an incentive to expand crop acreage (at the expense of forests) increasing the threat of global warming. Clearly, land use policies in the future will be influenced by new information about the causes and impacts of global warming.

Water Quality

Nonpoint source pollution will be the focus of environmental groups well into the next century. The Federal Water Pollution Control Act Amendments of 1972 and the Federal Insecticide, Fungicide, and Rodenticide Act Amendments of 1987 both focus on agrochemicals as a primary factor in degrading water quality and the human environment. Production agriculture and agrochemicals will continue to be a target of these and other major pieces of legislation. However, the scope of the water quality problem will continue to expand and include other potential farm problems such as contamination from sewage and dumpsites. Rural/urban conflicts will likely increase along with the increasing costs of water purification. Many large cities, seeking to reduce purification costs, have already begun to provide incentives or new regulations for upstream farmers to reduce runoff and/or chemical use.

Wildlife and Biodiversity

Efforts to increase wildlife populations have led to considerable conflict between various land users. Grassland and forestland leases currently favor the agricultural land user. While the current movement is to increase user fees on land leases, the leases may be put on the bidding block in the future. With the expected increases in the number of outdoor enthusiasts, new pressures to reallocate land for recreation will occur.

Property Rights

The ability to change farmer behavior through legislation will be limited in the short run by the current structure of property rights. However, persistent attempts to obtain environmental objectives with land use may change the structure of property rights. Property rights are defined in terms of the initial ownership (allocation) of rights, and rules under which they may be exchanged (Braden 1982). Agriculture has always enjoyed fee simple ownership of land and conditional access to entitlements.

The rules under which property rights may be exchanged can be summarized as three options: (1) property rules (2) liability rules, and (3) inalienability (Calabresi and Melamed 1972). With property rules, consent to an exchange of rights must be given in advance by all parties at an acceptable price, like market transactions. Under liability rules, prior consent to an exchange is not required, and prices are set by an objective third party. With inalienability, the exchange of rights is disallowed under some or all circumstances. These three delineations provide a reference point in the continuum of property rules, with most property rules being more easily defined as closest to one of the three.

A very subtle shift in the property rights continuum occurred in the conservation title of the 1985 act. Until then, exchange of property rights in agriculture was conducted under traditional property rules. Voluntary soil conservation programs provided cost sharing for farmers to reduce soil erosion. Thus, the right of the farmer to lose soil was inferred and the price of exchange was negotiated in terms of cost share and technical assistance. The conservation title of the 1985 act added further conditions on access to commodity program benefits which may be seen as a movement towards the use of liability rules rather than property rules to minimize environmental disturbances. Congress placed specific requirements on the management of highly erodible lands and wetlands. The penalty for failing to comply with these management requirements is the loss of all commodity program benefits. A further shift in the rules of exchange occurred under the 1990 act. The wetlands provision was changed to revoke commodity program benefits if wetlands are altered, rather than when an agricultural commodity is produced on altered wetlands.

Under the recent application of section 404 of the Water Quality Act of 1965, the Army Corp of Engineers assumed responsibility for agricultural wetlands, enabling the federal government to control any alterations or pollution in wetlands, thereby invoking inalienability as the rule of exchange.

The new conditions placed on access to commodity program benefits may be seen as a shift in the rules governing the exchange of land use rights from property rules towards liability rules for most agricultural property (e.g., land, soil). For others (e.g., wetlands) the shift may be far more significant, from property rules to inalienability.

Economic Consequences of Land Use Policies

A unique characteristic of land use policies is the geographical distribution of their impacts. The more targeted the land use policy the more likely the impacts will be concentrated in specific areas. Consider for instance, 4 land use policies implemented in 1987 under the 1985 act [Acreage Reduction Program (ARP), Paid Land Diversion Program (PLD), 0-50/92, and the Conservation Reserve Program (CRP)]. Together these programs idled more than 76 million acres in 1987. The geographical distribution of benefits for participating in these programs were considerably different from the geographical distribution of costs associated with the idling of cropland (Dicks and McCormick 1993). Thus, while the programs reimburse producers (directly or indirectly) for their foregone net returns, there is no guarantee of equitable distribution of these benefits (providing payments in proportion to foregone returns). Further, rural communities receive no direct reimbursement for the decline in economic activity associated with the reduced agricultural production. Too often rural development goals are attached to raising net farm income. However, total farm expenditures and volume of production are the stimuli for rural economic activity. Net farm income may be raised while total expenditures and volume of output are reduced. Thus, increasing net farm income may not stimulate rural economies.

Another often overlooked problem when considering land use policies to achieve agricultural and environmental objectives is the process by which these policies are implemented. Land use policies are most frequently implemented independently. That is, the agency responsible for implementing each policy will assign a unique group of employees to develop and carry out the implementation strategy of that policy. The combined and/or cross effect of several land use policies is rarely analyzed. The combined impact of all land retirement programs on non-targeted groups (e.g., rural communities) may be considerably larger and more adverse than that of any of the programs individually. Further, these impacts will increase with the increased dependence of the regional economy on agriculture. For example, the CRP legislation limited enrollment to 25 percent of a county's cropland acreage because of the potential adverse impacts expected with greater levels of enrollment on the economy of local communities. Some waivers were provided when the local community indicated that increased enrollment would produce minimal adverse impacts. However, the combination of the ARP, PLD, 0-50/92 and CRP idled more than 60 percent of total cropland in some counties. In areas where agriculture was a predominant industry, employment and income in the economy was reduced by as much as 20 percent (Hyberg, et al. 1991).

Multiple-objective policies force implementing agencies to consider the trade-offs between conflicting objectives such as environmental enhancement, supply management, and rural development. The difficulty with using multiple-objective policies comes from unweighted and/or unconsidered objectives. That is, a land use policy with an objective of controlling water pollution may also affect productivity and, in turn, economic activity. In this case, if the objective of rural development is not explicitly considered in determining an optimal implementation strategy, the land use policy could adversely impact rural communities. Further, even if these potential impacts are considered, but weights to indicate the relative importance of each objective are not provided in the legislation, the implementing agency may fail to consider the impacts on rural communities.

An important constraint to achieving agricultural and environmental objectives through land use policies is the ability of public and private institutions that provide the technical assistance necessary for land use changes to occur. For instance, legislating water quality requirements and then requiring that the Soil Conservation Service (SCS) ensure that these requirements are met is unlikely to produce the desired consequences without providing SCS with additional resources, retooling its current staff and/or hiring new staff trained in water quality management. SCS was created to reduce soil erosion and assist farmers and ranchers in maintaining and increasing the productivity of their land. Not until the 1980s was this agency assigned the role of managing farmland for water quality. There is no basis for assuming that the agency is able to provide the assistance necessary to *efficiently, effectively and equitably* obtain desired water quality levels.

The success of land use policies in obtaining the desired changes will always be constrained by the capabilities of the implementing and supporting institutions. Further, achieving the objectives of conservation compliance, sustainability, water quality, biodiversity, or averting global warming through land use policies will be difficult unless alternative land uses are available and proven. Too often, the perception of land use changes that should take place differs widely from land use changes that can take place.

The Conservation Compliance program provides an example of the vast difference between expectation and reality. This provision originally sought to influence farmers with highly erodible cropland to modify their production systems so as to hold erosion to less than the natural rate of soil regeneration. However, before county-level implementation, the rules were changed to influence farmers to use best available practices to reduce their current levels of soil erosion. The initial rules were modified because many farmers believed they could not meet the goals because of financial hardship, lack of available technology, or physical constraints.

The inability of programs that rely on incentives to meet their objectives is often used to support the call for regulation. Many of the current farm programs as well as nonagricultural programs that affect agriculture (e.g., Water Quality Act of 1965, Federal Insecticide, Fungicide and Rodenticide Act of 1947) rely on incentives to influence land use decisions. However, because the perceived benefits of these programs are sometimes slow to come, arguments to move to regulatory programs frequently arise.

Land Use Policy Trade-offs

Land use policies have played a central role in the agricultural policy agenda since our Nation's beginning. These policies began as simple single-objective instruments designed to influence the level of agricultural output. The primary role of land use policies developed in the 1930s was to stabilize prices and support farm income. Today land use policies are being used to achieve a broader range of agricultural, environmental, and development objectives. Policies have become more complex, with multiple objectives, multiple targets, and a combination of incentives and compliance provisions. Because a change in land use in a single location may affect both specific (e.g., productivity) and broader (e.g., global warming) concerns, greater effort must be placed on understanding the trade-offs between and amongst policy objectives and the array of concerns.

Should federal agricultural policy continue to retain more resources (e.g., land, capital) in agriculture than those required to meet the current levels of food and fiber demand? To the extent that these "reserve" resources may be needed in the future to compensate for unknown events such as adverse weather (either here or abroad), the maintenance of reserve resources will continue to provide for food security. How much reserve production capacity should be maintained continues to be one of the major questions that must be answered to establish omnibus food and agricultural legislation. While considerable budget savings and environmental benefits can accrue with reduced land use, the rate of response to exogenous events that adversely affect agricultural output is restrained, perhaps leading to higher consumer prices during the adjustment period. Stocks, on the other hand, afford instant adjustment to annual production shortfalls and food aid can be used to trim stocks to a manageable level. Thus, each of the policy tools designed to manage surplus production potential is necessary, the trick is to find the optimal level of each, faced with uncertainty about supply and demand each year.

Once the desired level and optimal allocation amongst alternative forms is established, an optimal allocation of the surplus production potential to be maintained under land use policies must be determined. Land use policy tools of the past have taken the form of annual land retirement (e.g., Acreage Reduction Program, Paid Land Diversion, 0-50/92), intermediate term land retirement (e.g., Conservation Reserve, Water Bank, Wetland Reserve) and long term land retirement (e.g., Conservation Easements). These programs take land out of production and thus hold excess capacity as the potential to produce rather than as surplus production. The longer land is idled, the less responsive potential supply will be to higher prices. However, longer term land retirement programs have a greater potential for obtaining environmental benefits.

While land use policies are normally thought of as land retirement, they could just as easily allow unlimited use of land but constrain the use of other inputs (e.g., fertilizer, pesticides) and perform the same level of control of excess capacity. Such an approach could provide considerable environmental benefits and enable quick supply response in the face of higher prices from increasing demand or short supply. Land use policies have also been used to enforce minimum acceptable standards for various management options such as tillage, conservation practices, or crop rotations (e.g., conservation compliance).

Because the land use policies now contain multiple objectives, it is important for Congress to explicitly stipulate the objectives and rank them in terms of priority, or, more preferably, assign a relative weight for each. To simply specify that a program should attempt to reduce the adverse affects of agriculture on the environment and control supply is insufficient to guarantee a successful program.

Implementing agencies can interpret objectives of a general directive as minimizing soil erosion, reducing surface water contamination, guaranteeing the pristine quality of groundwater, increasing biodiversity, reducing global warming, any number of other environmental goals, or none of these goals. Stating each specific goal and indicating its relative importance will assist the implementing agency in developing a program to match the impacts expected by Congress.

References

Braden, J. B. 1982. Some Emerging Rights in Agricultural Land. *American Journal Agricultural Economics* 64(1):19-27.

Calabresi, G., and A. D. Melamed. 1972. Property rules, liability rules, and inalienability: One view of the Cathedral. *Harvard Law Review* 85:1089-1128.

Cochran, Willard W., and Mary E. Ryan. 1976. *American Farm Policy 1948-1973*. Minneapolis, MN: University of Minnesota Press.

Dicks, Michael R., and Ian McCormick. 1993. Distributional Impacts of Federal Land Retirement Programs. Oklahoma Agricultural Experiment Station Working Paper AEP-111.

Hays, S. P. 1959. *Conservation and the Gospel of Efficiency*. Cambridge, MA: Harvard University Press. P 297.

Hyberg, Bengt T., Michael R. Dicks, and Thomas Hebert. 1991. The Economic Impacts of the Conservation Reserve Program on Rural Economies. *The Review of Regional Studies* 21:91-105.

Nash, R. 1968. *The American Environment: Readings in the History of Conservation*. Reading, MA: Addison-Wesley Publishing Company. P 236.

Paarlberg, Donald. 1980. *Farm and Food Policy*. Lincoln, NE: University of Nebraska Press.

Petulla, J. M. 1977. *American Environmental History: The Exploitation and Conservation of Natural Resources*. San Francisco, CA: Boy and Fraser Publishing Company. P 399.

Rasmussen, Wayne D. 1985. Historical Overview of U.S. Agricultural Policies and Programs. *Agricultural-Food Policy Review: Commodity Program Perspectives*. USDA Economic Research Service. Agricultural Economic Report No. 530. Pp 3-8.

7

Sustainable Agriculture

Glenn A. Helmers and Dana L. Hoag

Overview

Public interest in sustainable agriculture (SA) was expressed as long ago as 1580 (Gates 1988). Yet, it was not until 1985, through the passage of the Food Security Act, that SA was adopted into American food and agricultural policy. Subtitle C of Title XIV called for research and education to promote the development and adoption of low-input farming systems through low-input sustainable agriculture (LISA). The social commitment to SA was reaffirmed in the Food, Agriculture, Conservation, and Trade Act of 1990, and will likely lead to more policy changes in the anticipated 1995 food and agricultural legislation.

According to the 1990 act, SA is:

> An integrated system of plant and animal production practices having a site specific application that will, over the long term: satisfy human food and fiber needs; enhance environmental quality and the natural resource base upon which the agricultural economy depends; make the most efficient use of nonrenewable resources and on-farm resources and integrate, where appropriate, natural biological cycles and controls; sustain the economic viability of farm operation; and enhance the quality of life for farmers and society as a whole.

There are many definitions of SA, but none are very specific. The more specific definitions support a particular agenda. They may focus on the resources (e.g., soil), the people (e.g., small farmers), or the institutions (e.g., family farms) that should be protected, the inputs that should be used (e.g., nonchemical), the production methods that should be used (e.g., conservation tillage), or a combination of the above. One explanation for the difficulty in reaching a concise definition is that people feel more certain about what is *not* sustainable, i.e., "conventional agriculture," than what *is* sustainable. The public has become increasingly concerned about the environmental impacts of agriculture, the decline of family farming, a troubled rural economic community, preserving resources for future generations, and over-reliance on off-farm resources.

Perhaps the most intuitive, but still not very specific, definition is that future generations should have access to an equal or greater quality and quantity of goods and services than we have today (Dicks 1991, Victor 1992). This definition expresses the desire to leave future generations at least as well off as those now here, but does not provide guidance as to what future generations will consider equally well off. It is much easier to agree to *be* sustainable than it is to *achieve* sustainability. As a practical matter, trade-offs in sustainable objectives must be considered in order to determine if progress has been made. For example, is a soil conservation system that utilizes more chemicals more sustainable than a conventional tillage system?

The attention to SA has brought about questions concerning the role government has played in supporting economic, social, and environmental sustainability. The government establishes the institutional setting under which markets operate, which arguably has the greatest influence on sustainability. For example, property rights determine if environmental impacts are properly accounted for and if current consumption is imposing too great an expense on future generations. We turn our attention more narrowly to the impact of direct agricultural supports.

Government supports to agriculture are provided in three major forms: financial support (or penalties), research, and education. Many benefits have been derived from these programs, such as nearly tripling agricultural production since World War II. However, some people question if these benefits have cost too much by changing farm structure, fostering technologies that pollute and threaten human health, and by failing to accomplish, at reasonable cost, what they were intended to do. The purpose of this chapter is to examine how major farm policies may have affected SA in the past and how recent policy changes may alter those impacts in the future.

Policy Affecting Sustainable Agriculture

Federal, state, and local agencies, farmers, industry, and other participants in the agricultural sector would likely argue that they have always maintained the goals outlined in the Congressional definition of SA. However, policymakers and researchers, frequently in response to economic circumstances, have led farming technology in a direction that, some argue, has sacrificed certain parts of the definition, such as the environment, for short-run profits. The SA programs in the 1985 and 1990 food and agricultural legislation are part of an effort to "level the playing field" between conventional high-tech agricultural systems and systems that rely more on biological principles and better management skills. Practitioners of sustainable agriculture have had moderate success at convincing policymakers to tilt the playing field toward the SA systems.

The 1985 and 1990 food and agricultural legislation addressed the perceived inequity between conventional and SA systems in four major ways by: (1) stepping up research and education programs in SA, (2) changing certain commodity program rules on crop flexibility that may have discouraged SA systems in favor of monocultural farming and high input systems, (3) providing financial incentives for selected SA practices, and (4) regulatory action.

Research and Education

The 1985 act initiated funding for SA research and education in Subtitle C of Title XIV. However, it was not until 1988 that Congress appropriated any funds. The 1990 act was much more specific about research and education. Now called sustainable agriculture instead of LISA, the act divided research and education into three chapters under Subtitle B, Title XVI. These titles are (Brown 1992, Bird and Hubbard 1992):

Chapter 1: Best Utilization of Biological Applications
Chapter 2: Integrated Management Systems
Chapter 3: Technology Development and Transfer

By the end of 1992 only the first of these had received any funding. Chapter 1 is designed to "initiate projects that conduct mission-oriented research, obtain data, develop conclusions, demonstrate technologies, and implement educational activities that" (Bird and Hubbard 1992 p 4):

1. Reduce to the extent feasible and practicable, the use of chemical pesticides, fertilizers, and toxic natural materials in agricultural production.
2. Improve low-input farm/ranch management to enhance agricultural productivity, profitability, and competitiveness
3. Promote crop, livestock, and enterprise diversification.

Chapter 1 calls for joint projects, between farmers and ranchers and researchers and others, that will facilitate the goals stated above through exchange of information and experience, and through research. A program labeled Sustainable Agriculture Research and Education (SARE) has come from this Chapter. SARE activities are the focus of a later section of this chapter.

Chapter 2 calls for research and education in Integrated Crop Management and Integrated Resource Management. Up to $20 million was authorized, but thus far no funds have been appropriated.

Chapter 3 focuses on training. This chapter calls for technical guides and handbooks, national and regional training centers, and a system of state coordinators and specialists. No part of the 20 million dollars authorized for Chapter 3 has yet been appropriated. Provisions of this Chapter require that all Extension Service agents complete a training program by November 1995, and that after 1993, all new Extension agents must complete training within 18 months following employment. The act is unclear about how training is to be done, especially in view of no appropriations. In addition, mandatory training in the Extension Service is controversial since some assert that Extension clientele would be better served if Extension workers were left free to respond to clients' needs without a national mandate (Hoag and Pasour 1992).

Commodity Program Rule Changes

Another way that the 1990 act has affected SA is through changes in the commodity program provisions. Many provisions will have a direct or indirect effect on cropping systems that utilize sustainable agriculture concepts. The soil conservation provisions implemented in the 1985 act were strengthened and included new provisions for increased flexibility to rotate crops, cost-sharing for SA-type practices, and increased accountability for chemical use.

A number of studies have noted that commodity programs tilt the economic balance of agriculture toward monocultural farming (Cochrane and Runge 1992, Faeth et al. 1991, National Academy of Science 1989, Young and Goldstein 1987). Incentives to practice monocultural farming as opposed to the use of crop rotations are asserted to occur because of target prices for program crops, rigid program crop bases, and idling of cropland. This is suspected of increasing inputs used and associated environmental problems. Although empirical evidence on these issues is ambiguous, the 1990 act introduced increased flexibility into the commodity programs for food and feed grains, and for cotton. There are a variety of reasons for introducing more flexibility, but one hope is that producers can rotate program crops with legumes, grasses, and other crops to take greater advantage of "natural" pest controls and natural sources of fertilizer, thus reversing or offsetting incentives for monocultures.

Two major changes provide more "flexibility." The "Normal" flex acreage program reduces program payments on 15 percent of crop acreage, freeing the grower to produce almost any crop on that land. With the "Optional" flex acreage program of 10 percent a producer can increase flexibility to one-quarter of base program acreage.

Financial Incentives

The 1985 and 1990 food and agricultural legislation also provided new direct financial incentives for farmers to utilize SA systems. The Integrated Farm Management Program (IFMP) pays farmers to further reduce the fraction of program crops grown in order to produce resource-conserving crops (i.e. legumes and grasses). This program combines flexibility with cost-sharing to promote rotations with resource-conserving crops. Two other major cost-share programs were the Integrated Crop Management Program (ICMP) and the Water Quality Incentives Program (WQIP). These programs provide cost-sharing and other incentives for implementing "best-management" practices in a farm plan.

Regulations

For the most part, regulatory actions against farmers are carried out by state agencies or by federal agencies other than USDA. Nevertheless, the 1990 act did introduce some regulatory action aimed at SA. This act tightened up accountability in pesticide use by requiring pesticide record keeping for restricted-use pesticides. This requirement will provide more accurate and site-specific information about pesticide use and encourage producers to follow application directions more carefully. An organic certification program was also authorized. Consumers may benefit if they can get products grown with reduced, or no, chemicals, and growers can benefit if they receive higher prices or save on input costs.

Non-Federal Sustainable Agriculture Efforts

In addition to federal efforts to encourage SA, many state governments and national and regional nongovernmental organizations have instituted SA programs. No fewer than 224 sustainable agriculture programs have emerged across the country (Cuchetto et al. 1992). These include national programs such as the Alternative Farming Systems Information Center, the Center for Holistic Resource Management, and the Sustainable Agriculture Coalition. Universities, state agencies, industries, and private foundations and institutes have responded to the call for SA with cost-sharing, education, and research programs that may compare to federal efforts.

The Issues

The sustainability movement is a reaction to increasing understanding about the impact of human activity on the environment and concern for the socio-economic impacts of current production systems on producers and the people around them. The concept of sustainability has become a major force in world politics, to the point of being the focus of the well publicized Earth Summit, the United Nations Conference on Environment and Development, in Rio de Janeiro in 1992 (Goklany and Sprague 1992). In agriculture, there is concern that the sector relies too heavily on technology that minimizes human activity and the use of biological solutions to produce food and fiber. The reputed result is environmental pollution and a breakdown of the farm family and associated rural sector.

Research has increasingly identified ways in which farming and ranching activities have led to environmental degradation. Linked to the degradation is the intensification of agriculture by replacing labor and management with technology. Farmers contribute only 5 percent of the value of agricultural production compared to the contributions of the marketing and input sectors (Smith 1992). They rely increasingly on technology to become more efficient which leads to increased farm size and fewer farmers. One possible result is that such specialization leads to more pollution and a heavy impact on the rural farm sector. However, an alternative possibility is that fewer farms, even though they are more intensive, pollute less because they are more efficient and spread over fewer acres than would be required with less-intensive systems (Goklany and Sprague 1992, Tweeten 1991).

Some observers have noted that the research and education system, primarily USDA and the Land Grant Colleges and Universities, have a bias toward narrowly focused, high input technology over broader systems that utilize more management and labor (Schaller 1991, Smith 1992, National Academy of Science 1989). For example, chemical means to control pests are given more attention than biological controls such as crop rotations. Therefore, these studies note, research and education has either strengthened or created incentives for farmers to move to high-tech agriculture and thus led the agricultural production sector away from sustainability. Many dispute this claim, pointing to the many gains agriculture has made in sustainable technologies including soil conservation, nutrient management, and pesticide management (Tweeten 1991).

Congress passed Title XVI, subtitle B, in the 1990 act to encourage more research and education about sustainable systems. Research was to be encouraged through increased funding in the Sustainable Agriculture Research and Education (SARE) program. Education would come through SARE and a mandatory training program for Extension personnel.

Between 1988 and 1991, 1059 proposals were submitted to the SARE program requesting $91,093,406 (Bird 1992). Project funds were distributed between four regions, the Southern, Western, North Central, and Northeast. A total of 164 projects were approved for a total expenditure of $16,704,000, and $1,000,000 was contributed by the U.S. Environmental Protection Agency's Agriculture in Concert with the Environment (ACE) program. About 17 percent of the money went into education and demonstrations. Whole-farm research projects were given 31 percent of the funds and component research received 42 percent. The remaining 10 percent went to providing information and impact assessment.

Critics argue that the SARE money has been too little since only about 18 percent of proposals were funded. Nevertheless, the grant money provided has generated several new research and education programs, as well as national initiatives. Administrators have successfully organized the program to encourage diverse proposals. These projects have been spread across the country, over several disciplines, and have involved farmers, university researchers, state and federal agencies, and private groups. An abbreviated description of program results can be found in SARE's 1992 Annual Report (Bird 1992). These include innovations in livestock-crop systems, dryland cropping systems, vegetable production, low-input fruit systems, soil fertility management, weed control, water quality and global change, information management, and integrated pest management. In addition, economic studies are underway and several publications and videos have been developed.

A major hurdle in developing a successful research and education program in SA is its definition. The intentionally vague definition has advantages in promoting unity for a potentially controversial subject. Virtually everyone agrees that systems that are more profitable and better for the environment should be chosen when available. However, it is when trade-offs occur—which is the majority of the time—that the definition fails to convey guidance in developing a credible research and education program (Tweeten 1991). To some, sustainability means evaluating environmental and economic trade-offs; however, since the definition is vague and everyone assigns different weights to environmental and economic outcomes, some think that SA means the elimination of chemicals and other high-input technologies. The lack of agreement on specific identification of goals and objectives makes it impossible to determine what is being accomplished by SA and even where SA is being practiced.

The difficulty accompanying the definition is immediately obvious when addressing the situation of farm size. There is no doubt that new technology, primarily from USDA and Land Grant Colleges and Universities, has led to increases in farm sizes and reductions in farm numbers. This is unfortunate for those individuals who wish to enter farming, and to the rural communities dependent upon agricultural activities. However, the payoff is that farmers, in the past 40 years, have more than doubled production even though using fewer acres. For those farmers that remain, farm income has risen from 60 percent of nonfarm income in 1950 to 100 percent and above in the 1980's (Carlson 1992). In addition, real farm prices have declined by 50 percent. Which is preferred, the positive attributes of the rapid improvements in technology that have occurred, or the positive outcomes that we might still have without such rapid growth in technology?

Most likely, there is middle ground where certain labor- or management-saving technologies can be reduced or eliminated to create more value added by the farmer. The SARE and ACE programs will help bring more of these technologies to farmers; however, there will continue to be forces that limit the overall success of these programs. These include greater marginal returns to centralized research than to decentralized on-farm research, more flexibility with technological solutions, and the extent to which farmers avoid paying the costs of their pollution.

Contrary to some opinions, research and educational programs of universities and other public agencies have a rich history in SA systems. Some of the differences between conventional and SA systems are listed in Table 7.1. There is *nothing* in this listing of sustainable farming systems that has not already been the subject of major research. However, emphasis has not been placed on promoting the sustainable systems over the conventional systems. Economic forces, not biases against sustainable systems, have dominated research priorities, and these have led to the systems now currently used by farmers. Researchers put their efforts into those activities from which to achieve the greatest gain. By centralizing their efforts, they can concentrate on technologies (such as nitrogen fertilization) that benefit large groups of farmers at one time. Yield enhancements of equal magnitude might have been achieved by site-specific prescriptions of a biological technology, such as legumes in rotation. However, the optimal applications of this approach can show great variation from site to site. Therefore research cost, per unit of gain, is greater when compared to the above-described fertilization approach.

Gathering information is expensive. Therefore, a centralized system of research will show an inherent advantage over a decentralized system that is specific to individual farms unless the added costs of the customized information are exceeded by the gain. For example, agronomists have shown that nitrogen applications could be reduced by as much as 50 percent without reducing yield (Pesek 1991). However, this does not necessarily imply that farmers apply too much nitrogen. There is a degree of uncertainty when a farmer applies fertilizer. Babcock (1992) showed that nitrogen rates are 36 percent higher in corn when farmers include uncertainty in their decisions to fertilize. The fact that farmers apply the higher rates suggests that they think that the cost of determining the optimal rate exceeds the costs of "overfertilizing." "Systems research" in the current institutional structure is accomplished to some degree by the users of information themselves. Farmers and ranchers pick and choose results and integrate them into their own farming systems, making adjustments to account for their own individual circumstances.

TABLE 7.1 Conventional Versus Sustainable Crop Farming Systems.

Element of Farming System	*Conventional Farming*	*Sustainable Agriculture*
Crop rotation	Corn-corn-corn or corn-soybeans-corn	Multiple crops, emphasis on legumes and winter cover
Tillage practices	Minimum-till, no-till, or moldboard plowing	Mechanical cultivation and ridge till when possible
Fertilization program	Chemical fertilizers (NPK)	Legumes in rotation, organic N, less chemical N P & K use, soil structure regeneration
Pest control	Routine application of chemical pesticides	Low-input biological controls, mechanical cultivation, crop rotation, disease-resistant cultivars, scouting for pests and beneficials; applying chemical pesticides as a last resort

Source: Cooperative State Research Service 1991.

Reform of the current research and education system may not bring about a better representation of sustainable systems. More research money in programs like SARE and ACE will attract researchers into looking harder for sustainable systems, but solutions will likely always be limited by the current economic conditions. Most researchers and farmers conclude that sustainable systems *on average* are less profitable than conventional systems (Daberkow and Reichelderfer 1988, Lee 1992, Hoag and Pasour 1992). Let it be noted, however, that many sustainable practices such as conservation tillage or integrated pest management have proven profitable and were readily adopted. The U.S. General Accounting Office (1990 p 36) concluded that studies "on the economic performance of alternative agriculture, . . . are few, methodologically limited, and enterprise-specific." They could not conclude, based on the literature available, that sustainable systems were either more or less profitable than conventional systems but did conclude that farmers felt the systems were less profitable and that they "will naturally be reluctant to change existing production practices without convincing information regarding profitability" (U.S. General Accounting Office 1990 p 39).

Potential for reduced profits are not the only barrier. Labor- and management-saving technologies often associated with conventional systems may be preferred as they are easier and more adaptable to changing environmental or economic conditions. For example, relying on a crop rotation for pest management makes it difficult to respond to changing crop prices. Based on a farmer survey, The U.S. General Accounting Office (1990) found that farmers believe adopting alternative agriculture practices requires greater management skills and causes greater weed problems. A recent North Carolina study found that SA farmers have less farm experience, rely less on Extension agents for information, and have lower farm incomes. As Molly Anderson (1990), the author of the study, points out, if farmers' "production choices are rooted in values that are not shared by most farmers, making more technical information available or removing policy barriers will not be sufficient to effect widespread changes in farming practices."

Finally, farmers currently pay very little of their production costs that occur off their farm, such as soil erosion into lakes and streams. Many of the sustainable systems would be more attractive if off-site and long-run costs were made part of the decision-making set through government intervention or better assignment of property rights. However, government intervention may not be effective if goals are not clearly defined. For example, the existence of environmental damages from intensive, high-tech, agriculture might lead to efforts to make agriculture less intensive. Agricultural "extensification" would reduce pollution and has other desirable qualities. Nevertheless, it may be at cross purposes with environmental goals in some situations. Reduced land area required for high-tech systems may partially or fully offset per-unit increases in pollution from intensification. Gokany and Sprague (1992) calculated that four times more land would be required to produce the same quantity of food produced today if technology were held to 1910 levels. In addition, there may be lower per-unit costs for pollution-prevention technology in larger operations. Large confined animal operations often have better resources to prevent animal wastes from entering into streams than smaller operations. In addition, sulfony urea herbicides and pyrethroid insecticides developed by the agrichemical industries have dramatically reduced risk since they are used at very low rates, decay rapidly, and have low toxicity.

Commodity Program Impacts on Sustainable Agriculture

It has been suggested that a number of incentives and disincentives to sustainable agriculture reside in the traditional commodity programs. These include target prices, program bases and deficiency payments, set-aside requirements, the Conservation Reserve Program, and program yields. In the discussion that follows, we examine these observations and comment on conditions that validate or discredit them. Our observations are drawn from completed studies, perspectives of policy analysts, and the application of economic principles.

There are two ways in which commodity programs impact sustainable agriculture practices. The first is the impact of the program (or particular provisions of the program) on program-crop acreage. This impact is important because program crops tend to be high users of fertilizer and chemicals. The second is the resulting intensity issue affecting per acre application levels of fertilizer and chemicals. This impact is influenced by a resource substitution phenomenon. The issue of input substitution and low input agriculture has been discussed by Daberkow and Reichelderfer (1988). Together the acreage and intensity impacts determine the aggregate impact of a change in policy on fertilizer and chemical use.

Target Prices

Target prices function as the effective price received for program crops (on base acreage) and therefore appear to send alternative, and higher, price signals to growers than the market. Many unintentional and undesirable effects have been argued as having resulted from these price distortions. First it can be suggested that higher prices from commodity and price support programs raise overall crop production and input use. Output is increased by growers bringing back into production land that, with market prices, would be unprofitable. Further when fragile land is drawn into production, the potential for soil erosion increases significantly. Second, it has been argued that target prices, available only for specific commodities, encourage acreage and input use on those crops and those crops tend to be monoculture, high chemical-using crops. Further it is argued that, with "supported" crops, soil erosion increases due to increased intensity of production.

One solution to maintain the desirable income-enhancing elements of the programs, yet eliminate their undesirable input and output distortions, is to "decouple" income maintenance from provisions that influence crop choice. In the broadest sense, decoupling would provide an income support that is independent of crop acreage or production. Therefore, an income support of this nature would eliminate the above-described undesirable impacts.

The concept of total decoupling was very unpopular in the negotiations for the 1990 food and agricultural legislation. However, decoupling has been occurring slowly over the past several years. Target prices, for example, have already been largely decoupled from production decisions. Deficiency payments are received only on the effective base acres and on farms with program yields already established. Effective base acres can be increased only during years in which a farm is not participating in the program. Program yields have been frozen since 1986. Thus, for program participants, acreage decisions are determined predominantly by basic economic forces (enterprise cost and returns and any complementary relationships among enterprises) and program mandates (set-aside and base acreage requirements). Changes in target prices (ignoring the optional flex) have been shown to not affect cropping decisions of program participants (Helmers et al. 1993a).

While program participation involves set-aside and base acreage requirements, the direct effect from changes in target prices (deficiency payments) on acreage decisions accrue from participation decisions. Increases in the target price may lead to new entrants into the program but may also induce current participants to drop out, build base, and later re-enter the programs (Helmers et al. 1993b). Reductions in target prices may induce some participants to exit the programs. Nevertheless, the exit and entry costs are sufficiently high that changes in the target price do not translate directly into an equivalent price signal to farmers.

The indirect effect of changing or eliminating target prices is difficult to trace because, if not replaced by another incentive to retain participants, reduced or eliminated target prices result in less (or none if target prices are eliminated) program participation. Greater nonparticipation could result in increased acreages of program crops and decreased acreages of nonprogram crops. Thus, program crop prices tend to fall while nonprogram crop prices tend to rise. Assuming a reduction in target prices prompted the first-round impact of reduced program participation, increased (reduced) deficiency payments resulting from this decreased (increased) program crop price act as a second-round incentive to increase (decrease) program participation until a new equilibrium is reached.

The impact of the target price mechanism on marginal lands cannot be ignored. Marginal lands revert to noncrop use when target prices decline, unless replaced by a different incentive to encourage farmers to participate. This offsets to some degree the opposite impact on crop acreage caused by participants on nonmarginal land who no longer remain in the program.

The effect of target prices on input intensity appears to be negligible. This is because program production is sold at market prices, and program yields on farms can no longer be increased above the county yield average. Thus, a decision to intensity the use of fertilizers or pesticides is based upon the return from the program crop at the market price rather than at the target price. It is unknown, of course, if future program crop yields will be updated, and if in anticipation of this, farmers will increase input intensity.

The implication of the preceding discussion is that for program participants target price levels do not impact significantly what is produced on their farms or the resources (land, labor, machinery, and other purchased inputs) used in producing it. Target prices are only one part of the mechanism used to encourage program participation, and only in that more indirect context do they impact selection of crops grown and intensity of fertilizer and chemical use. Target prices currently are not indexed for inflation, and thus decline in a real sense under conditions of inflation. With the passage of time, other things equal, program participation will decline because the real deficiency payment diminishes.

The result of this is that target prices are often blamed for influencing cropping decisions when, in fact, the level of target prices only impacts program participation decisions; base acreage and set-aside mandates affect cropping choice. Perhaps many view target prices as equivalent to price supports that obviously do have the capacity to impact crop acreage decisions unless restrained by acreage base.

Analyses of the impact of program incentives on SA systems include a wide range of studies (Dobbs et al. 1988, Duffy and Chase undated, Faeth et al. 1991, General Accounting Office 1990, National Academy of Science 1989, Young and Painter 1990). These studies differ in their assumptions regarding program base, causality of target prices, and what constitutes a nonprogram environment. For some studies different cropping systems are compared with respect to profitability including a deficiency payment allocated to the program crop assuming a program base of 100 percent of a farm's cropland. In other studies program base is explicitly varied in proportion to the program crop in each cropping system with the result that cropping systems with greater proportions of the program crop fare better economically. Finally, for some studies cropping systems are compared under current program participation versus a setting in which no programs exist. In the latter studies disposition of former set aside land is crucial to assessing the impact of programs on program crop acres.

In these studies cropping systems defined as sustainable include a lower proportion of program crops, and therefore fare poorly economically, compared to conventional systems under current program participation. However, a comparable analysis using market prices usually (but not always) alters the relative economic advantage of cropping systems with high proportions of program crops because the target price benefit is removed. For this reason, it is sometimes suggested that target prices are the cause of the "distortion;" implying that if target prices were reduced, cropping systems defined as sustainable would be more competitive. However, once a farmer has established base, the target price does not affect cropping decisions unless he or she exits the program in order to make an adjustment in base acreage. Therefore, it is primarily the (indirect) influence of participation choice that target prices affect desired cropping system choices.

Program Bases and Deficiency Payments

In past commodity programs, eligibility for deficiency payments required producers to maintain program crop acreage at a base level, determined on the basis of historical crop plantings. Further, the deficiency payment was directly related to the level of the acreage base. In the 1990 act, a 15-percent mandatory (called "Normal") "flex" provision reduced payment acreage by 15 percent but allowed producers to plant any crop that they wished on this acreage. A second option, of 10 percent, was also made available. These flex provisions are a form of decoupling program payments from acreage requirements. From a sustainable agriculture standpoint, a flex provision of this nature is desirable if producers are being required to grow the base acres of program crops when alternative crops more appropriate to SA farming would be profitable.

In addition, two other program features are now available directed at past rigidities of program base requirements. These programs have been suggested as consistent with efforts to encourage sustainable agricultural practices. The first is the Zero Certification program, it permits producers to plant their crop acreage base to another crop without sacrificing base. Deficiency payments are not available under this program which makes it less attractive than the 0-92 alternative which still provides deficiency payments. The second program is the Integrated Farm Management Program (IFMP). Here resource conserving crops can be planted on 20 percent or more of the program crop base without loosing deficiency payments.

It has been argued that base-acreage preservation acts to encourage farmers to plant more program crops than they would under relaxed or eliminated base-acreage minimums (Cochrane and Runge 1992 p 70). Carlson (1992) says, however, that program participants are already rotating crops, suggesting that bases are not the dominant instrument in acreage-rotation decisions.

The direct impact on crop rotation of elimination of program acreage requirements depends upon the program base acreage of a farm and strength of the economic forces favoring joint production versus specialization on program or nonprogram crops. Assuming beneficial complementary relationships between program and nonprogram crops and no strong overriding profitability of either, farms with low bases tend not to flex their program acreage while farms with high program bases tend to flex (normal flex provision). Where no beneficial yield or cost interactions exist among crops, and nonprogram crops are more profitable than pro-gram crops, program crops will be grown at minimum base-acreage requirements. When program crops are more profitable, farmers can be expected to grow them at maximum levels. Elimination of base-acreage requirements would not be expected to significantly change acreage levels, but program crop acreage would increase slightly.

The linkage of the size of the farm base to the level of deficiency payments sometimes becomes confused as a target price issue. For example, suppose a certain cropping system becomes identified as an ideal system because it involves a higher percentage of nonprogram crops (particularly forages) than do conventional cropping systems. Also, assume the program crop is a higher-return crop (not including a "target price allocation" to that crop) than other rotation crops. For farms with a high program base, this ideal rotation can be grown only by sacrificing base to the level that matches the needs of the system, while loosing very sizable deficiency payments anticipated when using a conventional system. For farms with low bases, the sustainable rotation may be the rotation selected under participation, but the deficiency payment may be so low that remaining out of the program is preferable. Raising the target price in both cases results in the farm with a smaller base participating when it did not before (using the sustainable rotation) while for the farm with a larger base, a higher target price increases the sacrifice for not participating (growing the sustainable rotation) compared to participating.

In recent years questions have arisen regarding overall impacts of restricting fertilizer and chemical use, either totally (Knutson et al. 1990) or moderately (Helmers et al. 1990). One perspective of this is related to fertilizer usage. Given a fertilizer response function in which, over the range of very high application rates, there is a considerable degree of "flatness," it could well be asked how much is truly sacrificed by farmers as a result of partial curtailment of fertilizer use? Stated similarly, how much would target prices need to be increased to compensate farmers for cutting back on fertilizer applications? In a Nebraska setting, it was found that such target price increases would be only minor (Helmers et al. 1992).

The requirements of participation is the key to achieving a desired outcome. The current program has base-acreage requirements coupled to program payments. Were historical bases decoupled from base acreages and payments based on cropland, fewer differences in economic pressures toward participation versus nonparticipation among farms with different bases would be observed. Further, were the use of particular cropping systems made a condition for participation, the full removal of base rigidities would lead to fewer distortions in achieving participation among farms.

Set-Aside Acreage

Participation in the set-aside program requires producers to retire a proportion of their program base. This is done annually and the retired acreage is not required to be rotated. The impact of increases or decreases (including elimination) in set-aside levels on sustainable agricultural practices can be questioned. Two perspectives can be noted. The first is how annual changes in set-aside requirements affect sustainable practices. The second, and more-important, perspective is the effect of the overall institution of set-aside acres on sustainable practices. Annually changed set-aside requirements and the overall institution of the set-aside program involve several economic forces, some of which are indirect and not intuitively obvious. Part of the complexity of this issue occurs because set-aside land is a resource and a substitute for other resources, particularly fertilizers and chemicals.

At the firm level, other things equal, reducing set-aside requirements releases previously retired acreage to be used in program crops. Reduced set-aside requirements also indirectly impact program and nonprogram acreages due to increases in program participation. Together these direct and indirect effects, along with subsequent second-round impacts, generate a complex set of equilibria-related forces. Such impacts occur whenever set-aside requirements are changed.

A more systematic effect of the set-aside program is the long-run removal of an expected level of crop acreage from production, abstracting from year-to-year changes. In the long run, farmers cannot plan on the level of yearly set-aside; because these are unknown. This uncertainty, plus program cropping requirements on set-aside acreage reduces the possibility of using set-aside acreage directly in a long-term rotation. It can be suggested that set-aside can be useful as an "establishment year" for forage crops or as a green manure crop. It appears that little of this occurs. Perhaps, in addition to the above-mentioned uncertainty issue is whether set-aside land is significantly lower in quality than other cropland with less to be gained from rotation treatment.

A more subtle impact on chemical use in agriculture from the set-aside program lies in a substitution relationship. Here the issue is substitution of nonland inputs (particularly fertilizer and chemicals) to counter removal of cropland from production. The aggregate impact on fertilizer and chemical use caused by set-aside requirements depends upon the nature of substitution between land and other inputs, particularly fertilizer and chemicals, and the "slippage" (retired land may be less productive than nonretired land) involved. The net effect from a land set-aside provision involves both a substitution and an output effect. Often, it is perceived that a reduction in crop acreages simply reduces levels of "accompanying" inputs, but this view of the complementarity of inputs has a weak theoretical foundation. Recognizing that land retirement may significantly increase the intensity of use of fertilizer and agricultural chemicals needs greater emphasis in agricultural policy development.

Conservation Reserve Program

Retirement of large cropland acreages into the CRP has been justified for supply control, as well as for removing erosion-prone land from cultivation. First, intuition suggests the removal of cropland from production might be viewed as consistent with the general objectives of sustainable agriculture, in that the demand for fertilizer and chemicals might be expected to decrease along with reducing soil erosion and water runoff. However, as discussed in the last section, as cropland is retired, a proportional decrease in demand for fertilizer and chemicals is not likely to materialize if land and chemical inputs are substitutes for retired land.

Because of the size of the CRP, its impact on fertilizer and chemical use on remaining cropland needs serious consideration.

Program Yields

A criticism of past commodity programs is that producers could qualify for higher deficiency payments by securing higher proven yields, thus encouraging the use of fertilizer and chemicals to achieve those higher yields. Hence, since 1986, program yields on farms have been held constant. This would appear to eliminate the incentive to increase historical program crop yields. Unresolved, however, is the issue if and how program yields will be changed at some future date. Some farmers may perceive that yield increases will be rewarded by higher program yields at some future date. Therefore, program yields may not have been entirely decoupled from production decisions.

Conclusion

There is considerable need for a greater focus on the trade-offs involved under sustainable agriculture. Enough attention has not been directed to the varied consequences of widespread participation by farmers in practices which can be classified as sustainable. Sustainable agriculture is characterized as leading to (1) improved returns and reduced risk for farm operators, (2) increased long-run efficiency of agricultural production, (3) improved health of farm operators, (4) improved quality of soil resources and surface and ground water, (5) reduced dependence on "outside" inputs, (6) a stronger family-farm structure, (7) revitalized rural communities, and (8) improved food quality. Ignoring, rather than addressing, possible trade-offs frustrates the policy process in relation to sustainable agriculture.

There is likewise a need to identify the impacts resulting from differing policy provisions in terms of effectiveness in influencing sustainable agriculture. The types and magnitudes of the above-described trade-offs that exist when differing policy provisions are implemented need to be identified. Consensus about the types of benefits that arise from commodity program provisions—such as set-aside, CRP, target prices, acreage bases, and program-acreage flexibility—is rare. Other than acreage-flexibility alternatives and the "locking in" of program yields for farms, commodity programs have not been particularly directed to sustainable agriculture. Commodity programs can be designed to dramatically influence the use of sustainable agriculture practices. Should society wish to encourage a greater participation in what is perceived as sustainable agriculture practices, greater attention will need to be directed to commodity program evaluation.

References

Anderson, M. 1990. "Farming With Reduced Synthetic Chemicals in North Carolina." *American Journal of Alternative Agriculture* 5:60-68.

Babcock, B. 1992. "The Effects of Uncertainty on Optimal Nitrogen Applications." *Review of Agricultural Economics* (forthcoming, Vol. 14(2)).

Bird, G. 1992. *Sustainable Agriculture Research and Education Program: 1992 Annual Report*. Washington, D.C: Cooperative State Research Service Report No. 92-1.

Bird, G., and D. Hubbard. 1992. *Chapter 1 Guidelines: Sustainable Agriculture Research and Education Program, Best Utilization of Biological Applications*. Washington, D.C: Cooperative Extension Service, USDA.

Brown, W. 1992. "Sustainable Agriculture Research and Education Program, Southern Region." In Larry Johnson (ed). *Sustainable Agriculture: Enhancing the Environmental Quality of the Tennessee Valley Region Through Alternative Farming Practices*. Knoxville, TN: University of Tennessee at Knoxville, Department of Agricultural Economics and Resource Development.

Carlson, G. 1992. *Production Economics: Into the 21st Century*. J. W. Fanning Memorial Lecture. Athens, GA: University of Georgia, College of Agriculture.

Cochrane, W. W. and C. F. Runge. 1992. "Reforming Farm Policy—Toward a National Agenda." Ames, IA: Iowa State University Press.

Cooperative State Research Service. 1991. *The Basic Principles of Sustainable Agriculture*. A Policy Briefing Book, GA5534. Washington, D.C: SARE Program Director.

Cuchetto, A., A. Mansourov, C. Mizera, B. Petrucci, and L. Wettstead (eds). 1992. *Sustainable Agriculture Program Directory, 1992*. Washington, D.C: American Farmland Trust.

Daberkow, S., and K. Reichelderfer. 1988. "Low-Input Agriculture: Trends, Goals and Prospects for Input Use." *American Journal of Agricultural Economics* 70:1159-1166.

Dicks, M. 1991. "What Will Be Required to Guarantee the Sustainability of U.S. Agriculture in the 21st Century?" *Journal of Alternative Agriculture* 6:191-195.

Dobbs, T. L., M. G. Leddy, and J. D. Smolik. 1988. "Factors Influencing the Economic Potential for Alternative Farming Systems: Case Analyses in South Dakota." *Journal of Alternative Agriculture* 3:26-34.

Duffy, M., and C. Chase. Undated. "Impacts of the 1985 Food Security Act on Crop Rotations and Fertilizer Use." Staff Paper 213. Ames, IA: Iowa State University, Department of Economics.

Faeth, P. R., R. Repetto, K. Kroll, Q. Dai, and G. Helmers. 1991. *Paying the Farm Bill: Vs. Agricultural Policy and the Transition to Sustainable Agriculture*. Washington, D.C: World Resources Institute.

Gates, J. Potter. 1988. *Tracing the Evolution of Organic/Sustainable Agriculture: A Selected and Annotated Bibliography.* Bibliographies and Literature of Agriculture, No. 72. Beltsville, MD: USDA National Agricultural Library.

Goklany, I., and M. Sprague. 1992. "Sustaining Development and Biodiversity: Productivity, Efficiency, and Conservation." *Policy Analysis* No. 175. Washington, D.C: CATO Institute.

Goldstein, W. A., and D. L. Young. 1987. "An Agronomic and Economic Comparison of a Conventional and a Low-Input Cropping System in the Palouse." *American Journal of Alternate Agriculture* 2:51-56.

Hanson, J. C., S. E. Peters, and R. R. Janke. 1990. "The Profitability of Sustainable Agriculture in the Mid-Atlantic Region: A Case Study covering 1981 to 1989." *Northeastern Journal of Agricultural and Resource Economics* 19:90-98.

Helmers, G. A., A. Azzam, and M. F. Spilker. 1990. "U. S. Agriculture Under Fertilizer and Chemical Restrictions—Part 1." Department of Agricultural Economics Report No. 163.

Helmers, G. A., K. J. Bernhardt, and J. A. Atwood. 1993b. "Base Building Incentives and the Food, Agriculture, Conservation and Trade Act of 1990." Selected paper presented at American Agricultural Economics Association Meetings, Orlando, FL.

Helmers, G. A., K. J. Bernhardt, M. Spilker, and J. Atwood. 1993a. "Impacts of commodity Programs on Sustainable Agriculture." Selected paper presented at Western Agricultural Economics Association Meetings. Edmonton, Alberta, Canada.

Helmers, G. A., M. F. Spilker, A. Azzam, and J. E. Friesen. 1992. "Target Price Incentives to Reduce Nitrogen Use in Agriculture—First Round Impacts." Paper presented to American Agricultural Economics Association Meetings. Baltimore, MD.

Hoag, D. and Z. Pasour. 1992. Mandated Training in Sustainable Agriculture: A Dilemma for the Cooperative Extension Service. *Choices*, First Quarter, p 32-34.

Knutson, R. D., C. R. Taylor, J. B. Penson, and E. G. Smith. 1990. "Economic Impacts of Reduced Chemical Use." College Station, TX: Knutson & Associates.

Lee, L. K. 1992. "A Perspective on the Economic Impacts of Reducing Agricultural Chemical Use." *American Journal of Alternative Agriculture* 7(No. 1 and 2):82-88.

National Academy of Science. 1989. *Alternative Agriculture.* Washington, D.C: National Research Council, National Academy Press.

Pesek, J. 1991. "Alternative Practices for an Environmentally Benign Agriculture." In D. Avery (ed). *Global Food Progress.* Indianapolis, IN: Hudson Institute.

Schaller, N. 1991. *An Agenda for Research on the Impacts of Sustainable Agriculture, Assessment and Recommendations of a Panel of Social Scientists.* Greenbelt, MD: Institute for Alternative Agriculture.

Smith, S. 1992. "Farming—It's Declining in the U.S." *Choices* First Quarter, p 8-10.

Tweeten, L. 1991. "The Economics of An Environmentally Sound Agriculture (ESA)." ESO 1784. Columbus, OH: Ohio State University, Department of Agricultural Economics and Rural Sociology.

U.S. General Accounting Office. 1990. *Alternative Agriculture: Federal Incentives and Farmer's Opinions*. Washington, D.C: GAO/PEMD-90-12.

Victor, Peter A. 1992. "Indicators of Sustainable Development: Some Lessons from Capital Theory." *Ecological Economics* 4:191-213.

Young, D. L. 1988. "Economic Adjustment to Sustainable Agriculture: Discussion." *American Journal of Agricultural Economics* 70:1173-1174.

Young, D., and K. Painter. 1990. "Farm Program Impacts on Incentives for Green Manure Rotations." *American Journal of Alternative Agriculture* 5:99-106.

Young, D. L., and W. A. Goldstein. 1987. "How Government Farm Programs Discourage Sustainable Cropping Systems: A U.S. Case Study." Selected paper presented at Farming Systems Research Symposium, University of Arkansas, Fayetteville, AR.

Food and Food Aid

8

Domestic Food Aid Programs

Jean D. Kinsey and David M. Smallwood

Importance of Food Aid Programs

The goal of any country's food policy is to ensure an adequate, safe and nutritious supply of food to all its citizens at a reasonable price. Since prices that are reasonable to most households may not allow low income households to secure adequate nutritious food, special food aid programs are designed to prevent hunger and suffering and to invest in human capital. This investment supports the development of educated voters, a skilled labor force, and generally self sufficient households. Ensuring adequate supplies of food is related to national security as well as to feeding of people; and justifies food and agricultural policies that support farmer's incomes and enterprise.

The original objective of U.S. food aid programs, in the 1930s, was to provide a way to dispose of surplus agricultural commodities purchased by the government in order to stabilize farm prices and incomes. Even though an early characterization of the goals of food aid was that "there will be no starvation amidst plenty" (Black 1942), it was not until the late 1960s that food aid programs began to be treated as a way of meeting society's need for food and nutrition independently of farm price support programs.

Now, the most important purpose of domestic food aid programs is to alleviate hunger and improve the wellbeing of poor people. There are three primary mechanisms by which governments can try to alleviate hunger. One is to administratively lower food prices so the poor can afford to purchase enough food with whatever incomes they may have. This method is often used in developing countries. Another is to deliver food directly to the poor. The third is to provide the poor with adequate purchasing power to buy food at market prices. Food aid programs in the United States are a mixture of the last two, delivering food and/or purchasing power. The largest program, the food stamp program, raised the purchasing power and nutritional status of 25 million of the more than 40 million poor in the United States in 1992.

Food stamps and other food aid programs for large numbers of poor, mostly urban, consumers have been supported by rural politicians in exchange for urban politicians' support for farm programs for a small number of farmers. This political coalition has tied the welfare of urban consumers to the welfare of farmers and has dictated that the U.S. Department of Agriculture administer food aid programs.

In this chapter, we first present facts about various current food aid programs, their costs, the number of people they serve, and controversial policy issues surrounding the individual programs. Secondly, we treat estimates of the benefits to direct recipients, producers, and to society. Finally, we explore future expectations for food aid.

Food Aid Programs—Current Facts, Costs, and Controversies

Domestic food aid is largely financed through federal programs which are administered at the state and local levels. The overall cost of these programs in 1992 was $33.7 billion, about 55 percent of the 1992 federal budget for farm and nutrition programs ($60.75 billion). After adjusting for inflation, the overall cost of these programs increased 50 percent between 1980 and 1992 with the largest increases being in the food stamp and the Women, Infants, and Children (WIC) programs. Table 8.1 lists the cost of these programs in constant 1992 dollars and the percentage increase in these real costs since 1980. The cost increases are attributable to a larger number of participants, real increases in the value of benefits received, and rising administrative costs. For example, the average monthly benefit to a food stamp recipient increased from $35 in 1980 to $69 in 1992. This is a real increase of $9.50, or 14 percent, more than the change in the overall consumer price index.

Food Stamp Program

During the late 1930s when the food distribution programs were under attack for being inefficient and USDA was contemplating a controversial two-tier price system, the food industry proposed a food stamp program. In 1939, the first demonstration program was initiated. Participants were required to purchase orange stamps and received free blue stamps that could be used only for foods on the surplus list. The pro-

TABLE 8.1 Total Cost of U.S. Food Assistance Programs (1992 Dollars).[a]

Fiscal Year	*Food Stamps*[b]	*WIC*[c]	*Food Distribution*[d]	*Child Nutrition*[e]	*Total Costs*[f]
	------------------- million dollars----------------				
1980	14,170	1,146	267	6,693	22,417
1981	16,105	1,306	302	6,408	24,234
1982	15,308	1,352	609	5,373	22,758
1983	17,391	1,599	1,802	5,601	26,507
1984	16,412	1,897	1,905	5,646	25,969
1985	16,332	2,002	1,810	5,587	25,838
1986	15,842	2,064	1,687	5,670	25,360
1987	15,170	2,106	1,524	5,756	24,656
1988	15,419	2,144	1,155	5,274	24,558
1989	15,111	2,146	737	5,311	23,404
1990	16,885	2,249	689	5,509	25,430
1991	18,770	2,369	620	6,018	27,886
1992	23,425	2,627	584	6,760	33,576
Percent Change in Real Cost 1980-1992	60	133	118	1	50

[a] Child nutrition program costs deflated by the CPI for away-from-home food purchases. All other program costs deflated by the CPI for at-home food purchases.

[b] Includes State administrative matching funds and other program costs, plus Nutrition Assistance to Puerto Rico and the Northern Marianas.

[c] Includes all costs for WIC and the Commodity Supplemental Food Programs.

[d] Includes the Food Distribution Program on Indian Reservations and the Nutrition Program for the Elderly plus food distribution to summer camps, soup kitchens, food banks, and TEFAP.

[e] Includes all program costs--cash, entitlement, and bonus commodities, plus administrative expenses. Also includes Nutrition Studies and Education and Special Milk.

[f] Includes food program administrative costs and disaster feeding.

Source: Food and Nutrition Service, USDA, Program Information Division

gram was heralded for: (1) decreasing farm surpluses, (2) increasing grocery store sales, (3) relieving hunger and improving nutrition for the poor, and (4) improving economic conditions of the nation (Perkins 1940).

Currently, food stamps comprise the nation's major food and noncategorical income assistance program. They are sometimes referred to as

America's second currency (Senauer 1993). The Food Stamp Program (FSP) is designed to enable households without income to buy food comparable to the "Thrifty Food Plan" and is calibrated to cost about 80 percent as much as the low cost food plans developed by the Human Nutrition Information Service (HNIS) of the USDA. The "Thrifty Food Plan" provides virtually all of the recommended dietary allowances of nutrients, and costs 66 percent as much as the "moderate cost food plan." The Hunger Prevention Act of 1989 raised the maximum allotment to 103 percent of the cost of the Thrifty Food Plan by 1992. In 1992 the *maximum* allotment for a family of four, with no income, was $360 per month ($1.02 per meal). This maximum benefit is reduced for those with incomes over the maximum allowable deduction for their household size. For each additional dollar of net income, food stamp benefits are reduced by 30 cents. The income requirements, adjusted for household size, are that net cash income cannot exceed the poverty threshold and gross income cannot exceed 130 percent of the poverty threshold. In 1992 the poverty threshold income was $9,190 for a two-person household and $13,950 for a four person household.

In fiscal year 1992, the FSP provided about $22.6 billion in food benefits to about 25 million people; i.e.; one person in ten. Each recipient received an *average* of $826 for food during the year (about 75 cents per meal based on three meals a day).

The numbers of participants in the FSP has grown substantially during the last two decades, though the participation rate (defined as the number who participate as a percent of the poverty population), has remained fairly constant at about 60 percent. In other words, the number of persons participating in the FSP rises and falls at about the same rate as the number of people living in poverty. The program can adjust rather quickly to changes in economic conditions and provides a safety net for those with inadequate income.

The FSP embodies many of the principles of a good welfare program, i.e., uniform national standards administered by the states, universal coverage based on income and need, strong work incentives, and financed federally with automatic cost of living adjustments. It is an entitlement program but if appropriated funds run out and Congress fails to pass a supplemental appropriation, distribution of benefits will have to stop.

The Food Stamp Program for Puerto Rico was replaced by a cash block grant in 1982 called the Nutrition Assistance Program for Puerto Rico. Due to the extreme level of poverty in Puerto Rico, almost 43 percent of its population participates in the cash assistance program.

Food Stamp Controversies

Cashout or Coupons. Many economists argue that cash is the preferred form for providing food assistance to recipients. Several reasons are given: (1) lower cost of distribution; (2) programs can be combined to reduce administrative costs; and (3) benefits are valued more highly by recipients because they can purchase the goods and services most needed or desired. In contrast, public (taxpayer) and congressional support for the programs often depends on targeting both the types of benefits and the recipients.

In-kind benefit programs can ensure the delivery of food and nutrition, since it is more difficult to divert coupons or food to unintended uses. However, there are tradeoffs. The more highly targeted a program, the higher its administrative costs, including production, distribution, and authorization procedures (red tape). These higher costs are opposed by the taxpayers. Further, complicated procedures tend to cut down participation rates among the truly needy.

Currently, the Food and Nutrition Service (FNS) of the USDA is studying the effects of cashout in three states: Alabama, California, and Washington. Most former studies suggest that "cashing out" the food stamp program (providing an equivalent amount of cash as opposed to food stamps) would reduce food spending and nutrient intake (Mathematica Policy Review 1985, Senauer and Young 1986, Ranney and Kushman 1987, Fraker 1990, Devansy and Moffit 1991). But results are mixed; at least two studies have shown that the difference in food expenditures between cash and food coupons may not be very large or very significant (Hollonbeck et al. 1985, Levedahl 1991).

Electronic Transfers. In 1992, states received authorization to distribute benefits of the food stamp and other welfare programs electronically. Some pilot studies of this system are currently underway. The recipient presents an identification card to an ATM machine or a clerk in a grocery store. The card is run through an electronic terminal, the recipients identity is verified, and the benefits given in cash, food stamps, or food items. The benefits of this system include more services delivered for about the same costs to the government and lower costs to retailers and recipients. Recipients do not have to pay to have a check cashed or carry around cash or coupons that may be lost or stolen. In addition, there is less stigma attached to using a debit card. The government does not have the cost of printing, mailing and providing security for 2.5 billion coupons, some of which are lost in the mail each year ($16 million worth in 1990) (Minneapolis Star and Tribune 1991). Grocery stores save about 25 percent by not having to collect, tally, and turn in food stamps. The initial capital cost of the equipment required to handle debit cards is high, but is often shared by stores and the local government agencies. In some systems the same equipment will service debit cards of all customers in the local banking area.

The Federal Reserve Board, which oversees consumer protection in banking, proposed on February 8, 1993 that recipients of electronically transferred food stamps and other welfare benefits be afforded, under Regulation E, the same protection as other users of electronic banking services. The electronic transfer of food stamps was initially operating under a waiver from these regulations. Administrators of the FSP allege that such banking regulations would render the electronic method economically infeasible. In a compromise to save some costs, the Federal Reserve Board's proposal will limit the recipient's liability for losses to $50 if a lost or stolen card is promptly reported, as it does for all other users of electronic bank accounts. Also, government agencies or the banks would have to resolve reported errors within 10 days. But, since account balances are generally available at the time of each withdrawal and the recipients do not make deposits into the account, the requirement that a monthly statement be issued to the recipient would not be necessary, giving an estimated saving of about 32 cents to 75 cents per recipient per month. The final form of these regulations will be seen after a 90-day comment period (Federal Reserve Board 1993).

Fraud. Fraudulent sale or trade in food stamps is disturbing to everyone and stories about it in the press tend to outrage taxpayers. It represents, however, only a minor part of the value of food stamps. Fraud traced to recipients in 1990 amounted to about $36 million, 1.5 percent of the total program cost which is only about one-third the amount typically reported as fraud and waste ($1 billion). The difference is in stamps lost in the mail and certification errors, considered to be legitimate expenses of the program (Minneapolis Star and Tribune 1991).

There will always be some fringe activities by some unsavory characters. Except when proven otherwise, these costs should be treated as a "cost of doing business." A balance must be struck between the costs of tighter regulations and the benefits of decreasing fraud. Furthermore, the amount of money received by individual food stamp recipients (an average of $826 in 1989) is less than most people spend on food to eat at home (per capita expenditures for food consumed at home in 1989 was $961) (Senauer and Young 1986, USDL 1990). Consequently, there is little incentive for recipients to sell food stamps unless they have a desperate need for cash. The percent of such recipients is expected to be very small.

Expanded Uses. Since more than 40 percent of the food expenditures in the average American household is for food eaten away from home, it has been suggested that recipients of food stamps should be allowed to use food stamps for purchases of food in restaurants and fast food places. Although this would diminish food and nutrition purchased per dollar spent as well as reduce the impact on farm prices and incomes, it may increase the utility of food stamps to the recipients.

Child Nutrition Programs

The Child Nutrition Programs are comprised of the National School Lunch, School Breakfast, Child and Adult Care, and Summer Food Service Programs. An estimated 24.2 million children, in about 89,000 schools, are served daily through the National School Lunch Program. Some 47,600 schools also serve school breakfasts (Dixon 1992). Children from families which meet certain income requirements are eligible for free or reduced-price meals. Those below 130 percent of the poverty threshold are eligible for free meals and those between 130-185 percent of poverty are eligible for reduced-price meals. Those with incomes over 185 percent of poverty (52 percent of participating children) are also subsidized at a rate of 16 cents per lunch. In 1992, school lunch subsidies provided free lunches to 41 percent of children at a cost of $1.6625/lunch and reduced-price lunches to 7 percent of children at a cost of $1.2625/lunch. The daily cost is about $20.65 million.

The other food aid programs for children and their 1991 costs are listed in Table 8.2; the total 1992 cost is estimated to be $6,611 million, an increase of more than 8 percent over the 1991 level. These programs receive both cash and food commodities. The commodity value in each school lunch is about 14 cents.

School Lunch Controversies

Distribution of Benefits. Some questions have been raised about the distribution of the subsidy among income groups. For example, if the 7 percent of the children receiving reduced-price lunches were to be subsidized at the full price, program costs would increase by $122 million (40 cents x 306 million meals). This could be easily financed, however, by reducing the subsidy, from 16 cents to 10 cents, for the 51 percent who buy at the regular price, assuming that the relative number of participants in the groups remains unchanged.

The Special Milk Program. This program provides cash to states to subsidize milk served to children in eligible nonprofit schools, child care centers, and summer camps that don't participate in any of the other Child Nutrition Programs and in certain schools with split session kindergarten classes. About 1.6 million children per school day are served half pints of milk subsidized by this program at a total cost of about $178,333 per school day (based on a school year of 180 days).

TABLE 8.2 Child Nutrition Programs, 1991.

Program	Costs (millions)
School Lunch	$4,211.7
School Breakfast	685.1
Child and Adult Care	972.2
Summer Food Service	175.4
Total[a]	$6,096.8

[a]Includes Administrative costs, special studies, surveys and Nutrition Education and Training.

Source: USDA, Food and Nutrition Service, 1992, "Food Program Update." Program Information Division, Alexandria, VA.

Special Supplemental Program for Women, Infants, and Children (WIC)

The WIC Program is a targeted program for low income, pregnant, postpartum, and/or breastfeeding women, to infants, and to children up to 5 years of age who are found by health officials to be at nutritional risk. The program provides supplemental foods, nutrition education, and health care referrals. The program originated from the 1969 White House Conference on Food, Nutrition, and Health that recommended special attention be given to people in these groups. WIC authorization is separate from food and agricultural legislation. Like the child and elderly feeding programs, one of WIC's primary goals is investment in human capital.

In contrast to the FSP, WIC is not an entitlement program; participation is limited by the funds appropriated. The FNS program administrators have established a set of program priorities to target and ration these limited funds to those determined to be most at nutritional risk. Their highest priorities are infants and pregnant women with certain medical conditions and the lowest are children age one to five and some postpartum women. In 1990, it was estimated that 90 percent of the eligible infants, 45 to 85 percent of the eligible women, and 40 percent of the eligible children age one to five were covered. About 77 percent of all participants are infants and children.

WIC expanded rapidly from $10 million and 88,000 participants in 1974 to $2.6 billion and 4.9 million participants in fiscal year 1992. WIC provides the full cost of supplemental food packages for recipients; benefits are not adjusted for household income.

WIC is a highly targeted program. WIC provides food, food vouchers, or food checks so that eligible participants may supplement their diets with nutrients (such as iron, calcium, protein, and vitamins A and C) critical during pregnancy and early growth. The foods provided include milk, fruit/vegetable juice, infant formula, cheese, eggs, cereals, dried peas and beans, and peanut butter.

WIC Controversies

Infant Formula. The WIC program was subject to a controversy over infant formula during the early 1980s when formula prices increased much faster than other food prices and drove up the cost of the average WIC food basket. As a result, participation levels could not expand as fast as some would have liked. To help control formula costs, states initiated rebate contracts with formula producers in return for using particular brands of formula in the state's program. By 1991, rebates averaged about $1.50 per thirteen ounce can of formula concentrate. Some economists contended that these types of rebates raised the cost of formula for nonprogram participants and changed marketing strategies of formula producers. The top three formula companies account for about 95 percent or more of the infant formula market and the WIC program buys as much as 30 to 50 percent of all formula sold in the United States.

Breastfeeding. Many are also concerned that the infant formula distributed by the WIC program encourages formula usage at the expense of breastfeeding. In 1992, in an effort to increase breastfeeding, WIC initiated a special food package designed to provide additional nutrients needed during breast-feeding for post-partum mothers. In addition, WIC has undertaken special educational efforts to encourage more breastfeeding.

Commodity Supplemental Food Program

This program is an alternative to the WIC program and serves essentially the same people, but it also includes children up to age six and senior citizens living in project areas. The federal government provides actual foods to state agencies for distribution to program participants. The food items include those available through WIC and those acquired from farm surpluses as well as canned fruits and vegetables, canned meat, poultry, or tuna fish, dehydrated potatoes, and rice or pasta. As the name implies, the foods are meant to supplement those that are currently being purchased and not to provide a full replacement. In 1992, about 220,000 women, infants, and children and 110,000 elderly participated in the program. The benefit package cost between $20-$21.50 per person per month, depending on the age group.

Nutrition Education

The Expanded Food and Nutrition Education Program (EFNEP). Questions continue to be raised over how to best improve the nutritional status of low income households, renewing interest in nutrition education. The extent to which nutrition education can be used to leverage other food aid benefits, resulting in a healthier population, is unknown. EFNEP, conducted by the Cooperative Extension Services of the Land Grant Colleges and Universities, is the largest nutrition education program, but the FSP and WIC program also provide nutrition education as components of their programs.

EFNEP Controversies. The effectiveness of nutrition education on improved diets in low-income households is widely acclaimed, but many are uncertain whether formal programs are an economical way of improving nutrition and health. For example, the vast amounts of health and diet information in the news media may provide a more economical channel for educating the public, assuming the media news conveys sound nutritional information.

Food Distribution Programs

Section 416 of the Agricultural Act of 1949 made certain food commodities, acquired through farm price-support operations by the Commodity Credit Corporation (CCC), available for distribution to the poor. The intent was to use, without detriment to the farm price-support program, foodstuffs that might be lost because of deterioration and spoilage before they could be moved through normal domestic channels.

Surplus CCC stocks are often donated to domestic food assistance programs. Donations include cheese, butter, nonfat dry milk, cornmeal, flour, honey, rice, and wheat. During the 1980s, domestic food assistance programs, especially the Child Nutrition Programs and the Temporary Emergency Food Assistance Program (TEFAP), benefitted greatly from these donations.

The Food Distribution Program on Indian Reservations. This program operates as a substitute for the FSP on or near Indian reservations. The program allows tribal organizations to run commodity distribution programs in lieu of receiving food stamps.

Nutrition Program for the Elderly. This program is administered by the Department of Health and Human Services. Originally developed to provide nutritious foods to senior-citizen meal sites and meals on wheels, it has evolved into a mostly cash subsidy program. In 1992, about 92 percent of all benefits were distributed as cash.

Donations to Charitable Institutions and Summer Camps. Commodities acquired by the USDA are distributed to charitable institutions serving needy persons and summer camps for children. These include soup kitchens, some hospitals, the meals-on-wheels program, and orphanages that do not participate in other Child Nutrition Programs. Next to the school feeding programs, these are the largest recipients of commodities distributed by USDA. They provide an outlet for continued distribution during the summer when many schools are not in operation.

Temporary Emergency Food Assistance Program (TEFAP). In 1981 dairy product surpluses were at an all-time high. The Agricultural and Food Act of 1981, Section 1114, required that price-supported commodities "not likely to be sold by CCC or otherwise used in programs for commodity sale or distribution" be made available to food banks and nutrition programs providing food service. This was the beginning of the TEFAP. In a few years the CCC ran out of surplus stocks of some commodities, primarily cheese, honey, and nonfat dry milk. Pressures to continue the TEFAP program resulted in the Hunger Prevention Act of 1988, authorizing more than $120 million per year to purchase commodities for distribution. This action effectively "delinked" TEFAP from fluctuations in farm surplus stocks. The benefits of TEFAP are that, by supporting local, voluntary food distribution agencies (Ballenger and Harold 1991) it helps to feed hungry people who do not participate in the food stamp program.

Commodity Distribution Controversies

Fluctuating Benefits. Food available from the CCC stocks fluctuates with supply, and there is often great pressure to continue benefits when surpluses are gone. This causes fluctuations in the nonfederal cost of programs and influences the kinds and amounts of commodities bought with appropriated funds. Historically, the National School Lunch Program and the TEFAP program grew out of such pressures to keep food coming after surpluses ran out.

Higher Food Prices. Farm price-support programs raise food prices. Poor households are most affected by rising food prices because they spend almost 50 percent of their income on food, nearly five times the percent spent on food by the average household. Food price increases also raise the cost of food assistance, since most programs are indexed to the cost of food.

Slippage. Commodity distribution programs cause "slippage". That is, they substitute for some of the food that consumers would otherwise purchase, altering traditional marketing channels. They partially offset some of the farm price supports resulting from commodity removal programs. USDA studies of TEFAP commodity distributions found that, on average, each pound of cheese distributed displaced about one-third of a pound of commercial cheese sales and that butter donations displaced margarine sales pound for pound.

Inefficient. Price-support and food distribution programs are not an economically efficient means of providing food and nutrition assistance to the needy. The costs of procurement, distribution, storage, and management are higher than for many alternative forms of assistance, such as cash or food stamp programs.

Health Concerns. A controversy has evolved recently over the assumption that the foods available for distribution are healthy for the recipients. This is a particularly sensitive issue for high fat foods like cheese, whole milk, butter, and canned pork. As a consequence of serious questions about excessive fat in school lunches, USDA is revising school meal standards. The Special Milk program has also been criticized for having to offer whole as well as the lower fat milks. Children typically can choose the type of milk they prefer; about 31 percent of the milk distributed is whole milk, 64 percent is lowfat, and 4 percent is skim.

Food Aid Program Benefits

Food Stamp Program Benefits to Recipients

Most studies show that food stamps increase food expenditures for individual households, increase the nutrition available to household members, and increase overall demand for food. The maximum annual benefit for a household of four persons with no income is $4,320. Most studies have shown that food stamps increase food expenditure about two to three times that of an equivalent amount of money income. Each additional dollar of cash income increases food spending by about 5-10 cents whereas a dollars worth of food stamps increases food spending between 20-30 cents (Senauer and Young 1986; Fraker 1990). Another way to look at this is that it is consistent with a slippage factor of about 70 to 80 percent. That is, every food stamp dollar frees up 70-80 cents of cash previously spent on food (Fraker 1990).

Evidence suggests that not only do program participants spend more money for food than similar nonparticipants, they also get more nutrients in each dollars worth of food (Devansy and Moffit 1991). Table 8.3 summarizes findings from the USDA Survey of Food Consumption in Low-Income Households, 1979-80. All reported nutrients are consumed in larger quantities by food stamp households than by nonparticipating households with similar incomes.

TABLE 8.3 Household Nutrient Availability as a Percentage of the RDA for Persons Eating in the Household.

Nutrient	*FSP Participant*	*FSP Nonparticipant*	*Difference*
		----------Percent of RDA-------------	
Food energy	139	121	18
Protein	232	203	29
Calcium	119	111	8
Iron	151	137	14
Magnesium	134	123	11
Phosphorous	202	183	19
Vitamin A	213	178	35
Thiamin	194	165	29
Riboflavin	204	180	24
Vitamin B6	132	114	18
Vitamin B12	235	191	44
Vitamin C	290	264	26

Sources: Fraker, Thomas 1990. USDA Survey of Food Consumption in Low-Income Households, 1979-1980.

WIC Benefits to Recipients

WIC has been one of the most popular and successful food assistance programs. This is due, in part, to the well targeted nature of the program both in terms of the benefit package and the target group of participants. The program also yields cost savings for government by reducing health care costs. On average, every government dollar spent through WIC on pregnant women saved between $1.77 and $4.75 in medicare costs for new born children and their mothers with about half of that saving occurring in the first year of the baby's life. Women who had participated in WIC's prenatal program made greater use of prenatal care and had improved pregnancy outcomes (Devaney, B., L. Bilheimer, and J. Schore 1990; Minneapolis Star and Tribune 1992).

Producer Benefits

The impact on producer income and prices is small. In 1991, every dollar spent on food aid was estimated to increase farm sales by nine cents which translates into an increase in gross farm cash income of $3 billion, or 1.8 percent more than would otherwise have been available. Farm prices increased by less than 1 percent (Martinez and Dixit 1992).

For every dollar spent on the school lunch and breakfast programs total food expenditures increased by 66 cents. The comparable figure for the Food Stamp program is 33 cents. This is in spite of the fact that school meals substitute for home food expenditures at the rate of about 34 cents for every government dollar invested (Martinez and Dixit 1992). Overall, government allocations to food aid (minus administrative costs) increased total food expenditures by 40 percent in 1989 (Martinez and Dixit 1992). In 1992, if the same ratio holds, expenditures on food of $31.5 billion should translate into increased food expenditures of $12.6 billion. This is, however, less than a 3 percent increase in total expenditures for food at home.

General Benefits

As Figures 8.1 and 8.2 illustrate, the bulk of the dollars spent in food aid programs goes for the Food Stamp Program. Table 8.4 summarizes the benefits received by individual recipients and the number of recipients. The administrative costs are also listed so one can tell the relative amount of cost that goes directly to the recipients. Though the aggregate amounts listed on Table 8.4 seem large, individuals receive a quite modest sum. For example the FSP provides the average recipient with about $69 per month, about half of the average monthly food expenditure of the typical American consumer (U.S. Department of Labor 1990). On average, children in school each receive lunch subsidies valued at $153 per year; $299 if they receive a free lunch, $227 if they receive a reduced-price lunch and $29 if they pay the regular price for their school lunch, based on a 180-day school year. One must temper the observation about modest benefits, however, by recalling that many of the recipients also qualify for other food aid and welfare programs. Their cumulative benefits are likely greater than indicated by Table 8.4.

Future Expectations

The number of people who will need food aid benefits in the future and the costs of food aid programs will depend on the number of people living in poverty, the unemployment rate, and food prices. Resources to provide food aid will depend on fiscal policy and tax revenue as well as on the growth of productivity and income in the economy. Projections for future economic activity in the United States are that it will grow, but slowly. The food stamp program has become a substitute for a national-income maintenance program, partly because it automatically grows when employment and wages decline. It is one of the automatic stabilizers

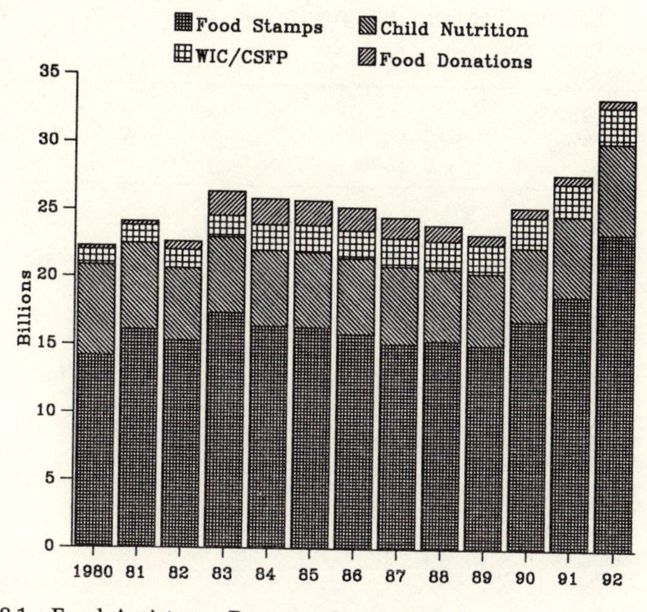

FIGURE 8.1 Food Assistance Program Costs, 1980-92.

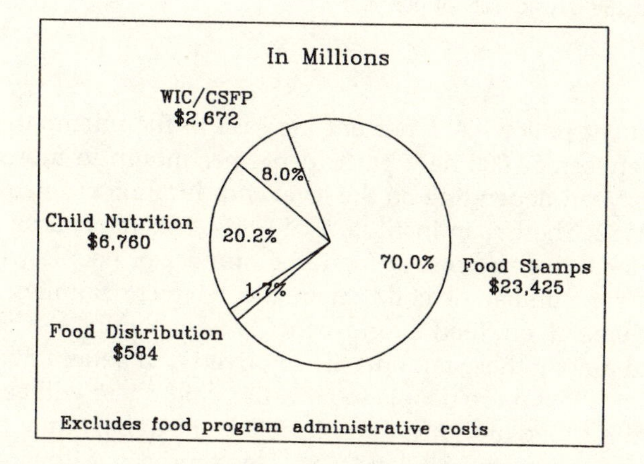

FIGURE 8.2 Federal Food Assistance Budget, 1992.

TABLE 8.4 Estimated Average Annual Total Food and Nonfood Costs and Benefits Received by the Individual Recipients of Food Aid Programs.

Program	*Total Food Benefits*	*Individual's Benefits*[a]	*Average Number Served*	*Nonfood Costs*[b]	*Percent Nonfood Costs*
Food Stamps	$20.9 bil	$822.36	25.4 mil	$1,500 mil	6.7
WIC	$ 2.6 bil	$362.00	5.4 mil	$ 604 mil	30.6
CSFP	$66.8 mil	$207.72 (Women/Children)	222.5 mil	$ 18 mil	21.2
		$171.36 (Elderly)	120,000 (Elderly)		
School Lunch[b]	$ 4.5 bil	$153.55[c]	24.2 mil		
Special Milk[c]	$32.1 mil	$ 20.00	1.6 mil children		

[a]Recipients estimated annual average benefit per individual recipient.

[b]Nonfood costs are for the whole program. School lunch does not have a separate nonfood cost item. The nonfood costs are included in the price of the meal subsidy.

[c]All items calculated from other data. School lunches and special milk programs are based on 180 days of school per year.

in government policy. A 1 percent increase in the unemployment rate leads to between 50,000 new participants per month to more than one million per year, depending on the level and duration of unemployment (Senauer 1982, The Urban Institute 1985). Its costs depend on a number of factors including the cost of food, the number of people who participate, and the number of children in the recipient families.

The demand for food stamp benefits is generally less among the elderly and among those temporarily in poverty. If better education and higher skills allow more people to be better paid, there will be an easing of the need for food aid. Most predictable is the aging of the population and the relative wealth of the elderly. This should put downward pressure on food stamps from that part of the population. Likewise, a decline in fertility rates means a smaller number of children in the population which could ease the overall need for child nutrition program expenditures. At the same time, however, the proportion of children living in poverty is climbing with the number of immigrants and the number of single parent households. Childhood poverty, under educated, and under employed adults will all act to increase the demands for food aid.

If government acquisition of surplus stocks of food under the Commodity programs are diminished, direct food distribution will not be an attractive way to alleviate hunger. Pressure to move towards a cash or income maintenance plan will probably increase in any case. It will be seen as more efficient, more equitable, more desired by recipients and easier to administer.

References

Ballenger, N., and C. Harold. 1991. "Revisiting Surplus Food Programs After Surpluses: The TEFAP and its Role in the District of Columbia." Washington D.C.: National Center for Food and Agricultural Policy, Resources for the Future, Discussion Paper Series No. FAP 91-01.

Black, John D. 1942. *Parity, Parity, Parity*. Cambridge, MA: The Harvard Committee on Research in the Social Sciences. P 309.

Devaney, B., L. Bilheimer, and J. Schore. 1990. *The savings in Medicaid costs for Newborns and Their Mothers from Prenatal Participation in the WIC Program*. Washington, DC: MPR.

Devansy, B., and R. Moffit. 1991. "Dietary Effects of the Food Stamp Program." *American Journal of Agricultural Economics* 73:202-211.

Dixon, J. 1992. "Schools Serve More Breakfasts, but Probably Millions Miss Out." *The Oregonian*. Portland, OR: (October 21 p A9).

Federal Reserve Board. 1993. *Press Release,* Washington D.C.: Board of Governors, Federal Reserve Board, February 8, 26p.

Food and Nutrition Service. 1992. *Program Information Report*. Washington, D. C.: USDA, July.

Fraker, Thomas. 1990. "The Effects of Food Stamps on Food Consumption: A Review of the Literature." *Mathematica Policy Research*. Washington, D.C.: USDA, FNS.

______. 1992. *Food Program Update*. Washington D.C.: USDA.

Hollonbeck, D., J. C. Ohls, and B. Posner. 1985. "The Effects of Cashing Out Food Stamps on Food Expenditures." *American Journal of Agricultural Economics* 67:609-613.

Levedahl, J. William. 1991. *The Effect of Food Stamps and Income on Household Food Expenditures.* Tech Bull. No. 1794, Washington DC: Economic Research Service, USDA.

Martinez, S. W., and P. M. Dixit. 1992. "Domestic Food Assistance Programs," *Measuring Benefits to Producers*. Washington D.C.: USDA, ERS, Agricultural and Trade Division, July.

Mathmatica Policy Research. 1985. "Evaluation of the Nutrition Assistance Program in Puerto Rico." Vol. II. Washington D.C.: USDA, FNS.

Minneapolis Star and Tribune. 1991. "Cost of Mistakes, Fraud Placed at $1 Billion for Food Stamp Program in '90." December 15, p 18A.

______. 1992. "Prenatal Aid to Poor Pays in Long Run, Study Finds." May 6, p 7A.

Perkins, Milo. 1940. "Thirty Million Customers for the Farm Surpluses." Pp 650-651 in *Farmers in A Changing World*. Yearbook of Agriculture, USDA.

Ranney C., and J.E. Kushman. 1987. "Cash Equivalence, Welfare Stigma and Food Stamps." *Southern Economic Journal* 53:1011-1027.

Senauer, Benjamin. 1993. "America's Second Currency." *The Region* 7:1, Minneapolis, MN: Federal Reserve Bank of Minneapolis. Pp 4-11.

______. 1982. "The Current Status of Food and Nutrition Policy and the Food Stamps." *American Journal of Agricultural Economics* 64:1009-1016.

Senauer, Benjamin and Nathan Young. 1986. "The Impact of Food Stamps on Food Expenditures: Rejection of the Traditional Model." *American Journal of Agricultural Economics* 68:37-43.

Urban Institute. 1985. *The Effects of Legislative Changes in 1981 and 1992 on the Food Stamp Program*. Washington, D.C.: Final Report to Congress, May.

U.S. Department of Agriculture. 1982. *Survey of Food Consumption of Low Income Households: 1979-1980*. Washington D.C: HNIS Preliminary Report No. 13.

U.S. Department of Labor, Bureau of Labor Statistics. 1990. "Consumer Expenditures in 1989." *NEWS*. Washington, D.C: USDL Newsletter No. 90-616, November 30.

9

Overseas Food Aid Programs

Mark E. Smith and David R. Lee

The Food and Agriculture Organization (FAO) estimates that in developing countries about one person in every five (about 780 million individuals) suffers from chronic malnutrition (Food and Agriculture Organization of the United Nations 1992). For perspective, this total is about three times the population of the United States. How the United States responds in terms of providing food aid will be part of the 1995 agricultural and food policy debate. U.S. food aid issues will likely center not on drastic changes to existing policy, but rather on evaluating the effects of the Food, Agriculture, Conservation, and Trade Act of 1990, which mandated some of the most substantive changes in U.S. food aid in years. Additional changes may represent a response to continuing uncertainties about food needs in countries undergoing political and economic restructuring such as the former Soviet Union, and food needs in chronically food-deficit countries, such as in Sub-Saharan Africa. Given growing global food aid needs, and severe budgetary restrictions in the provision of U.S. food aid, two key issues will likely be how to improve the effectiveness of food aid and the amount of food aid the United States is willing to provide.

U.S. Food Aid in Perspective

Food Aid in Global Terms

Food aid from all developed country donors has played a declining role in assistance to developing countries. Since the mid-1970s, food aid as a proportion of official development assistance (ODA, the total amount of concessional flows from developed to developing countries), has fallen from 12 to 15 percent of total aid to 6 percent in the early 1990s. While food aid has generally increased over the period, ODA has grown about twice as fast. In the early 1990s, donors' food aid amounted to about $3.4 billion, compared to approximately $56 billion in overall ODA. Donors generally focus on channels other than food aid to provide assistance to developing countries.

Cereals account for more than 90 percent of world food aid. The aggregate volume of cereals food aid has been erratic, but trended upward after a steep decline in the mid-1970s (Figure 9.1). Between July-June trade years 1987 and 1991, global food aid in cereals averaged nearly 12.2 million tons (about 70 percent wheat, 22 percent coarse grains, and 8 percent rice) (Food and Agriculture Organization of the United Nations 1993). Other food aid, comprising 5 to 10 percent, has included vegetable oil, dried skim milk, and pulses (Food and Agriculture Organization of the United Nations 1991).

Food aid accounts for a declining share of world trade. Cereals aid accounted for slightly more than 5 percent of the world cereal trade in the early 1990s, compared to about 10 percent in the early 1970s. Food aid as a proportion of imports by low-income, food-deficit countries has generally fallen from approximately 30 percent in the early 1970s to about 15 percent in the early 1990s. However, in Sub-Saharan Africa (and to a lesser extent Latin America), growth in food aid is outpacing total cereal imports (Food and Agriculture Organization of the United Nations 1991).

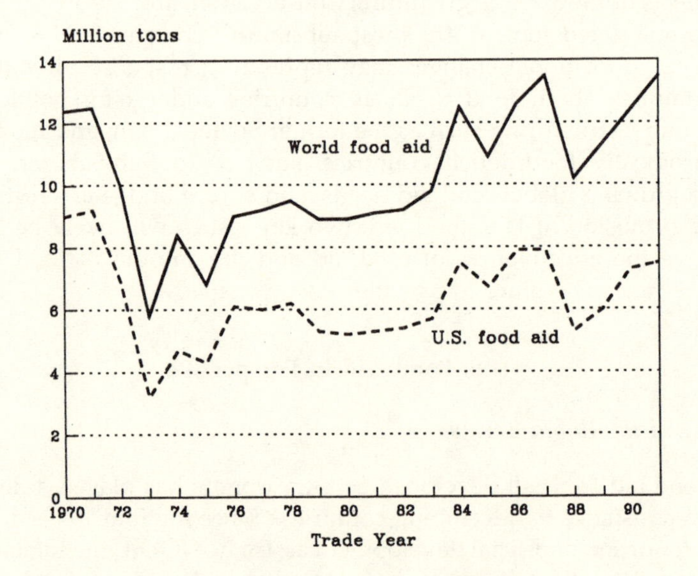

FIGURE 9.1 World and U.S. Cereals Food Aid, 1970-91.
Source: Food and Agriculture Organization of the United Nations.

U.S. Food Aid

The United States remains by far the world's major food aid donor, committing between $1.6 and $2.1 billion annually since the mid-1980s. This compares to about $8 to 10 billion in other forms of U.S. foreign economic assistance. In nominal terms, the value of U.S. food aid exports has been relatively steady since the late 1950s, although it has risen from the mid-1970s, especially in response to the Ethiopian famine of the mid-1980s (Figure 9.2). However, in real terms (measured in 1987 dollars), the level of U.S. food aid has experienced a long-term decline since the 1960s. This reflects not only smaller volumes provided since then, but also lower real commodity prices.

The United States provided close to 60 percent of global cereal food aid in Trade Years 1987 to 1992. Compared to the late 1950s, when the United States provided 95 percent of global food aid, or even the early 1970s, when the U.S. share was 70 to 75 percent, the U.S. position as a global food aid supplier has diminished substantially as other donors have increased their assistance (Hopkins 1990).

Although the United States continues to play an important role in providing global food aid, food aid as a share of both U.S. foreign economic assistance and U.S. agricultural exports is small and declining. For example, in fiscal years (FY) 1990-92, food aid represented about 15 percent of foreign economic assistance, compared to about 33 percent in the early 1970s (Figure 9.3). The share of U.S. agricultural exports shipped as food aid has dropped even more dramatically as commercial exports have increased. In recent years, U.S. food aid accounted for less than 5 percent of U.S. agricultural exports, compared to 25 to 30 percent in the late 1950s and early 1960s. In general, food aid is still a relatively important source of U.S. foreign assistance but much less so for domestic U.S. farm assistance. However, it is still important for some commodities. For example, more than half of U.S. soybean oil exports shipped in the early 1990s went as food aid.

The United States provides a wide array of commodities through its food aid programs. In fiscal year 1992, for example, grains constituted the majority (56 percent) of food aid, but additional commodities included dairy products and vegetable oils (12 percent each), processed cereal products (e.g., wheat flour and corn meal, 11 percent), and miscellaneous products (e.g., cotton and beans, about 10 percent).

U.S. food aid in fiscal year 1992 was programmed to 87 countries. African nations received 45 percent of the total, Asian countries 18 percent, Latin American nations only slightly less, and European countries nearly 15 percent. Near East recipients received about 5 percent. While the primary recipients of U.S. food assistance have

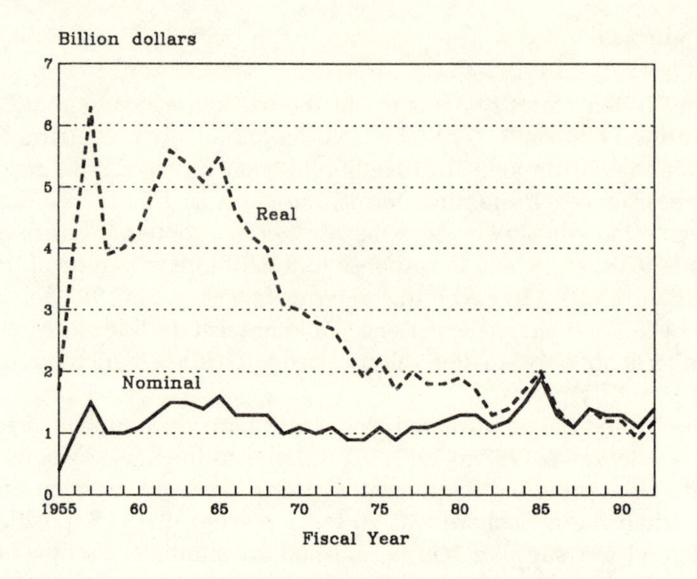

FIGURE 9.2 U.S. Food Aid Exports, Real and Nominal Value.
Source: U.S. Department of Agriculture, Economic Research Service.

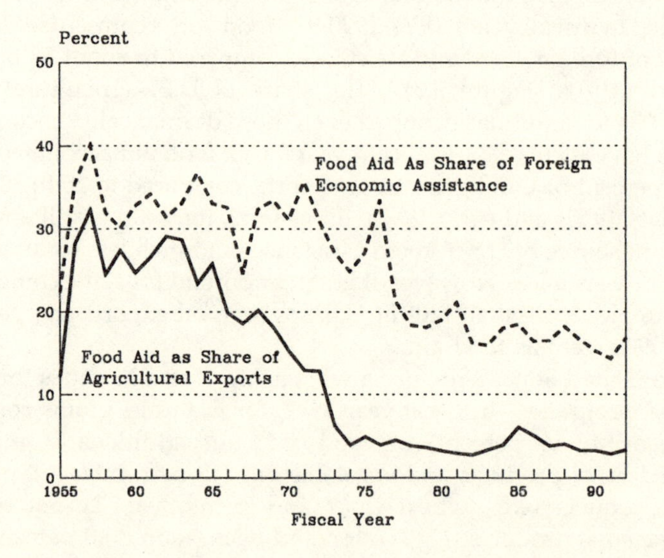

FIGURE 9.3 Role of Food Aid in U.S. Foreign Economic Assistance and Agricultural Exports.
Source: U.S. Department of Agriculture, Economic Research Service and Agency for International Development.

included some notably food-insecure countries (Ethiopia, Sudan, and Bangladesh, for example), much U.S. food aid has at times been directed toward countries of geopolitical importance, such as Egypt, Morocco, Jordan, and El Salvador. With major changes in the 1990 legislation, this may be less the case in the future, although not enough time has elapsed to measure its effects on this score.

The composition of U.S. food aid by program source is shown in Figure 9.4. U.S. food aid is authorized mainly by Public Law (P.L.) 480, originally enacted in 1954 and otherwise known as the Food for Peace Program. It consists of three main titles. Title I provides food aid on a concessional, long-term credit basis. It has traditionally been the mainstay of U.S. food aid, accounting for two-thirds or more of the U.S. total as late as fiscal year 1983. Title II authorizes food aid donations for "humanitarian" purposes—combatting famine and malnutrition—and also for promoting economic development such as through food-for-work projects. Until it was revised by the 1990 Act (discussed below), Title III had often been considered with Title I since Title III authorized concessional credit but also included specific debt forgiveness provisions.

Smaller than and separate from P.L. 480, food aid authorized under Section 416(b) of the Agricultural Act of 1949, as amended, consists of donations from surplus stocks of the Commodity Credit Corporation (CCC). However, the mix and amount of commodities available under this program are dependent upon inventories of surplus CCC stocks. Since volume and composition of these stocks are closely tied to developments in domestic U.S. farm commodity programs and world markets, Section 416(b) food aid has fluctuated from year to year.

The Food for Progress Program, originally authorized by the Food Security Act of 1985, allows food aid provision to assist agricultural policy reforms in recipient countries. It can draw upon the resources of P.L. 480 Title I, Section 416(b), or other CCC funds.

Food Aid Analysis

Food aid in general can be analyzed from different perspectives. From a donor's perspective, in addition to providing humanitarian relief, food aid can be viewed as a boost in demand for donor-provided commodities, thus increasing farm prices to the benefit of producers and stimulating trade in the donor country. This, in turn, benefits processors, handlers, exporters, and shippers. To the extent that commodity prices are increased, domestic farm support payments may be reduced, benefiting taxpayers. For example, depending upon assumptions, U.S. food aid shipments are estimated to have boosted domestic U.S. wheat prices by about 1 to 5 percent in the mid- to late 1980s (Price et al. 1992).

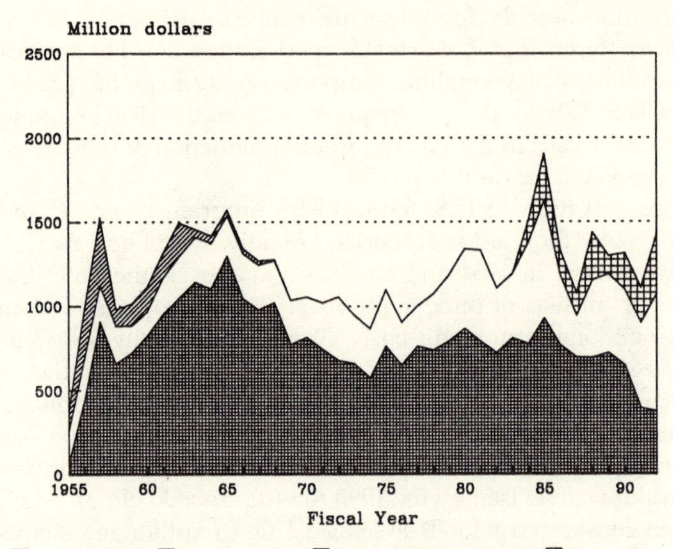

Title III included with Title I from fiscal years 1978-90 and with Title II from fiscal years 1991-92. Sec. 416 (b) donations were re-authorized in 1982.

FIGURE 9.4 U.S. Food Aid Shipments, by Program, 1955-92.
Source: U.S. Department of Agriculture, Economic Research Service.

Higher farm prices meant lower deficiency payments. Considering these savings, the estimated net U.S. taxpayer cost of a dollar's worth of wheat provided as food aid in that period ranged between 40 and 75 cents. It is worth noting that, a donor's benefits from food aid can be diminished to the extent that food aid displaces commercial exports.

From the recipient's perspective, food aid is seen as an increase in supply. Most importantly, food aid can alleviate hunger and malnutrition, especially in emergencies when the recipient's food supply is short because of natural or man-made reasons. Food aid can also provide balance-of-payments relief since foreign exchange freed by food aid can be used to import other goods, ideally those needed to generate development. This effect is maximized when food aid displaces commercial sales. Evidence suggests that food aid commodities sold in the recipient's domestic market substitutes for commercial imports to a significant extent (Saran and Konandreas 1991).

Despite food aid's benefits, potential problems that may accompany its provision have been identified. Primary among these are several disincentive effects. In a recent review, Maxwell (1991) identifies four significant potential disincentive effects of food aid: (1) a *price disincentive* effect on the recipient's producers who may suffer lower prices due to additional supply; (2) a *policy disincentive* effect in which food aid may lead recipient governments to neglect agricultural production and food security priorities; (3) *demand disincentives* for locally-produced foods, caused by changing consumer preferences for imported aid commodities (though this could also be called successful market development); and (4) *labor market disincentives*, caused by food-for-work programs which may disrupt labor markets, discouraging labor migration, and depressing output elsewhere in the country or economy.

While the literature on these and other criticisms of food aid is vast, it is often contradictory. A consensus, as summarized by Maxwell, is that while "food aid has the capacity to cause disincentive effects, . . . these can be and often are avoided by government policy" (p 70). These policies can include stimulating demand through food distribution programs or providing production incentives to domestic producers. Food aid from one donor will by itself likely have little leverage in influencing a recipient's policies. This is one reason for incorporating food aid into an overall development assistance package through which donors can discuss appropriate policies with recipients.

Food aid may also be an inefficient form of resource transfer compared to the provision of cash assistance. In other words, the benefits of a dollar that a donor spends on food aid may not be worth a dollar to recipients, perhaps because of high administrative or transport costs. If food aid is provided primarily because recipients lack purchasing power to buy food, then recipients may prefer a cash, rather than commodity, transfer. This would leave recipients with flexibility to purchase those goods most desired. Further, assistance in the form of cash transfers may be rapidly provided and can create a "demand-pull" effect which may regenerate the trade, transport, and marketing sectors of an economy (Sen 1990). In situations where these attributes are important, where aid recipients desire principally nonfood aid commodities, or in which direct food distribution is operationally difficult, then cash disbursement may generate significant efficiency gains over food aid. On the other hand, in situations where cash assistance would not be forthcoming, where the aid commodity is desired by recipients, or where local markets are unlikely to respond rapidly and sufficiently to food needs, then the potential efficiency gains of cash transfers may be less important.

Major Changes in 1990 Food Aid Legislation

The Food, Agriculture, Conservation, and Trade Act 1990 incorporated some of the most significant changes in U.S. food aid policy in decades. In part, these changes were in response to criticisms of U.S. food aid policy over the years. A principal criticism has been that among many, often conflicting, goals of food aid programs, political or marketing objectives have often superseded humanitarian and development goals, as reflected by the high proportion of food aid that has gone to countries of primarily strategic interest. Also, the many objectives of U.S. food aid policy have created bureaucratic hurdles, inefficiency, and inflexibility, and these have hindered food aid management and delivery.

Recognition of these (and other) problems prompted debate among food aid constituencies, including farm groups and private voluntary organizations (PVOs), about how to improve the effectiveness of food aid programs prior to passage of the 1990 legislation. The legislation, as finally adopted, affirms that U.S. policy is to "promote the foreign policy of the United States by enhancing the food security of the developing world" (Sec. 1512). A secondary goal includes the multiple objectives of (1) combatting world hunger and malnutrition, (2) promoting broad-based equitable and sustainable development, (3) expanding international trade, (4) expanding U.S. export markets, and (5) fostering private enterprise and democracy in developing countries.

Although the multiple objectives of U.S. food aid were retained, the 1990 legislation reflected a major modification in policy. Of the major themes that emerged from the new legislation, a key one is the greater emphasis on food security and development concerns in low-income countries. Prior to 1990, U.S. policy (as stated in P.L. 480) had been first to "expand international trade; [and] to develop and expand export markets for United States agricultural commodities . . ." The 1990 Act placed explicit attention on development (versus political and marketing) objectives. Priority countries to receive Title I assistance are redefined to include those which have demonstrated food needs, are adopting policies to promote food security and agricultural development, and are potential commercial markets. Title II supports humanitarian and development initiatives undertaken by PVOs, cooperatives, and multilateral aid organizations such as the World Food Program. New Title III "Food for Development" provisions authorize bilateral agreements providing food aid to least-developed countries as grants based on explicit poverty and malnutrition-based criteria. These changes in legislation should gradually shift food aid from countries primarily of political importance to those with pressing food needs.

A second theme of the 1990 legislation involved administrative changes intended to reduce bureaucratic impediments to effective food aid implementation and delivery. The Secretary of Agriculture was granted sole authority for Title I programs, while the Administrator of the Agency for International Development (AID) received sole authority for operating the Title II and III donation programs. This represented a major bureaucratic streamlining compared to the past when as many as five principal agencies—the Departments of Agriculture, State, and Treasury, AID, and the Office of Management and Budget—played detailed roles in food aid programming decisions, extending the time required to make food aid allocation decisions. This change also made U.S. food aid procedures more similar to those of other donors, where foreign aid ministries play a significant role in food and operations.

A related theme of the new legislation was increased program flexibility in providing food aid as humanitarian and development assistance. For example, Title II now provides for a set level of funds to be made available to the PVOs. Title III is now a grant program rather than a concessional loan program with debt-forgiveness provisions. Currencies generated under Title III may be used to purchase commodities produced in recipient countries to meet urgent food needs in the recipient or neighboring countries. For least-developed countries pursuing market-oriented economic reforms, certain Title I debt repayments may be waived by the President.

Issues for the 1995 Agricultural and Food Legislation

Enhanced Humanitarian and Development Assistance

One of the principal changes in the 1990 Act was its increased focus on alleviating hunger and its causes. It is likely that demands for greater operational flexibility and otherwise increasing the developmental impact of food aid will continue to be heard (Food Aid Management 1991, Fletcher 1992, Clay and Stokke 1991). An issue is the extent to which the enhanced flexibility provided in the 1990 Act has actually improved the delivery of effective assistance.

Monetization. The selling of food aid commodities to generate revenues, called monetization, is one of the means Congress has authorized to allow greater flexibility of food aid. Monetization helps overcome the inefficiencies of providing aid as commodities rather than as cash. Congress has set a 10 percent minimum for the monetization of commodities distributed under nonemergency Title II programs. In fiscal 1992, about 15 percent of Title II non-emergency resources were monetized. An issue in the 1995 policy debate will likely be the effectiveness of monetization in meeting program goals and the extent to which commodities should be monetized. A question is whether the benefits of activities pursued with funds generated by monetization exceed the benefits of providing commodities to hungry people.

Monetizing commodities donated by U.S. taxpayers can help meet program goals. For example, PVOs are able to generate funds to help meet transport costs to distribute remaining commodities inland to areas where needs may be greatest. Alternatively, with revenues generated by monetizing commodities at the port of entry, the PVO may purchase locally produced commodities inland, thus providing foods perhaps more desirable to recipients and also strengthening local markets. Revenues generated through monetization may be used to fund immunizations and health education at the site of food aid disbursement, as well as other development activities.

However, monetization does have drawbacks. As mentioned, monetized food aid may displace commercial exports. Some feel that since PVOs are given the commodities, they should themselves generate the funds for their delivery. If commodities are to be monetized, then perhaps aid in cash rather than in commodities may be more appropriate.

The chief beneficiaries under Title II monetization programs are the receiving nations and the PVOs. As such, they would bear the effects of increased or decreased monetization. Recipients benefit to the extent food aid provision is improved. PVOs costs of distribution or enhancement of the effectiveness of food aid are eased when they have the ability to monetize commodities. U.S. producers and exporters do not benefit from monetization, and may be hurt by monetized sales interfering with commercial exports. Monetization yields no direct benefits to U.S. taxpayers.

Local Purchases and Triangular Transactions. Another means with potential to enhance the developmental impact of food aid is local purchase or triangular transaction. Local purchases involve the distribution of commodities purchased elsewhere in the recipient country. Triangular transactions involve the purchase of food in one country for distribution in another. In the process, desired foods are provided to recipients and local markets and interregional trade may be strengthened.

During the 1980s, a growing share of world food aid was provided through local purchases and triangular transactions. Japan purchases commodities for its food and programs overseas (wheat from the United States, rice from Pakistan and Thailand, and at times maize from Zimbabwe) and ships them to the recipient nations. The European

Economic Community also distributes a sizable amount of food aid through such procedures. The United States, however, has provided relatively little aid through such transactions.

There is no direct benefit to U.S. taxpayers, producers, or exporters when the United States procures aid commodities overseas. Foreign recipients may benefit by receiving a more appropriate (perhaps indigenous) commodity and also from strengthening local markets while addressing the primary goal of hunger alleviation. However, such assistance is a complicated means of providing aid and is at odds with one of the aims of P.L. 480, (i.e., to expand export markets for U.S. agricultural commodities). U.S. producers and exporters are harmed by the lost opportunity to provide U.S. commodities as food aid.

Local purchases and triangular transactions, if desired, should be pursued with caution. Research has shown that potential benefits from local purchases and triangular transactions may not necessarily be realized (Clay and Benson 1990). In some cases, aid commodities acquired and shipped from a developing country were no more timely than aid shipped from the donor country itself.

Cargo Preference

The Merchant Marine Act of 1936, as amended in 1985, requires that at least 75 percent of U.S. food aid shipments be transported on vessels of U.S. registry. Some suggest that these "cargo preference" provisions should not apply to food aid. The U.S. General Accounting Office (GAO) reports that ships of U.S. registry is generally more expensive than shipping on vessels of other fleets, and that PVOs believe cargo preference provisions have resulted in less timely lifting of aid (U.S. General Accounting Office 1990). Currently, the U.S. Department of Agriculture (USDA) pays cargo preference costs on 50 percent of the food aid volume while the Department of Transportation pays the costs on an additional 25 percent.

Given a fixed food aid budget, there is a trade-off between the amount of commodities that the USDA may provide as food aid and its support for the U.S. merchant marine. Taxpayer support for the U.S. merchant marine through food aid programs reduces the volume of aid provided.[1] The owners, operators, and crew members of the relatively small number of U.S. vessels that transport food aid commodities are the

---

[1]However, if food aid were exempted from cargo preference rules, food aid volumes would not necessarily increase if savings were instead used for deficit-reduction purposes.

beneficiaries. However, the U. S. General Accounting Office (1990) cited a Department of Transportation report stating that "federal programs, including cargo preference rules, have not kept the U.S. merchant marine viable and competitive in world trade."

This discussion emphasizes how the greater the share of food aid funds devoted to noncommodity costs (e.g. transportation or processing of commodities), the lower are the benefits received by U.S. producers (in terms of higher market prices) and by taxpayers (through reduced CCC deficiency payments).

Growing Global Food Aid Needs

The 1989 National Research Council (NRC) report on global food aid projections for the 1990s suggests that by the end of the century global food aid needs will rise sharply. Using projections from six sources, estimates of needs by the year 2000 generally ranged from 30 to 50 million metric tons. (This compares to estimated needs in 1992/93 of 16 to 27 million tons when donors provided about 13 million tons.) The NRC aid estimates may be too low because they did not consider emergency needs and also because of two other major global developments.

First, it may take considerable time before some of the economies of the former Soviet Union will be able to fully provide for their own food needs, either through domestic production or commercial imports. These countries may represent an additional demand on global food aid supplies that was overlooked when the NRC estimates were made.

Second, an agreement in the Uruguay Round of the General Agreement on Tariffs and Trade (GATT), mandating multilateral reduction of protectionist trade barriers and policies in agriculture, presents a potential for two adverse effects on global food aid: price increases, as supports for internationally traded commodities are reduced, and reduced surplus stocks of selected commodities. Trade negotiations on the food aid provisions of GATT have highlighted the continuing need, under trade liberalization, for *bona fide* food aid to reinforce the food security objectives of developing countries. The challenge is doing so without increasing the potential for trade distortions. Suggestions include increased use of cash grants, expanding market purchases of food aid, and increasing the provision of multilateral aid (Ballenger and Mabbs-Zeno 1992).

The ability of the United States to respond to increasing food aid needs in the 1990s will be constrained. Current U.S. budgetary constraints will likely continue into the foreseeable future, limiting the potential for a significant rise in real (and perhaps even nominal) U.S. food aid expenditures. There is a secondary budgetary effect which reinforces the first: tight budgets limit the generosity of domestic U.S. farm commodity programs, diminishing the prospect of the CCC accumulating large stocks which could be used for food aid programs. As food aid becomes viewed as foreign aid rather than in terms of support for U.S. producers, it is difficult to foresee any significant expansion. Indeed, the President's proposed program level for P.L. 480 in fiscal year 1994 was about 5 percent less than that of fiscal year 1993.

Summary

As the time for debate over 1995 agricultural and food legislation nears, two likely issues will be: further flexibility in U.S. food aid for humanitarian and development assistance, and the applicability of cargo preference rules to food aid. The amount of food aid the United States will provide is an important, related issue. New flexibility in food aid procedures, authorized in the 1990 Act, was intended to achieve greater impacts with U.S. food aid. Issues to consider are the extent to which new legislative authorities have been used, and the extent to which program changes have improved program operation and outcomes. The issue of the degree of support to the U.S. merchant marine through food aid programs will also likely arise. However, given growing and changing food needs, the most important issue confronting policymakers when revisiting food aid programs may not be altering legislation dramatically, but rather determining how much additional food aid U.S. taxpayers are willing to provide.

References

Agency for International Development. *U.S. Overseas Loans and Grants and Assistance from International Organizations*. Washington, DC (various issues).

Ballenger, N., and C. Mabbs-Zeno. 1992. "Treating Food Security and Food Aid Issues at the GATT." *Food Policy*, August, pp 264-276.

Clay, E., and O. Stokke. 1991. *Food Aid Reconsidered: Assessing the Impact on Third World Countries*. London: Frank Cass & Co., Ltd.

Clay, E., and C. Benson. 1990. "Aid for Food: Acquisition of Commodities in Developing Countries for Food Aid in the 1980s." *Food Policy*, February, pp 27-43.

Fletcher, L.B. 1992. "Food Aid as a Development Resource: Performance, Potential and Prospects," in L.B. Fletcher, ed., *World Food in the 1990s: Production, Trade, and Aid*. Boulder, CO: Westview Press.

Food Aid Management. 1991. "Monetization Seminar," mimeo. Washington, D.C., November.

Food and Agriculture Organization of the United Nations. 1992. *Nutrition and Development—A Global Assessment*. Rome: Food and Agriculture Organization and World Health Organization.

______. 1991. "Prospects for Food Aid and Its Role in the 1990s." Rome: Committee on World Food Security. January.

______. 1993. *Food Aid in Figures*. Rome.

Food Policy. 1992. "Monitor: Deliberations in the WFP on Food Aid Issues." August, pp 308-309.

Hopkins, R. 1990. "Increasing Food Aid: Prospects for the 1990s." *Food Policy*, August, pp 319-327.

Maxwell, S. 1991. "The Disincentive Effect of Food Aid: A Pragmatic Approach." Pp 66-90 in E. Clay and O. Stokke (eds). *Food Aid Reconsidered: Assessing the Impact on Third World Countries*. London: Frank Cass & Co., Ltd.

National Research Council. 1989. *Food Aid Projections for the Decade of the 1990s*. Washington, D.C.: National Academy Press.

Price, M., M. Smith, and G. Mathia. Undated. "The Domestic Effects of U.S. Overseas Wheat Aid," USDA, Economic Research Service unpublished manuscript.

Saran, R., and P. Konandreas. 1991 "An Additional Resource? A Global Perspective on Food Aid Flows in Relation to Development Assistance." Pp 37-65 in E. Clay and O. Stokke (eds). *Food Aid Reconsidered: Assessing the Impact on Third World Countries*. London: Frank Cass & Co., Ltd.

Sen, A. 1990 "Food Entitlements and Economic Chains." In L.F. Newman (ed). *Hunger in History: Food Shortage, Poverty and Deprivation*. New York: Blackwell.

U.S. Department of Agriculture. 1992. "Backgrounder: Export Credit Guarantees and Food Aid to Russia." Office of Public Affairs, Oct. 9.

______. "U.S. Government Concessional Agricultural Exports Database." Economic Research Service.

U.S. General Accounting Office. 1990. *Cargo Preference Requirements Their Impact on U.S. Food Aid Programs and the U.S. Merchant Marine*. NSIAD-90-174. Washington, DC.

10

Food Quality: Safety, Nutrition, and Labeling

Carol S. Kramer and Julie A. Caswell

The Issues: Quality Concerns over Food Safety and Nutrition

An expanding array of food quality, safety, and nutrition issues has concerned and challenged the U.S. public and policymakers in recent years. As societies become more affluent, food concerns shift from quantity towards quality. Consumers' desires supersede observable characteristics such as color or visible spoilage, and experiential characteristics such as taste and texture, to encompass credence characteristics which are those reliant on consumer trust. Because key credence characteristics such as food safety or nutritional quality cannot be known by consumers through their senses, appropriate policies must establish effective methods of quality assurance.

The numerous dimensions of food quality are defined by the attitudes, cultural practices, technology and circumstances of a country. In the United States, food safety and nutrition concerns have grown in importance. The primary food safety concern is that of consuming a particular product which causes either acute or chronic harm to health. Potential food safety hazards include disease-causing microbes, naturally occurring toxins, environmental contaminants, pesticide or animal drug residues, and harmful food additives. Nutrition-related concerns reflect knowledge or beliefs that consuming particular products may contribute to a diet harmful to health. Causes range from over-consumption of overall energy or of particular nutrients such as fats and sodium, to under-consumption of valuable food components such as fiber. For many Americans, the major challenge is consuming less fat, sodium, sugar, alcohol, and tobacco and consuming more fiber, fruits, and vegetables.

Many surveys indicate, and experts agree, that U.S. consumers' knowledge of and interest in food quality issues increased in the 1980s and early 1990s. In addition, portions of the public express concern about the uses of new technologies in food production and processing. Examples are the development and use of anabolic steroids in meat production, bovine and porcine growth hormones, genetically engineered organisms (such as the "ice minus" bacterium), and irradiation. Increasingly, consumers appear to care about *how* food is produced in addition to what is produced. Thus, concerns about animal welfare, or the ecological damage associated with pesticide use or, even the effects of biotechnology on the family farm may increasingly figure into food consumption decisions.

The Central Food Policy Challenge

The central underlying food quality policy issue is assuring that consumers can purchase acceptably safe and nutritious foods. In other words, how and to what standard should the government attempt to ensure food quality, given the U.S. economic system and consumers' freedom to choose among food products. The experts and the public often disagree on the relative risk or acceptability of risks from different hazards (e.g., risks from microbial versus pesticide sources) and, therefore, the order of food safety policy priorities. An important aspect of establishing food quality standards concerns whose risks or needs should count the most, in other words, the distributional impact of standards. Should standards be set to protect the most sensitive consumer or the average consumer, and who should pay?

Experts realize that delivering food quality—such as lower levels of microbiological contamination or pesticide residues or enhanced nutrient profiles—is an often complex process requiring extensive coordination among producers, handlers, processors, distributors, and retailers. A relevant policy issue is how the necessary coordination can best be achieved.

Policymakers, as they fashion federal and state legislation and design and implement programs, respond to changing consumer and industry demands and attempt to exploit an evolving knowledge base. Improved understanding of pathogens and foodborne diseases therefore creates opportunities to implement new policies and programs. In the nutrition area understanding of the links between nutrition, other risk factors, and disease continually evolves. Finally, food safety and nutrition policies reflect the dramatic changes in food production, marketing and distribution, and consumption practices revolutionizing food markets in recent years. These changes include globalization of food markets and the growing affluence of consumers in developed countries.

Major Food Quality Policy Issues

Major food quality policy issues discussed here include:

- avoidance of microbial hazards by revamping and improving food inspection systems;
- reconciliation and rationalization of pesticide laws to better protect consumers (particularly children) farmers and farm workers, as well as the environment;
- development of new policy approaches for assuring food quality, including labeling and education;
- design, implementation, and evaluation of programs intended to assist consumers in improving their dietary habits; and
- appropriate monitoring of the development and use of new technologies to assure their safety.

The "Facts"

The major food safety issue now confronting federal policymakers is making sure that consumers can purchase foods acceptably free of microbial contaminants and of hazardous chemicals—including pesticide residues, natural toxins, environmental contaminants, animal drugs, or harmful additives. While these concerns are not new (recall Upton Sinclair's exposé *The Jungle*), new dimensions to the problems continually challenge regulators. The piecemeal development, over time, of regulations to assure food quality has resulted in diverse risk standards being applied in different areas. Rationalization of these standards is central to a rationalization of the system.

Microbial Contaminants

Microbial problems in the food supply may occur in a variety of products, originate anywhere from the farm through food service or the consumer's kitchen, and can be potentially reduced, avoided, or eliminated using alternative means of varying, often uncertain, cost-effectiveness. While problems with *Salmonella* or *Campylobacter* contamination in eggs and poultry products have gained wide notoriety and are symptomatic of a variety of concerns, red meat safety has recently been in the news.

One of President Bill Clinton's early acts was to dispatch newly appointed Agriculture Secretary Mike Espy to Olympia, WA, to learn more about the deaths and illnesses linked to contaminated hamburgers.

Between mid-January and March 1993, approximately 500 persons in the states of Washington, Idaho, California, and Nevada reported cases of hemorrhagic colitis associated with *E. coli 0157:H7* (U.S. Department of Agriculture 1993). Nationally, it is estimated that from 6.5 to 81.0 million cases of foodborne illnesses from microorganisms occur each year, causing approximately 9,000 deaths annually (Centers for Disease Control 1990).

Scientists continually discover new pathogens, which may imply that the figures cited above are underestimates. For example, the strain of E. *coli* pathogen implicated in the 1993 Pacific Northwest outbreak (*E. coli 0157:H7*) was not discovered until 1982 (*Food Chemical News* 1993).

The recent problems in the Northwest have focussed attention once more on questions of the safety of livestock-derived products and how best to assure that safety. Since at least 1985, when the National Academy of Sciences published its ground-breaking report on establishing a scientific, risk-based, meat and poultry inspection system, many of the issues—the need for systems design, risk assessment, rapid tests for microbial contaminants, microbial standards, and traceback authority—have been with us. Enhanced efforts to "optimize" the current system in the short run and subsequently to make more major changes toward a risk-based, prevention-oriented inspection system based on risk analysis were announced by USDA in early 1993 (U.S. Department of Agriculture 1993).

Increasingly consumers buy food in restaurants or in-store delis, that is, food produced and prepared away from home. With more meals being purchased or eaten away from home, food safety in food service situations has become increasingly important. In addition, microwave cookery, while revolutionizing food preparation and distribution possibilities, raises new challenges. Development of information and inspection and quality control systems that make optimal use of new technologies and appropriate sampling methods is a pressing challenge.

Protection from Chemicals

Among the chemical issues, the safety of pesticides used in food production is perennial; it grows in importance each year as the use of pesticides remains high and as legislative inactivity persists. Major public concerns center on the adequacy and consistency of *safety standards*, the *quality of information* for regulators and the public, *how risks and benefits should be compared* (if at all), and *the backlog of pesticide registration decisions* at the EPA. In addition, *establishing safety standards for pesticides vis á vis the diets of infants and children* is addressed in an important study released in June 1993 by the National Academy of Sciences (National Academy of Sciences 1993).

Perhaps the prime pesticide policy issue is establishing an appropriate, consistent safety standard for the carcinogens, but also for reproductive hazards, neurological toxins, and chemicals with other adverse effects. The various Delaney clauses to the Federal Food, Drug and Cosmetic Act of 1938 set zero risk as the standard for chemicals classed as carcinogens and used as food or color additives or new animal drugs (including pesticide residues concentrating in processing). Present assay methods can detect residues in parts per billion, so the Environmental Protection Agency (EPA) had attempted to effectively establish a floor under the Delaney Clause by adopting what was called a *de minimus* standard of negligible risk. Essentially, the EPA interpretation expressed the notion that the law does not bother with "trifles," in this case residue levels resulting in fewer than one in one million cancers over lifetime exposure.

In recent months a coalition of consumer and labor groups challenged the EPA in court. The U.S. Court of Appeals decided that EPA should not continue its *de minimus* interpretation of Delaney. In February 1993 the U.S. Supreme Court let stand the lower court decision. The implications of the decision could result in the loss of some 35 currently registered pesticides that have shown carcinogenic properties, but whose risk the EPA had considered negligible or insignificant. The Court decision focuses attention once more on the need to revisit safety standards for pesticides. Whether the Congress can reach agreement remains to be seen.

In June 1993 the National Academy of Sciences released a long-awaited study of pesticides in the diets of children. This study focused attention on appropriate methods for assessing risks to children from dietary exposure to pesticides, given the facts that children's diets vary less than those of adults, and that children consume proportionately more foods such as fruits and vegetables per unit of body weight than do adults. The study points out that children are not merely "little adults" but, depending on their stage of development, experience qualitatively different reactions to chemicals. Hence, the issue of improving risk assessment and setting conservative standards to protect the most vulnerable consumers has been placed squarely before both Congress and the Administration.

A corollary to safety standards is the issue of which decision rule is most appropriate. How should economic and other benefits of pesticides be weighed in approval decisions? Methyl bromide, an extensively used soil, postharvest, and quarantine fumigant provides a current case in point where both risks (to the ozone layer) and benefits (to producers in semitropical climates and consumers) were both considered in decisions to phase out the pesticide.

Pesticide regulatory decisionmaking has been slow at best. As the base of toxicological information expands each year so does the need to revisit pesticide studies that were originally used to substantiate registration of now older pesticides. In 1972 and again in 1988, the Congress amended the Federal Insecticide, Fungicide, and Rodenticide Act of 1947 (FIFRA) to require EPA to reevaluate registered pesticides under more current scientific and regulatory criteria. To accelerate the re-registration review process, time frames that would result in the completion of most pesticide re-registration decisions by 1997 were set. Nevertheless, the mandated EPA re-registration process lags far behind schedule. More than 50,000 pesticide products have been registered since FIFRA was enacted in 1947, and some 17,000 (containing about 676 distinct pesticide active ingredients) are subject to re-registration. As of July 1992, the EPA had reached final determinations on only two (U.S. General Accounting Office 1992, p 2).

While pesticide issues receive the most prominent attention regarding the safety of chemicals found in the food supply, other concerns exist and may become the focus of increased public policy activity. Among these are residues of animal drugs in food products, particularly in milk (U.S. General Accounting Office 1990, 1992), and the potential effects of various biotechnologies on the chemical composition of foods.

Nutrition and Nutrition Labeling

As noted, public health experts recognize the importance of good nutrition and the links between nutrition and many of the leading causes of death and illness in the United States. Public opinion surveys indicate that public health recommendations and consumers' concerns are converging. Many Americans now recognize the importance of nutrition and are concerned about the amounts of fats, sodium, fiber, and other nutrients in their diets.

Unfortunately, concern about diet has not, in many cases, brought about implementation of improved dietary patterns. Changes in American consumers' diets have not kept pace with knowledge, and there is some evidence of backsliding in some areas (Putler and Frazao 1991).

The new mandatory nutrition labeling effective in 1994 will give consumers a powerful tool for following dietary recommendations (Caswell and Padberg 1992). This tool is in the form of a label that makes it easier to place foods in the context of a healthy diet, and in regulations that ensure the validity of claims such as "light" and "free". Critics argue that the labeling regulations, while imposing costs on the food industry, will not significantly better inform consumers or change their dietary habits. Critics say that the labels contain too much information and are too complicated for the average consumer to use effectively. However, after intense contention, the regulations are now in place. If, and under what circumstances, consumers will use them to improve their diets remains an empirical question.

An important determinant of the new labels' success may be the degree to which public and private educational programs explain their purpose and encourage their use. Planning for these types of educational programs is underway.

For the large majority of Americans, the major nutritional issue is limiting consumption of fats, sodium, sugar, alcohol, and tobacco, and increasing consumption of fiber, fruits, and vegetables. Consensus regarding these dietary recommendations has been painstakingly built over the last 20 years (U.S. Department of Health and Human Services 1988, National Academy of Sciences 1989). In 1992, USDA incorporated the recommendations in its "eating right" pyramid, a key component in future nutrition education programs. The recommendations themselves are based on expanding medical knowledge of the links between diet and ten of the leading causes of death and illness in the United States.

Currently, the most pressing challenge is to identify ways to help consumers modify their own behavior, with expanded nutrition labeling and nutrition education the major approaches underway. In addition, many in-consistencies exist in public policies and programs affecting nutrition. Consumer groups such as Public Voice for Food and Health Policy have criticized what they consider the high fat content of government donations to school lunch programs, for example (*Washington Post* 9/17/92, p 19).

New Technologies

New technologies present opportunities and challenges for food quality. Some new technologies facilitate food safety assurance by replacing older, less safe ones or by making safety monitoring easier. However, other technologies such as irradiation or the use of bovine somatotropin (bST) are often challenged, themselves becoming quality issues for consumers. As an example, irradiation has been approved for some uses to control development of microbial pathogens, but consumer opposition has been strenuous.

At the food product level, biotechnological techniques may lead to enhanced shelf life, nutritional and/or mineral fortification, or other "designer" foods. Finally, because convenience counts large in many consumers' utility functions, precooking or packaging practices that lend to convenience are popular but may also have implications for safety or quality.

Policy Background

Administrative Responsibility

Responsibility for regulating food quality extends over several federal agencies in a patchwork of programs. The Food and Drug Administration (FDA) has prime responsibility for safety and nutritional labeling of processed foods. Meat and poultry are primarily regulated by USDA. Nutrition education has also been the bailiwick of USDA. Other governmental agencies also have roles in ensuring food quality.

In general, the federal policy approach to new technologies has been to attempt to evaluate them under existing laws. Thus, the safety of biotechnological procedures for developing food commodities or products with enhanced or selected quality attributes is regulated by various agencies under several different statutes.

Food Safety

Since 1906, the foundation of food safety law has been the Federal Food, Drug, and Cosmetic Act (FFDCA) as amended, and associated regulations found in the Code of Federal Regulations. Under its various sections, the FFDCA addresses microbial contamination, pesticides and animal drug residues, naturally occurring toxins, environmental contaminants, and unsafe food and color additives. In addition, a series of laws such as the Wholesome Meat Act of 1967 and the Poultry Products Inspection Act of 1957 elaborate legal jurisdiction for specific product groups. For particular issues (e.g., seafood safety), legal and regulatory responsibility is unclear.

Federal pesticide policy, which has given rise to much controversy in recent years, is defined principally by the Federal Insecticide, Fungicide, and Rodenticide Act of 1947 (FIFRA) administered by EPA, but food additive provisions of the Federal Food, Drug, and Cosmetic Act of 1938 also apply. Thus, residue safety enforcement falls under the FFDCA and to the FDA. Since 1970, under FIFRA, EPA has been charged with determining the legal uses of pesticides and establishing the maximum residue levels, or tolerance levels, permissible in or on foods. EPA is charged, in a process called "Special Review," with weighing a pesticide's benefits against its risks as it decides whether to register, suspend, or cancel registration of a particular pesticide for use on a given commodity (Reichelderfer 1990). However, in what has been called the "Delaney Paradox," if a pesticide demonstrating carcinogenic properties is shown to concentrate in processed foods, then a tolerance may not be granted (National Academy of Sciences 1987). In essence, different safety standards apply to processed and unprocessed foods (Reichelderfer 1990; National Academy of Sciences 1987).

Policy Instruments

In general, the U.S. food safety policy approach combines *final product standards* setting minimum acceptable quality, some specification of *permitted production and processing methods,* and some *monitoring via government inspection systems.* In addition, the government provides *information* in the form of research on the development of good manufacturing practices, and mandates that firms provide specific information, either to the public in the form of ingredient labels, or to government agencies to meet requirements for premarketing product approval.

Safety standards vary over products or with respect to different food constituents, creating problems of inconsistency. Food and color additives, as well as pesticide residues that concentrate in processed foods, are subject to the no-risk cancer standard contained in the Delaney Amendments to the FFDCA. Other foods or constituents may meet alternative, less stringent safety standards, such as the "Generally Recognized as Safe" (GRAS) standard, and are not required to meet the no-risk Delaney standard.

It is especially important to note conflicts in the decision rules regarding pesticide safety between FIFRA and FFDCA. FIFRA permits the use of carcinogenic pesticides provided their risk to consumers is not "unreasonable." FFDCA on the other hand prohibits approval of any carcinogenic pesticide in or on foods that leaves residues that concentrate in processing.

Nutrition

Beginning in the mid-1980s in response to growing consumer interest and demand and in conjunction with the growing consensus on the importance of diet, food processors began to vigorously promote the nutritional content of their products. In the laissez-faire regulatory environment, some label and advertising messages were unsubstantiated or false. Potential health gains from improved diets and strong consumer interest in nutritional contents of foods along with some market abuses, led to calls for a general overhaul of food labeling requirements (see for example, U.S. Department of Health and Human Services 1990; National Academy of Sciences 1990, 1991). The Nutrition Labeling and Education Act of 1990 directed the FDA to make extensive changes in the labeling requirements of foods it regulates. Changes include *mandatory labeling, standardization of serving sizes,* strict *regulation of descriptors* (for example, "free", "less"), stringent *limits on health claims*—such as messages linking particular nutrients to specific health conditions or diseases (e.g., linking fiber consumption and colon cancer)—and the inclusion of *daily reference values*, which inform consumers of the percentage of the average daily requirement of a nutrient (e.g., fat) contained in a serving of a particular food product. By agreement, the USDA will apply nearly identical standards to labels for processed meat and poultry products it regulates. Final regulations take effect in May and July 1994.

With the recent revisions in federal food labeling policy—now a major focus of nutrition policy—changes in the foreseeable future are likely to be refinements only. Public and private educational programs to accompany the introduction of the new labels are expected to leverage their impact on consumers. One important unresolved side issue is whether nutrition-based advertising claims, under Federal Trade Commission jurisdiction, will be regulated in a manner consistent with the labeling standards.

Economic and Political Considerations

Each of these food policy areas—safety, nutritional quality, labeling, and technology—has both economic and political dimensions.

Contributions of Economic Theory

The rationale for government involvement in safety assurance and market regulation flows from experience with problems in the operation of unregulated markets. Welfare economists term these problems manifestations of market failure: unregulated food markets often produce suboptimal levels of food quality.

Some problems are driven by the fact that consumers or buyers throughout the system lack adequate information. An unregulated food market typically does not offer the producer incentives to generate and provide information about quality. This information frequently contains a public-good component, meaning in part that many who receive it have no incentive to pay for it. Thus, the producer who invests in information or enhanced quality may not recover the costs of doing so.

Throughout history, governments have attempted to reduce the prevalence of "caveat emptor" ("let the buyer beware") in food markets and have actively regulated the sale of food and drink. How policies and regulations are formulated affects not only the costs and profitability of producing and distributing food—that is, the supply side of the market—but also the quantity, quality, and prices of foods demanded by both U.S. and foreign consumers. The economic impacts of food safety or nutrition policies can be analyzed in the context of welfare economics using benefit cost analysis or related techniques such as risk benefit analysis or cost effectiveness analysis.

From the policy standpoint, where does the demand for food quality originate? Private parties (consumers) and public sources or society as a whole demand food quality (Kramer 1991a). Consumers wish to avoid foodborne risks and the costs associated with them, be they illness, death, or simply loss of peace-of-mind (van Ravenswaay 1992). There is public-good value associated with general confidence in the food supply. In addition, public demand exists for food quality because of the societal costs or externalities associated with acute foodborne illnesses such as salmonellosis, or chronic diseases such as arterio-sclerosis (Caswell 1990). Social costs of foodborne illness include both medical costs and productivity losses to the economy.

Benefits and costs associated with regulation, taxes, or information requirements and the relative responsiveness of supply and demand to changes in costs, prices, and quality changes—affect social welfare as well as the distributional consequences of policy changes. To the extent that consumers suffer illness or death related to food safety or nutritional characteristics, policy alternatives affect public health and costs associated with foodborne illness and these become part of the economic calculus.

From a public finance or public policy perspective, costs associated with assuring food quality can be weighed, both by consumers and by policymakers, against the entire range of alternative uses for dollars that may improve social welfare.

Political Dimensions

In addition to the economic dimension are political concerns: the distributional effects of policy changes on relative costs or benefits have repercussions on clientele groups in the political process. From public-choice theory comes the insight that small, well-organized producer groups experience greater economic incentives to express political preferences about a policy change than do large groups of consumers. Each individual consumer might benefit only slightly in relation to the transactions costs of taking action, though the aggregate benefits for all consumers of a policy change might be great and social net benefits positive.

Alternative Policy Approaches

Ensuring food quality can be reduced to two major questions: What standards should be set and how should those standards be implemented

(Caswell 1990). The first question involves measurement of risks (risk assessment) and judgements about what levels of risk are acceptable in particular situations (and to whom). Standards are set by combining this information with cost information for attaining various levels of risk reduction. As noted above, no one would argue that the federal government's current acceptable risk standards are consistent. Ultimate coherence in food quality programs relies on consistent standards, although this is likely unattainable. Meanwhile, the policy process often proceeds on a piecemeal basis, examining particular risks and risk standards singly.

Under current policy, federal risk standards provide a key benchmark or floor in all food quality areas. Private parties (or the states) may choose to adhere to stricter standards, as happens, for example, when producers, distributors, and consumers choose organic foods. As tolerance for foodborne risks by U.S. consumers seems to be decreasing, two major policy alternatives arise.

Update Risk Standards

One policy approach is to update product risk standards so as to bring them in line with consumers' acceptable risk levels, although this raises questions about whose risk tolerance should be respected, the most concerned consumer, an average consumer, or that of the experts in the field? Updating (often tightening) risk standards has the advantage of preserving the traditional high floor under food quality and the "every food a safe food" approach but the disadvantage of introducing higher costs into the food marketing system.

Encourage Private Market Approaches

The second major policy option is essentially to leave current standards in place and encourage the development of private markets to serve consumers who prefer foods that meet more stringent final product standards. From the standpoint of food safety, this represents a significant policy departure. The two options discussed here align with alternative approaches to risk standard implementation. These two approaches can be referred to as the "banning or minimum standard" and the "information" approaches (Zellner 1988, Caswell 1990). Under the banning or minimum standard approach, government sets the product standards that serve as a floor and any product that does not meet the standard is considered illegal for sale. Under the information approach, government standards serve only as a benchmark and regulation focuses on making sure consumers have reliable information that will allow them to make informed choices. Obviously, the easier it is to inform consumers and the better they are able to judge product quality, the more attractive are information strategies.

In virtually all areas of food quality assurance, the federal government mixes minimum final product standards/banning and information strategies. However, it is very important to understand that, to date, the United States has mostly chosen a banning/minimum standard approach to food safety, and accepted and institutionalized an information (labeling) approach to nutritional content. It is equally important to understand that both regulatory approaches are applicable to both quality issues in certain cases. A major source of regulatory innovation in the future may come from applying the approaches in new areas.

Since a regulatory policy of information (labeling) and education in regard to nutritional content is firmly entrenched, food safety is the major area in which alternative policies may be considered in the near future. The alternatives include modification or further development of minimum standards related to safety, adoption of safety labeling, or a combination of the two.

A labeling approach to food safety requires the development of markets for food products with varying degrees of safety, markets which have been slowly developing. Formal labeling regulations may encourage this development, just as national standards for organic products are expected to facilitate their marketing. The major trade-off to be considered is the value to consumers of being able to choose products with varying safety levels versus an "all food is safe food" policy. There is also some doubt whether markets will really support products of varying safety or if only the safest product will ultimately survive in the market.

An important upcoming policy issue is whether products produced with new technologies (e.g., biotechnology) will be required to be labeled. Opponents of such technologies argue that labeling preserves consumers' freedom to choose without stifling innovation. They suggest that markets for products free of specific technologies will develop.

Differences exist between two labeling policies: 1) *requiring* that products using the technology be labeled as such, and 2) *allowing* products that do *not* use the technology to be labeled as such. Whereas food producers, processors, and retailers can usually use the second approach without restriction, the first approach is more powerful, because the label requirement tends to suggest to many consumers that questions about the technology's safety may exist. Given that most new technolo-gies face some organized opposition, a major issue will be whether, in the political process, labeling of a technology will become its price of admittance to the market.

Challenges for Change

As preparations begin for formulation of the 1995 food and agriculture legislation, it is useful to consider the manner and extent to which concerns over food quality have (or have not) been incorporated in past iterations of the bill. A brief review of the Food, Agriculture, Trade, and Conservation Act of 1990 suggests that while this legislation affects domestic and foreign food consumers in numerous important ways, for the most part major issues of food quality, safety, and nutrition are treated elsewhere. The 1990 Act does of course exert numerous direct and indirect effects on food availability, food costs, food safety and food quality, marketing rules, food assistance and nutrition programs, and the public research agenda. It also imposes costs on consumers as taxpayers to fund provisions of the legislation.

To provide some perspective, the 1990 act contains 25 titles in a 5-year framework for agricultural and food policy. The first 11 titles deal with specific commodity programs and general commodity provisions. Subsequently, the mix of titles includes forestry; conservation; agricultural trade; credit; research; rural development; food assistance; fruits, vegetables, and marketing; grain quality; and organic certification (Kramer 1991b, p 914).

In the end, what did the 1990 act mean for consumers in terms of impacts on food quality, safety, nutrition, or labeling? Perhaps most symbolically important were some of the newer "green" measures included: to develop national organic certification standards, to require that farmers keep records of their pesticide applications, to establish new water quality measures, to undertake research into the effects on pesticide use and consumer demand of 'cosmetic' grade and quality standards for fruits and vegetables, and to redirect some research expenditures toward food safety, sustainable agriculture, and environmental improvement.

More tangible and established in some senses, the 1990 act also reauthorized the major food assistance programs without significant changes, and in an era of recession, incorporating some program operation changes to make these programs more accessible to eligible recipients (Kramer 1991).

Thus far U.S. food policy, as opposed to agricultural policy, has never been developed in a systematic manner—either in the context of food and agricultural legislation or in any other comprehensive legislative vehicle. Indeed, this has been one criticism consistently leveled by food policy analysts over time. What are the chances for a change in the future, perhaps with food quality, safety, and nutritional issues integrated into broader food and agriculture legislation? If one considers the rather surprising success of environmental groups who instigated the design and incorporation of conservation and other environmental measures, first in the Food Security Act of 1985 and, subsequently, in the Food, Agriculture, Conservation, and Trade Act of 1990, it encourages intriguing speculation as to how attempts to rationalize agricultural and food policies might be pursued and how measures akin to the cross-compliance and cost-sharing employed by the environmental lobby might be incorporated.

If future food and agricultural legislation developers wished to actively incorporate food quality measures in the legislation, they would have the choice of developing distinct food safety or nutrition content titles, or incorporating measures, where appropriate, in the existing commodity or research titles.

There may be some value in considering other instances where multiple objectives have been combined in provisions of food and agricultural legislation like that of the 1990 act. For example, it might be convincingly argued that both domestic and foreign food assistance programs have always represented a marriage of convenience between objectives of making food available (sometimes with explicit nutritional objectives) to the needy and expanding agricultural sales. Perhaps the food assistance programs, more than any other in the food policy area, represent an operational melding of competing objectives. Although controversial in the sense that surplus commodities are not always among those highly rated by nutritionists, the food assistance programs, nevertheless, have passed the test of political survival.

Why have most of the major food quality and labeling issues been handled outside of the food and agricultural legislation framework, and is this likely to be a permanent situation? Asking the question another way: Is there a role for future food and agricultural legislation to explicitly promote food quality, safety, and nutrition, and at the same time promote consistency among the objectives of the commodity and conservation titles? The probable answer, "it depends," reflects not only the yet to be determined imagination and cooperation of policymakers, but the political organization of Congressional responsibility for the various jurisdictional areas and the legislation involved. In addition, the responsibility for oversight of various administrative agencies *and* the potency of public and constituency concerns at the time of legislation development will shape the political will and feasibility of modifying business as usual.

References

Caswell, Julie A. 1990. "Food Safety Policy Fights: A U.S. Perspective." *Northeastern Journal of Agricultural and Resource Economics* 19:59-66.

Caswell, Julie A. and Daniel I. Padberg. 1992. "Toward a More Comprehensive Theory of Food Labels." *American Journal of Agricultural Economics* 74:460-468.

______. 1990. *Food Safety and Quality: FDA Surveys Not Adequate to Demonstrate Safety of Milk Supply.* Washington, D.C.: GAO/RCED-91-K26.

Kramer, Carol S. 1990. "Food Safety: The Consumer Side of the Environmental Issue." *Southern Journal of Agricultural Economics* 22.

______. 1991a. "Health Risks as a Cost and Driving Force in Public Food Policy Decisions." *Understanding the True Cost of Food: Considerations for a Sustainable System.* March Proceedings of Institute for Alternative Agriculture, Eighth Annual Scientific Symposium, Washington, D.C.

______. 1991b. "Impact of the 1990 Farm Bill on Consumers," *American Journal of Agricultural Economics* 73:913-916.

National Academy of Sciences, National Research Council. 1985. *Meat and Poultry Inspection: The Scientific Basis of the Nation's Program.* Prepared by the Committee on the Scientific Basis of the Nation's Meat and Poultry Inspection Program, Food and Nutrition Board. Washington, D.C.: National Academy Press.

______. 1987. *Regulating Pesticides in Food: The Delaney Paradox.* Prepared by the Committee on Scientific and Regulatory Issues Underlying Pesticide Use Patterns and Agricultural Innovation, Board on Agriculture. Washington, D.C.: National Academy Press.

______. 1989. *Diet and Health: Implications for Reducing Chronic Disease Risk.* Report of the Committee on Diet and Health, Food and Nutrition Board, Commission on Life Sciences. Washington, D.C.: National Academy Press.

______. 1993. *Pesticides in the Diets of Infants and Children.* Committee on Pesticides in the Diets of Infants and Children, Board on Agriculture and Board on Environmental Studies and Toxicology. Commission on Life Sciences. Washington, D.C.: National Academy Press.

National Academy of Sciences, Institute of Medicine. 1990. *Nutrition Labeling: Issues and Directions for the 1990s.* Report of the Committee on Nutrition Components of Food Labeling, Food and Nutrition Board, Institute of Medicine. Washington, D.C.: National Academy Press.

______. 1991. *Improving America's Diet and Health: From Recommendations to Action.* Report of the Committee on Dietary Guidelines Implementation, Food and Nutrition Board, Institute of Medicine. Washington, D.C.: National Academy Press.

Putler, Daniel S. and Elizabeth Frazao. 1991. "Assessing the Effects of Diet/Health Awareness on the Consumption and Composition of Fat Intake." *Economics of Food Safety*, ed. Julie A. Caswell, pp 247-270. New York, NY: Elsevier Science Publishing Co., Inc.

Reichelderfer, Katherine H. 1990. "Pesticide Regulatory Policy: Pros, Cons, and Prospects for Change." Paper prepared for Resources for the Future briefing on Pesticides and Food Safety, January 16. Mimeo.

U.S. Department of Agriculture. 1993. Briefing book. Prepared for American Meat Institute Workshop, Chicago, Il.

U.S. Department of Health and Human Services. 1988. Food and Drug Administration. *The Surgeon General's Report on Nutrition and Health.* DHHS (PHS) Publ. No. 88-50210. Washington, D.C.: Government Printing Office.

______. 1990. Food and Drug Administration. "Food Labeling: Definitions of the Terms Cholesterol Free, Low Cholesterol, and Reduced Cholesterol; Tentative Final Rule." Federal Register 55, No. 139, 19 July, 29456-29473.

U.S. General Accounting Office. 1992. *Food Safety and Quality: FDA Strategy Needed to Address Animal Drug Residues in Milk.* Washington, D.C.: GAO/RCED-92-209.

van Ravenswaay, Eileen O. 1992. "Public Perceptions of Food Safety: Implications for Emerging Agricultural Technologies." In Volume 2: A New Technological Era for American Agriculture—OTA Commissioned Background Papers: Part E: Food Safety and Quality/U.S. Congress, Office of Technology Assessment, National Technical Information Service, Springfield, VA.

Zellner, James A. 1988. "Market Responses to Public Policies Affecting the Quality and Safety of Food and Diets." *Consumer Demands in the Marketplace: Public Policies Related to Food Safety, Quality, and Human Health,* ed. Katherine L. Clancy, pp 55-73. Washington, D.C.: Resources for the Future, National Center for Food and Agricultural Policy.

International Agriculture

11

International Trade Agreements

C. Parr Rosson III, C. Ford Runge, and Dale E. Hathaway

The United States is one of 23 charter members of the multilateral trade organization known as GATT, or the General Agreement on Tariffs and Trade, founded in 1948. Prior to the 1980s, the United States depended upon GATT for increased market access and resolution of trade disputes. During the second half of the 1980s, however, the United States began to move toward preferential bilateral or regional trading arrangements known as International Trade Agreements (ITAs). ITAs provide for the elimination of tariff and nontariff barriers (import quotas and licenses) to trade. Investment laws, transportation regulations, and mechanisms for resolving trade disputes are often included.

In most cases, ITAs are used to strengthen existing trade ties among nations, creating opportunities for additional employment and stimulating investment. More recently, concerns about air and water pollution, worker safety, and unfair labor regulations have led to the inclusion of environmental and labor issues in the negotiation of ITAs.

Although ITAs are relatively new in U.S. trade policy, many nations have used and participated in various forms of preferential trade for decades. Currently there are more than 23 forms of preferential trading arrangements among the 119 countries that account for 82 percent of world trade (Fieleke 1992). The use of ITAs to achieve both domestic and international trade policy objectives is clearly increasing.

U.S. International Trade Agreements

Since 1985, the United States has negotiated ITAs with Israel and Canada, and a regional agreement with Mexico and Canada. It now appears poised to begin trade talks with Chile. The outcomes of the latter two agreements are at the moment undecided, but public concerns regarding the ability of the United States to compete with low wage economies that produce low-cost goods for export have been raised. The loss of U.S. jobs to countries such as Mexico, the environmental consequences of more-open common borders and increased industrialization in Mexico, and the prospects of the world economy fracturing into openly hostile trading blocs are major public policy issues related to the use of ITAs as trade policy tools. Some writers doubt that free trade is achievable, particularly in highly protected sectors such as agriculture.

The Reciprocal Trade Agreements Act of 1934 (RTA) authorized the President to fix tariff rates, and resulted in a liberal tariff stance for the United States. From 1934-47, in fact, the United States negotiated bilateral trade agreements with 29 nations. However, when the GATT emerged as the primary trade forum, the RTA has declined in importance as a mechanism for trade liberalization.

In September, 1985, the United States and Israel entered into a bilateral agreement to reduce tariff and nontariff trade barriers in merchandise trade. Services trade was liberalized and provisions were made to protect intellectual property rights.

The Canada-U.S. Trade Agreement (CUSTA) went into effect on January 1, 1989. It provides for the elimination of both tariff and nontariff barriers to trade between the two countries, specifies rules governing trade in services, and reduces restrictions on investments by both nations over a 10-year transition period.

The North American Free Trade Agreement (NAFTA) was signed by President Bush on December 17, 1992, clearing the way for its submission to the U.S. Congress for approval or refusal sometime in 1993. NAFTA calls for the elimination of tariffs and nontariff barriers to trade by commodity sector over a 5-, 10-, or 15-year period, depending upon the economic and political sensitivity of the particular sector.

Forms of Economic Integration

For the most part, ITAs agreed to by the United States have taken the form of *free trade areas* consisting of provisions to remove barriers to trade with members, while retaining barriers with nonmembers. However, *free trade areas* represent but one form of economic integration.

A *customs union* requires that members eliminate import tariffs with each other and takes integration a step further by establishing identical barriers against nonmembers, most often in the form of a common external tariff. The European Community (EC), created by the Treaty of Rome in 1957, is probably the world's most well-known *customs union*. It includes a common agricultural policy (CAP) for all 12 member states. CAP is characterized by high internal support prices, a variable levy system to limit imports, and export subsidies to move surplus production onto the world market.

A *common market* takes integration yet further by permitting the free movement of goods and services, labor, and capital among member nations. In January, 1993, the EEC moved closer to full common-market status by implementing the Single European Act. It calls for elimination of barriers to trade in goods and services, labor, and capital.

Finally, an *economic union* completes the economic-integration process by harmonizing, or even unifying, the monetary and fiscal policies of member nations. An *economic union* represents the most advanced form of economic integration and has been achieved by Benelux, a union created following World War II by Belgium, the Netherlands, and Luxembourg.

These various forms of economic integration have all occurred within the guidelines of the General Agreement on Tariffs and Trade (GATT) to foster liberalization of world trade. Although Article I of GATT prohibits the use of preferential tariff rates, an important exception is allowed by Article XXIV. The latter Article specifies the conditions under which signatories may form and legally operate international trade agreements. These conditions are: (1) trade barriers are eliminated on substantially all trade among members; (2) trade barriers remaining against nonmembers are not more severe or restrictive than those previously in effect; and (3) interim measures leading to the formation of the agreement are employed for only a reasonable period of time. When these three conditions are met, international trade agreements do not violate any of the GATT Articles and may exist legally under the GATT framework.

The Single European Act, passed in 1987, will attempt to further unify the EEC by creating a single, barrier-free market for most goods and services. Customs documents and procedures, value-added taxation, and nontariff barriers to trade have been harmonized to facilitate the movement of goods among EEC member nations. However, more contentious issues, such as developing a common EEC currency, may delay the ratification of the Treaty of Maastricht, the document which completes the economic integration of the EEC.

It has been possible to rationalize, if not justify, the proliferation of ITAs under the GATT. Vague phrases such as "substantially all trade" and "reasonable time period" have allowed much latitude for interpretation of the GATT articles. Since its inception, GATT has been notified of over 70 ITAs, some establishing interim agreements with no dates for completing the *free trade area* specified. None have been formally disapproved (Jackson 1989).

Gaining or Losing from International Trade Agreements

With the prevalence of ITAs for liberalizing trade, at least among member nations, a question of critical importance is whether or not the United States will become better off as the economies of North America become more closely integrated. Each case of ITA creation must be examined empirically and on its own merits. It is impossible to assert unequivocally that ITAs may be more efficient than free trade among all nations. Given that trade is less than free, the creation of an ITA may or may not result in welfare gains. Recent advances in international trade theory allow some conclusions to be made about likely outcomes of ITA creation. Viner (1950) and Johnson (1965) were among the first to explain the costs and benefits of ITAs through the example of a *customs union*. These writers focused primarily on trade creation and trade diversion, along with total welfare gains. Later efforts by Salvatore (1993) focused on the dynamic effects of ITA creation.

Static Effects: Trade Creation and Trade Diversion

The static effects of creating an ITA are measured by trade creation and trade diversion. Trade creation occurs as some domestic production of a member nation is replaced by lower-cost imports from another member nation. Assuming full employment of domestic resources, trade creation increases the economic welfare of member nations because it leads to greater specialization in production and trade, based on comparative advantage. ITA members will import from one another certain goods not previously imported at all due to high tariffs. The trade-creation effect results in efficiency gains for member nations because some members shift from a higher-cost domestic source of supply to a lower-cost foreign source. A trade-creating ITA may also increase the welfare gains of nonmembers, since some of the increase in its economic growth will produce real increases in income that will in turn translate into increased imports from the rest of the world.

Trade diversion occurs when lower-cost imports from nonmember nations are replaced by higher-cost imports from members. This occurs naturally with an ITA, due to the preferential trade treatment provided member nations. Alone, trade diversion reduces global welfare since it shifts production from more efficient producers outside the ITA to less efficient producers within the ITA. The international allocation of resources becomes less efficient and production shifts away from comparative advantage.

In reality, most attempts to create ITAs contain both trade creation and trade diversion and can increase or decrease member welfare, depending on the relative strength of the two opposing forces. ITAs will most likely lead to trade creation and increased welfare of member nations under the following conditions:

1. High pre-ITA trade barriers increase the probability that trade will be created among members, rather than diverted from nonmembers to members.
2. The more countries included in the ITA and the larger their size, the more likely that low-cost producers will be found among its members.
3. An ITA formed by competitive, rather than complementary, economies is more likely to produce opportunities for specialization in production and trade creation.
4. When member nations are in close proximity to one another, transportation costs become less of an obstacle to trade creation.
5. If pre-ITA trade and economic relationships among members are strong, greater opportunities for welfare gains can be expected.

Dynamic Effects of ITAs

The formation of an ITA can be expected to have major dynamic benefits that should be considered important to the participating nations. In fact, it has been recently estimated that the dynamic gains from forming an ITA often exceed the static or welfare gains by a factor of five or six (Salvatore 1993). The more important dynamic gains include: increased competition, economies of scale, stimulus to investment, and more efficient use of economic resources.

Increased Competition. Possibly the most important single gain from an ITA is the potential for increased competition. Producers, especially those in monopolistic and oligopolistic markets, may become sluggish and complacent behind barriers to trade. With formation of an ITA, trade barriers among members are eliminated, and producers must become more efficient to effectively compete. Some may merge with other firms; others will go out of business. The higher level of competition is likely to stimulate the development and adoption of new technology. These forces combined will likely reduce costs of production and in turn consumer prices for goods and services. Importantly though, the ITA must ensure that collusion and market-sharing arrangements are minimized if these competitive forces are to operate efficiently.

Economies of Scale. Another major benefit of trade agreements is that substantial economies of scale may become possible with the expanded market. If firms were serving only the domestic market, the expanded market with the ITA will likely create substantial export opportunities, resulting in more output, lower costs per unit, and greater economies of scale. For instance, it has been determined that many firms in small nations such as Belgium and the Netherlands before joining the EEC, were comparable in size to U.S. plants, and thus enjoyed economies of scale by producing for the domestic market and for export. However, after becoming members of the EEC, significant economies of scale were gained by reducing the range of differentiated products manufactured in each plant, thereby gaining from increased specialization and greater reliance on comparative advantage.

Stimulus to Investment. The formation of an ITA is likely to stimulate outside investment in production and marketing facilities to avoid the discriminatory barriers imposed on nonmember products. Further, in order for firms to meet the increased competition and take advantage of the enlarged market, investment is likely to increase. In most cases, investment is an alternative to exports of goods—a benefit provided to ITA members. The large investments made by U.S. firms in Europe after the mid-1950s and in the years and months leading up to the Single European Act were fostered by their desire not to be excluded from this large potential market; investing ensured that their products would not be restricted by tariff and nontariff barriers.

Efficient Resource Use. Finally, if the ITA is a *common market*, the free movement of labor and capital is likely to stimulate better use of the economic resources of the entire community. Efficiency of industries and individual firms will likely increase with increased access to lower-cost capital and additional labor. Lower consumer costs and higher real incomes should follow. This argument, however, is less important when considering the benefits of a *free trade area* or a *customs union*.

Provisions of the North American Free Trade Agreement

The proposed North American Free Trade Agreement (NAFTA), involving the United States, Mexico, and Canada creates separate bilateral agreements in agricultural trade, one between the United States and Mexico and the other between Canada and Mexico. Because of the low level of Canada-Mexico agricultural trade, this discussion focuses primarily on the U.S.-Mexico bilateral agreement in agriculture. Its general provisions include:

1. Immediate elimination of all import tariffs on a broad range of agricultural products already facing low or negligible duties. About one-half of current U.S.-Mexico bilateral agricultural trade will be duty free when NAFTA takes effect. These commodities represent about $1.5

billion in current U.S. exports to Mexico and $1.6 billion in Mexican exports to the United States.

2. Systematic reduction of all remaining tariffs on agricultural trade between the United States and Mexico. Some commodities will be covered by special safeguard provisions. A small share of trade (about 10 %) will be liberalized over a 5-year period. These products were deemed too sensitive for immediate liberalization but not sensitive enough to require more than 5 years for transition to free trade.

3. Elimination of tariffs for most sensitive products over a 10- or 15-year transition period. Some of these products will be eligible for special safeguards in the form of tariff-rate quotas (TRQs) during the transition period. TRQs call for a low or zero duty on a specified quantity of imports. Imports in excess of the specified TRQ quantity will face a higher tariff (the current or original).

Both the within-quota and the over-quota tariffs will decline to zero during the specified time period. Initial TRQ quantities will be determined by recent average trade levels and will expand at a 3 percent annual compounded rate over the transition period. The United States will use 10-year TRQs entirely for selected fruit and vegetable imports from Mexico currently valued at $330 million. Mexico will apply the 10-year TRQs on $155 million in imports from the United States, mainly hogs, pork, potatoes, and apples.

4. Elimination of tariffs on a few selected fruits and vegetables over a 15-year period. A 15-year period with TRQs is provided for products that are economically and politically sensitive, including U.S. imports of sugar, peanuts, and frozen orange juice. Mexico will employ a 15-year transition with TRQs for corn, dry beans, and nonfat dry milk.

5. Elimination of nontariff barriers over specified transition periods. Mexico will eliminate its import licensing requirements on U.S. products. The United States will exempt Mexico from its Meat Import Act. The United States will also replace Section 22 (Agricultural Adjustment Act of 1933) quotas on imports from Mexico with TRQs during specified transition periods.

6. Provisions provide that complaining parties in disputes involving environmental measures or sanitary/phytosanitary regulations bear the burden of proof. This is in contrast to GATT, where the burden of proof rests with the defending party. Dispute settlement may rely on scientific proof of a country's regulation, and not on speculation as to the particular merit of the regulation.

7. A working group will be established to study ways to avoid using export subsidies. However, the United States may use export subsidies on products shipped to Mexico to counter subsidized competition from non-NAFTA countries such as the EEC.

Negotiation of side agreements related to the enforcement of environmental laws, labor regulations, and protection against surges in imports began March 17, 1993. Some members of U.S. Congress have indicated that a trinational commission appointed to oversee the implementation of side agreements should have the authority to impose punitive duties on NAFTA partners if needed to enforce domestic laws. Canada and Mexico alike, however, strongly oppose any such supranational entity. It may prove difficult to develop effective side agreements which will satisfy U.S. concerns about enforcement of existing environmental and labor laws, while at the same time meeting Mexican and Canadian concerns about giving up sovereignty over some domestic issues.

Reasons for International Trade Agreements

The emergence of the United States as a major player in the formation of international trade agreements has raised questions about why ITAs have become such a popular policy tool and why the United States is aggressively seeking to participate. Several reasons may be cited:

1. To provide for special trading arrangements with countries that are economically or politically important to the strategic interests of the United States.
2. The slow progress of multilateral trade negotiations, under the auspices of the GATT, to achieve timely, substantial reduction in barriers to trade, particularly agriculture, intellectual property rights, services trade, nontariff barriers, and dispute-settlement procedures.
3. To counter the economic and political power created in Europe by further integration of the EEC and the prospects for trade and economic cooperation with former East bloc nations, Central and South America, and Asia.
4. To reduce, by stimulating economic growth and development in Mexico, the flow of illegal immigrants from Mexico and other costly side-effects of trade barriers.
5. To foster political stability and economic prosperity in Mexico, thereby supporting the continuation of the democratic process in Mexico and reducing the likelihood of Mexican political and social disruption.

International Trade Agreements as a Second Best Policy

The GATT was created in 1947, along with the World Bank and the International Monetary Fund, to prevent the reemergence of trade protectionism and the slow economic growth which preceded World War II. After more than 6 years in the current (eighth) round of trade talks that began at Punta del Este, Uruguay in 1986, no full compromise agreement has yet emerged. Although the GATT was successful at reducing tariffs in manufacturing from 50 percent in 1947 to about 5 percent in the mid-1980s, trade in agriculture, services, intellectual property rights, and trade-related investment were not effectively dealt with during previous GATT rounds (Runge 1992). The Uruguay Round was convened as an attempt to liberalize trade in these more difficult sectors.

Rising farm program costs, the inability to separate international agriculture from domestic farm programs, and trade distortions caused by government intervention in agriculture all contributed to the complex set of policy issues facing trade negotiators as the Uruguay Round opened. As Runge (1988) points out, "If farm policy is to be liberal, rather than protectionist, it must reflect greater market orientation both at home and abroad, and reduce the distortions that now separate the domestic from the global markets."

Agriculture has emerged as the key to completion of a successful Uruguay Round. In 1985, the United States and the EEC were engaged in an all out fight to gain control of the world grain trade through the use of export subsidies. Caught in the middle were developing countries such as Argentina and Brazil, who depend upon agricultural exports as a key source of foreign exchange earnings and income. Other developed nations, such as Australia and Canada, were adversely affected by United States/EEC subsidies because world grain prices were forced lower.

Despite progress by the EEC to undertake reform of the Common Agricultural Policy by reducing price supports, the stalemate in GATT has continued. The "Blair House" agreement between the EEC and the United States calls for the EEC to reduce support of oilseed production, curtailing its output to 9.5 to 10.0 million tons. Even so, it appears that the Uruguay Round could still fail. Because of the slow progress and possibility of failure, the United States, Mexico, and Canada have adopted a "second best" solution and negotiated an agreement to liberalize trade which does not provide for a substantial reduction in domestic support to agriculture.

Counter Economic and Political Power in Other Parts of the World

The proposed NAFTA will create the world's largest *free trade area,* consisting of 360 million people, $6.2 trillion in economic output, and more than $1.2 trillion in trade. The EEC has 340 million people, $4.5 trillion in economic output, and trade of $2.4 trillion. However, the EEC already has extended preferential market access to Poland, Hungary, and the Czech and Slovak Republics and will likely form a larger *free trade area* with some or all of the members of the European Free Trade Association.[1] As economic and political integration in Europe continues, it is likely that the United States will experience more pressure to form larger and economically stronger ITAs in the Western Hemisphere.

In 1991, Argentina, Brazil, Paraguay and Uruguay initiated the Southern Cone Common Market, MERCOSUR, to foster economic growth and mutual interests among the member nations. Although smaller in economic strength than NAFTA, MERCOSUR has stimulated interest and concern among many publics. Some analysts speculate that the Enterprise for the America's Initiative, designed to stimulate investment and reduce debt in the Caribbean and Latin America, may be a key first step to linking the economies of the America's through trade.

Though less-well organized than the EEC or NAFTA, five Asian nations are attempting to strengthen trade ties. Japan, Singapore, Hong Kong, South Korea, and Taiwan have developed plans to extend preferential trading terms to each other. While this does not constitute a formal trading relationship, it may create the economic incentive for the formation of a *free trade area* or other trading arrangement in the near future.

Economic and political interests of the United States may dictate that as economic blocs emerge around the world, steps be taken to ensure that existing and future trading relationships are maintained, and even strengthened, by participation in one or several forms of economic integration. As Salvatore (1993) points out, any *free trade area* or *customs union* acting as a single unit in international trade negotiations will likely possess much more bargaining power than the total of its members acting separately. Further, reaching a compromise with a unified group of nations may sometimes be easier than dealing with each nation individually.

Reduce Costly Side Effects

ITAs may be used to encourage member nations to undertake desirable economic, social, or political reform. For instance, the proposed NAFTA, if effective, will create an economic incentive for Mexican workers to seek employment in Mexico, rather than in the United States. It is doubtful that any other policy action taken by Mexico could or

---

[1]The European Free Trade Association (EFTA) was formed in 1960 as a free trade area among Austria, Finland, Iceland, Liechtenstein, Norway, Sweden, and Switzerland. The EEC and EFTA created an industrial free trade area in 1972.

would have been as instrumental in stemming the flow of the more than 300,000 illegal Mexican immigrants annually to the United States. This could have other positive consequences for the United States such as reducing the poverty in border *colonias* and eliminating the strain on border health care, water, sewage, and other facilities required to accommodate a near doubling of population over the last decade.

Other "backdoor" reforms may include an increased effort on the part of the Mexican government to control air and water pollution in its major cities. Already there is evidence that the government shut down the state-owned and -operated petroleum refinery in Mexico City to indicate to the United States that Mexico was serious about environmental degradation and its consequences. Meat packing facilities are being relocated outside of cities in order to improve air quality and reduce problems with disposal of animal waste.

Foster Political Stability and Economic Prosperity

Probably an overriding goal of the United States in liberalizing trade with Mexico is to foster continuation of the democratic government and political stability of the past few decades. More open markets and the ensuing economic growth may be one of the most effective means of improving the standard of living in Mexico and ensuring that economic prosperity leads to a stable political environment.

Although various reasons may cause the United States to participate in ITAs, it is likely that strong economic and political interests will be the driving force behind such efforts. It appears likely in the future that leverage may be needed to negotiate continued or additional access to major markets. Negotiating from the strength of an economic bloc may prove to be the most effective alternative.

Summary

International trade agreements have been growing in importance in recent years. Presently, about two-thirds of world trade is accounted for by "trading blocs." Trade among members has grown at the expense of the rest of the world. By their very nature, ITAs discriminate against nonmembers, favoring trade with member nations.

ITAs function legally within the framework of existing GATT rules. Because of the vague nature of these rules, nations have been able to discriminate rather freely against nonmember nations with little fear of retaliation.

Trade theory provides only ambiguous conclusions regarding the consequences of ITA formation. Free trade is certainly more efficient than discriminatory trade, but in a world of less-than-free trade, ITAs may permit major economic gains under certain conditions. It is likely that dynamic gains, such as increased competition, economies of scale, and greater investment will far exceed static gains from trade creation or trade diversion.

If current efforts to liberalize world trade through the GATT are unsuccessful, it appears likely that ITAs will take on added importance as policy tools for negotiating fewer barriers to trade among groups of nations. Although many reasons may cause the United States to participate in the continued formation of ITAs, it is likely that strong economic and political interests will be the driving force behind these actions. In the future, it will likely become more important to negotiate continued or additional access to major markets using the additional leverage provided by nations acting as a single unit.

References

Fieleke, Norman S. 1992. "One Trading World or Many: The Issue of Regional Trading Blocs." *New England Economic Review*, May/June.

Jackson, John H. 1989. *The World Trading System: Law and Policy of International Economic Relations.* Cambridge, Massachusetts: The MIT Press.

Johnson, H.G. 1965. "An Economic Theory of Protectionism, Tariff Bargaining, and the Formation of Customs Unions." *Journal of Political Economy*, September.

Runge, C. Ford. "The Assault on Agricultural Protectionism." *Foreign Affairs*, No. 67108, Fall 1988.

______. 1992. "Economic Stability Rides on Success of GATT Negotiations." *Feedstuffs* (Special Report)64:28.

Salvatore, Dominick. 1993. *International Economics*, 4th Edition. New York, NY: Macmillan Publishing Company.

Viner, Jacob. 1950. *The Customs Union Issue*. New York, NY: The Carnegie Endowment for International Peace, (Reprint).

12

U.S. Agricultural Trade Policy

Ronald G. Trostle, Karl D. Meilke, and Larry D. Sanders

Trade policy, particularly agricultural trade policy, has been controversial since at least England repealed its Corn Laws. Although the words of Sir Robert Peel, Great Britain's Prime Minister, are now nearly 150 years old, one can easily imagine a current U.S. government official making similar arguments in favor of trade liberalization (Stigler and Friedland 1989):

> In ingenuity—in skill—in energy—we are inferior to none. Our national character, the free institutions under which we live, the liberty of thought and action, an unshackled press spreading the knowledge of every discovery and of every advance in science—combine with our natural and physical advantages to place us at the head of those nations which profit by the free interchange of their products. And is this the country to shrink from competition? Is this the country which can only flourish in the sickly, artificial atmosphere of prohibition? Is this the country to stand shivering on the brink of exposure to the healthful breezes of competition?
>
> -Sir Robert Peel, Feb. 9, 1846

Trade policy is controversial because border controls are seldom put in place, except to underpin domestic programs and policies. Hence, trade liberalization, which dismantles border protection, threatens domestic programs and the producers they were designed to protect.

While trade barriers for manufacturing goods have been greatly reduced since World War II, this has not been the case for agricultural product trade. Among the industrial market economies (IMEs) trade in agricultural products, through mutual agreement, has remained largely outside, or exempt, from the international rules that apply to manufactured product trade. This is partly a reflection of historical concerns about food security, particularity in Europe and Japan, and also of the political power of farmers.

While it is tempting to classify countries as either "free traders" or "protectionists," the composition of these groups has changed over time. Many of the trade barriers the United States would now like to see eliminated were allowed to exist at the earlier insistence of the United States. In some cases a country's trade position will change depending on the commodity under discussion, and even those countries whose rhetoric is most strongly protrade, typically will have some highly protected commodity sectors.

The Uruguay Round of General Agreement on Tariffs and Trade (GATT) negotiations is the latest attempt to "normalize" trading relationships for agricultural products. While the final result will fall far short of complete trade liberalization, it is now clear there will be no Uruguay Round agreement without some progress on agriculture.

Whether the Uruguay Round is concluded successfully, or is written off as a failed attempt, it has implications for domestic agricultural policies. This is because, although bureaucrats continue to discuss the costs and benefits of freer trade, in reality trade flows have a major influence on farm prices, farm incomes, commodity stocks, and government expenditures for farm programs—in the United States and around the world.

Periods of strong export growth tend to coincide with higher prices and cash receipts from the sale of farm products (Figure 12.1). When exports of program commodities decline, commodity inventories rise and market prices decline toward support price levels because domestic demand cannot absorb the excess supply at these prices. The lower market prices reduce farm cash receipts and trigger increases in government expenditures to support the sector.

Given the dependence of U.S. agriculture on foreign markets, government policies, both domestic and trade, are critical to the health of the sector. These policies play an important role in determining trade flows, in the plans of farmers and agribusinesses, as well as in influencing the long-term competitiveness of the agricultural sector. They also have a major influence on the budgetary costs of farm income support.

Figure 12.2 shows the relationship between exports and one measure of support for agriculture. Between 1981 and 1986, U.S. agricultural exports declined from $45 billion to $28 billion.[1] During the same period

---

[1] FAO's definition of agriculture (see footnote a, Table 12.1) is used for statistics on trade.

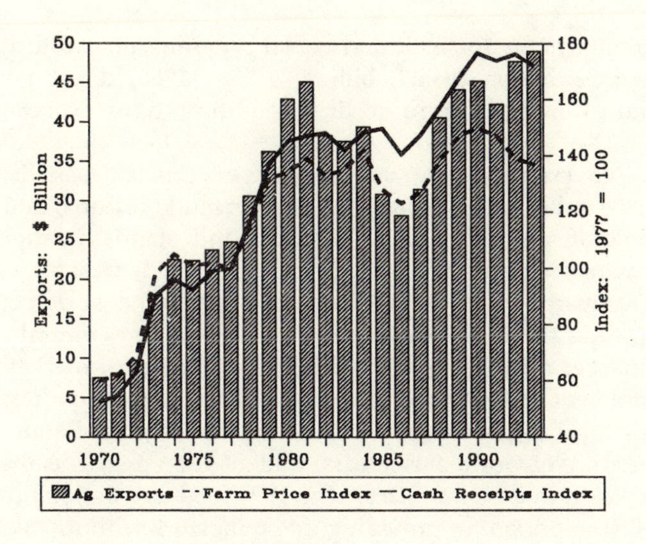

Author's forecasts for 1993.

FIGURE 12.1 U.S. Farm Prices, Cash Receipts, and Agricultural Exports, 1970-1992.

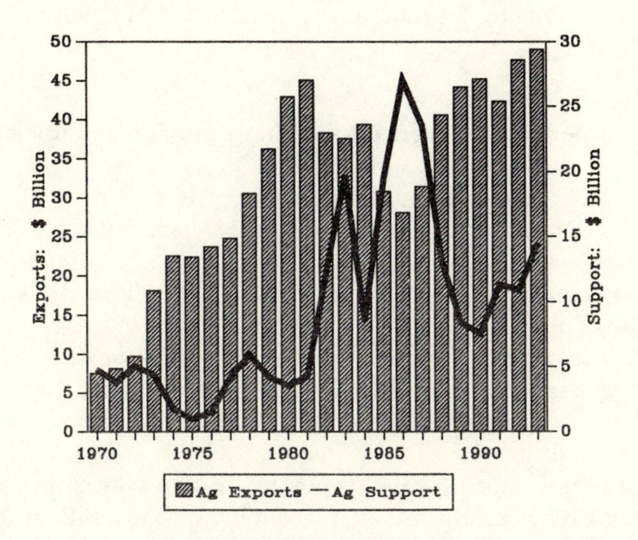

Support = CCC net outlays for domestic and export programs, minus disaster payments.

Authors' forecasts for 1993.

FIGURE 12.2 U.S. Government Support for Agriculture and Agricultural Exports, 1970-92.

CCC net outlays for domestic and export programs related to agriculture rose from $8 billion to $30 billion. After 1986, direct government expenditures on agriculture declined significantly as exports began to rise.

Some government programs facilitate exports and benefit both U.S. agriculture and foreign consumers. These include research and development, food inspection services, grading and standardization, market development programs and some forms of food aid. Other programs are more controversial, both domestically and within the world community. These include the major elements of traditional agricultural and trade policy: market price supports, stabilization programs, marketing loans, land retirement schemes, stockholding policies, import quotas, export subsidies, tariff escalation, and provisions for trade retaliation.

The nature of world agricultural trade is changing, as is the environment in which trade takes place. The 1995 food and agricultural legislation will be important in providing the framework within which the Administration and the farm sector can effectively respond to these changes. The fundamental issue underlying the broader agricultural policy debate is the degree and nature of government intervention. For trade policy, the essence of the debate is between advocates of freer trade, and those in favor of maintaining or increasing current levels of support and protection.

Importance of Agricultural Trade and Trade Policies

U.S. agricultural exports increased from less than $8 billion in the early 1970s to $48 billion in 1992 (Table 12.1). In recent years (1988-1991), about 17 percent of all farm output was exported. For the average U.S. farm whose gross farm receipts total $76,500, the farm value of exports equals $13,000. For crops, the production from one of every three acres is sold to consumers in other countries, but for some commodities this percentage is much higher. For example, exports account for more than one-half of the U.S. output of wheat, cotton, soybeans, rice, almonds and sunflower seed.

Agricultural exports have important linkages to employment and business activity in the general economy. Each billion dollars of agricultural exports supports about 27,000 jobs; 11,200 of them off-farm. A dollar increase in agricultural exports generates $1.60 in supporting business activity; $1.34 of it in nonfarm industries (Edmondson and Robinson 1991).

Imports, which are approaching $30 billion, also play a significant role in the U.S. markets for a variety of agricultural commodities and

TABLE 12.1 Value of World and United States Agricultural Trade, 1970-1992[a].

Year	*World Exports ($ bil.)*	*U.S. Exports ($ bil.)*	*U.S. Share (Percent)*	*World Imports ($ bil.)*	*U.S. Imports ($ bil.)*	*U.S. Share (Percent)*
1970	53.5	7.4	13.8	56.5	6.3	11.2
1971	55.2	8.1	14.7	60.9	6.3	10.3
1972	65.6	9.7	14.8	71.3	7.0	9.8
1973	95.3	18.1	19.0	101.5	9.1	9.0
1974	117.6	22.5	19.1	127.8	11.0	8.6
1975	123.2	22.3	18.1	137.1	10.2	7.4
1976	132.6	23.6	17.8	145.1	11.8	8.1
1977	152.5	24.8	16.3	167.1	14.1	8.4
1978	172.4	30.6	17.7	188.4	15.7	8.3
1979	204.3	36.2	17.7	224.4	17.8	7.9
1980	233.1	42.9	18.4	254.6	18.1	7.1
1981	233.4	45.0	19.3	253.6	18.4	7.3
1982	213.5	38.2	17.9	234.4	16.9	7.2
1983	208.9	37.6	18.0	229.2	17.7	7.7
1984	220.9	39.4	17.8	240.4	20.9	8.7
1985	209.0	30.8	14.7	233.9	23.1	9.9
1986	229.2	28.1	12.3	252.4	24.5	9.7
1987	252.8	31.4	12.4	279.0	24.1	8.6
1988	287.5	40.6	14.1	313.6	24.7	7.9
1989	301.9	44.2	14.6	328.4	25.4	7.7
1990	326.0	45.2	13.9	352.9	27.1	7.7
1991	328.5	44.6	13.6	352.4	28.7	8.1
1992	na	48.3	na	na	29.3	na

[a] FAO's definition of agriculture is broader than that used by USDA. FAO includes manufactured tobacco, distilled alcoholic beverages, various feed additives and several other minor items. USDA's 1990 data shows U.S. exports and imports to have been $5.8 billion and $4.3 billion lower, respectively, than that reported by FAO.

na = not available.

Source: FAO. *Trade Yearbook.* Rome: (various issues).

products (Table 12.2). Many imported items are tropical products that add variety to the American diet and do not compete with U.S. production. Other imported products do compete with U.S. production, although sometimes only seasonally. Intra-industry trade is important in many commodity groups. In particular, red meats, fruits and nuts, and vegetables are broad commodity groupings within which the United States is both a significant importer and exporter (Table 12.2).

Developments in U.S. and World Agricultural Trade

The U.S. share of world agricultural exports declined from 18.5 percent during 1980-82 to 14.0 percent between 1989-91 (Table 12.1). However, the U.S. share of world exports varies widely across commodities. Its export shares of corn (64%) and soybeans (59.9%) are high, while its export shares for shell eggs (0.5%) and live pigs (0.7%) are small. In 1981, the value of world agricultural trade grew to $233.4 billion and U.S. exports reached $45 billion—a record world trade share of 19.3 percent. Between 1981 and 1985, the value of world trade dropped 10.5 percent while U.S. exports fell even more—31.6 percent. In 1986, the U.S. share of world trade dropped to 12.3 percent, the lowest level in the last three decades. The United States responded by adjusting its domestic and trade policies to make its products more competitive in export markets.

The United States has traditionally maintained a surplus in its agricultural trade account. In the early 1980s the surplus averaged over $20 billion, but with the decline in exports in the mid-1980s the surplus dropped to less than $4 billion in 1986. Since then, the trade surplus has grown and is now again approaching $20 billion.

Distortions in World Agriculture

Experience has shown that government intervention in industrial market economies (IMEs) generally boosts domestic prices to well above world price levels. Higher prices simultaneously depress domestic demand and stimulate production. For commodities where production exceeds demand at the domestic price, governments are forced to use export subsidies to sell the excess product on world markets, thereby depressing world market prices for foreign producers. Imports become more competitive relative to domestic production and trade barriers must be erected to keep out cheaper foreign supplies (U.S. Department of Agriculture 1989).

U.S. intervention in agricultural markets can be characterized as domestic market price supports or deficiency payment programs, sometimes supplemented with supply controls, maintained with import restrictions and export subsidies for some commodities. Intervention by the European Economic Community (EEC) is more comprehensive, with high support prices and import restrictions applying to nearly all commodities. Export subsidies are used to dispose of an increasingly large fraction of

TABLE 12.2 U.S. Agricultural Trade Value and Share of World Trade, by Commodity Group, 1991.

Commodity	Exports		Imports		Net Exports
	$ Bil.	*Percent of World*	*$ Bil.*	*Percent of World*	*$ Bil.*
Total agriculture	44.6	13.6	26.7	7.6	18.0
Cereals & preps	11.1	25.2	1.1	2.3	10.0
Wheat & flour	3.5	21.9	.1	.4	3.5
Rice	.8	17.4	.1	1.9	.7
Coarse grains	6.0	45.2	.2	1.5	5.8
Corn	5.1	57.7	--	.4	5.1
Oilseeds	4.3	42.6	.2	1.3	4.2
Soybeans	4.0	65.9	--	.4	4.0
Soybean meal	1.1	21.3	--	.1	1.1
Soybean oil	.2	12.7	--	.4	.2
Livestock & products	7.8	8.7	5.6	6.0	2.1
Live animals	.7	7.4	1.2	12.4	-.5
Meat & products	3.7	10.0	3.1	8.0	.5
Beef & preps	2.2	13.3	2.2	12.4	--
Pork & preps	.5	4.0	.9	7.0	-.4
Poultry meat	.6	17.5	--	--	.6
Dairy products	.3	1.3	0.5	2.2	-.2
Hides & skins	1.4	23.3	0.2	2.3	1.2
Cotton	2.5	30.8	--	--	2.5
Tobacco[a]	6.0	31.4	1.3	8.0	4.8
Beverages	.8	3.5	3.9	16.8	-3.1
Sugar	.5	3.5	1.3	8.7	-.8
Fruit & vegetables	5.8	10.7	6.2	9.9	-.5
Fruit & prods	3.1	12.1	3.8	11.7	-.6
Vegetables & prods	1.8	7.8	1.9	7.2	-.1
Tree nuts	.9	26.0	.4	11.1	.5
Coffee, tea, cocoa & spices	.5	2.5	3.6	16.1	-3.1

[a] Includes manufactured tobacco.

-- = negligible; less than $50,000 or one-half percent.

Source: FAO. *Trade Yearbook*. 1991. Rome.

of EEC output. Japan supports high internal prices behind a wall of import restrictions. In all cases, distortions occur in prices, incomes, and domestic resource use; as well as in foreign trade.

The cost of government interventions to support agriculture have risen in the United States as well as in most other major industrial nations. The Organization for Economic Cooperation and Development (OECD) estimated that the industrialized nations of the world provided $320 billion in support and protection to their agricultural sectors in 1991, up from $295 billion in 1986. This was equivalent to $16,000 per full-time farmer (Table 12.3). These figures illustrate why trade liberalization, which would remove some support and protection, is so difficult to negotiate.

Because of government intervention, world prices have been driven so low that unsubsidized, though efficient, producers have difficulty competing. World price declines below free trade price levels resulting from support and protection in the industrial market economies (IMEs) is estimated to be 50 to 60 percent for dairy products and sugar, 25 to 35 percent for wheat and coarse grains, 10 to 20 percent for meat, and 5 to 10 percent for other crops (Roningen and Dixit 1989).

TABLE 12.3 Total Transfers (in U.S. dollars) Associated with Agricultural Policies for Selected Countries, 1991.

Country	*Total Transfers (PSE)*[a] ($ bil.)	*PSE as Percent of Cash Receipts* (percent)	*Transfers related to*		
			Population ($/capita)	*Farmer*[b] ($/farmer)	*Acres* ($/acre)
Australia	1.2	15	70	5,000	7
Canada	9.5	45	353	22,000	309
EC-12	141.8	49	409	13,000	1,937
Japan	63.2	66	510	17,000	20,810
United States	80.8	30	318	22,000	242
OECD	320.7	45	na	16,000	432

[a] The Producer Subsidy Equivalent (PSE) measures gross income transfers to producers due to agricultural policies. As such, it excludes some of the transfers captured in the total transfer figure. The percentage PSE is calculated as the ratio of total transfers to total cash receipts (including government transfers).

[b] Full-time farmer equivalent.

Source: Organization for Economic Co-operation and Development. 1992. *Agricultural Policies, Markets and Trade: Monitoring and Output 1992.*

World prices have become so distorted that approximately 40 percent of the support and protection provided to agriculture in the IMEs is necessary simply to offset the price-depressing effects of this very same support. While all IMEs contribute to distortions in world agricultural markets, nearly 75 percent of the depression in world prices is the result of policies in the European Economic Community (EEC) and the United States (Roningen and Dixit 1989). The EEC accounts for nearly one-half of world price distortions, and is an important actor in the market because of its size and because it uses policy instruments that typically distort both production and consumption. In the United States, consumer distortions are far less important, but U.S. policies affecting producers do distort world prices.

Efforts to Reduce Trade Distortions

The focus of the Uruguay Round negotiations in GATT and of the North America Free Trade Agreement (NAFTA) negotiations is on trade liberalization. However, during the six years of sluggish GATT negotiations, there has been a slow worldwide movement toward liberalization and policy reform. Many countries have moved unilaterally towards more trade-friendly policies. Significant changes have been made in Australia, Brazil, Canada, China, Japan, Mexico, New Zealand, Poland, Sweden, and the EEC, as well as in the United States (McClatchy and Warley 1991).

In the United States, sugar quotas have been replaced by tariff rate quotas, crop yields used to determine deficiency payments have been frozen, short-term acreage reduction programs have been scaled back, and flexibility provisions have all combined to reduce trade distortions. In the EEC, reform of the Common Agricultural Policy (CAP) will reduce intervention prices and require larger farms to idle crop land. As a result, the use of export subsidies should diminish, perhaps significantly (Helmar et al. 1993, Josling and Tangermann 1992). In Japan, import quotas for beef and some other products have been replaced by tariffs. In Canada, a net income stabilization program has been introduced to replace more-distorting forms of commodity-specific protection.

Other factors contributing to the trend toward unilateral liberalization include budgetary constraints, macroeconomic costs, distributional and environmental concerns, World Bank/IMF conditionality, GATT decisions, and bilateral and regional trade arrangements. There is also a general shift in focus away from "farm policy" toward more broadly defined agri-food and rural policy in which nonfarm interests have a stronger voice. New policies, introduced in response to these forces, may be less distorting of global agricultural resource use, production, consumption, and trade than the policies they replace. It is the goal of the Uruguay Round of GATT negotiations to cement these positive steps, to discourage new forms of protectionism, and to provide impetus for further liberalization.

Public Concerns

There are several major concerns with U.S. agricultural and trade policies. First, in these days of large budget deficits, the cost of domestic agricultural and associated trade programs is cause for alarm. Second, despite large expenditures for agricultural and export assistance programs, farmers do not seem to be much better off and U.S. agriculture does not seem to be much more competitive in international markets. Even with the use of aggressive export subsidy programs the United States has been unable to win back much of its lost export market share. This loss of export market share would not be of great concern if world trade values had been growing rapidly, and had the United States shared in this growth. However, this was not the case. The nominal value of U.S. exports has only recently equalled the values obtained in the early 1980s.

Third, the equity consequences of farm program benefits that flow disproportionately to large farms are being questioned. Although program benefits are often justified on the grounds of helping the "family farm", the benefits are closely tied to production. In the United States, 30 percent of farms receive nearly 90 percent of direct government payments to agriculture. These farms, compared to the other 70 percent of farms and to nonfarm families, have adequate incomes and substantial net worth. In 1990, only 15 percent of U.S. farms had sales exceeding $100,000. Farms in the $100,000 to $250,000 sales class had net cash incomes of $70,378, of which $15,055 was in government payments (U.S. Department of Agriculture 1991). These farms had an average net worth of $681,000.

Fourth, the protectionist policies of the industrial market economies (IMEs) have exacerbated the income problems of agricultural exporters in developing countries. These countries cannot compete with subsidized exports from developed countries, nor do they have the economic power to successfully negotiate unilateral reductions in import barriers.

Trade Policy Orientation: Freer Trade or Protection and Intervention

The failure of the Uruguay Round to conclude in 1990, an increasingly competitive trade environment, and budget deficit concerns, have caused agricultural trade policies to become more contentious. There are a variety of specific policy issues concerning export subsidies, import quotas, tariffs, and trade remedy laws associated with particular commodities. However, the essence of the trade policy debate is between advocates of trade liberalization and those who promote the subsidization and protection of domestic agricultural producers.

Advocates of trade liberalization seek to reduce distortions in domestic and international markets caused by government intervention. The alleged benefits from reduced protection and intervention include more efficient allocation of resources, faster increases in productivity, enhanced competitiveness, higher and more stable world market prices, lower consumer prices for some commodities, fewer opportunities for domestic monopoly pricing, and reduced taxpayer costs.

Proponents of continued government protection and intervention often rely on the "level playing field" analogy, pointing, for justification, to the large sums spent by competing nations to enhance their agricultural exports and to protect their own domestic markets. Some will go further and argue that the production of food and fiber is too important to be left to market forces, and each country has the right to feed its own domestic consumers.

At the extreme, those favoring government intervention in agriculture argue for international market management through the use of commodity agreements and producer cartels.

The Protection and Intervention Case

There are numerous groups, each with its own agenda, that oppose trade liberalization. The most politically powerful group consists of highly protected agricultural producers, in the United States and abroad, who have benefitted from restrictive trade practices and who favor continued support. While some of the arguments of this group are self-serving, others deserve careful consideration.

The objectives of this group are to increase and stabilize domestic prices and farm incomes by supporting prices, controlling domestic production, and preventing low-cost imports. They are concerned that trade liberalization will cause farm income and asset values to decline if lower support prices and payments are not fully offset by price increases on the world market. Other concerns include: expected declines in output in the most heavily protected sectors as some domestic production is replaced by imports; reduced domestic price stability; and reduced income security for smaller, less efficient, less competitive U.S. producers.

Advocates of continued protection of the agricultural sector see the effects of protection and subsidy policies of other countries, and are concerned about the economic effects of liberalization. They are skeptical about other nations strictly adhering to a multilateral agreement. They fear that trade liberalization could result in a surge of imports, without appreciably increasing access to foreign markets, and that an international agreement would sanction higher levels of support in foreign nations than in the United States.

Trade liberalization studies show income losses for producers of some commodities. Irrespective of large gains for U.S. taxpayers, producers worry about the sustainability of fully transparent taxpayer-funded nontrade-distorting compensation, particularly, when this compensation cannot be linked to production. However, without a compensation program, some producers will suffer economic losses in a free trade environment. Thus, although U.S. agriculture would become more efficient and competitive, more-rapid structural adjustment could cause hardship in some segments of the rural economy.

Opponents of liberalization also cite a number of "nontrade" concerns that may arise as a result of trade liberalization. First, a more rapid decline in rural infrastructure and amenities as the loss of less efficient producers speeds up the depopulation of rural areas. Second, potential adverse environmental effects as producers engage in unsustainable production practices as a result of short-run income declines. Third, the perception that although there may be short-run problems, U.S. agriculture, working within its traditional policy framework, has provided the nation with food security, and an overwhelming supply of low-cost, wholesome food and fiber. Fourth, agriculture has important linkages with downstream input suppliers and upstream processors whose contributions to employment and the economic well-being of the country are much larger than the size of the primary production sector would suggest. Finally, there is a great reluctance to constrain domestic policy choices and options in order to comply with international trade agreements.

The Trade Liberalization Case

All studies of agricultural trade liberalization conclude that reducing the trade distorting features in current U.S. agricultural policy will benefit U.S. taxpayers. Depending on the exact form of liberalization, there may be either small gains or losses for consumers. Producers of highly protected commodities would lose, but if other countries likewise reduce the trade-distorting features embodied in their policies, world prices and economic activity would increase, thereby easing farm sector adjustments. Hence, the case for multilateral rather than unilateral reform.

Gains from trade arise from three primary sources. First, trade allows countries to separate their production decisions from their consumption decisions. Second, it allows countries to specialize in producing products for which they have a comparative advantage, and trade for products for which they have a comparative disadvantage. Third, it inhibits uncompetitive pricing behavior in the domestic market.

Compare, for example, the production of food and television sets in the United States and Japan. The United States, because of its vast land resources compared with Japan, has a comparative advantage in producing food. If the United States is unable to trade with Japan, the price of food, relative to the price of television sets, will be low in the United States and high in Japan. Allowing Japan and the United States to trade, even if they produce the same quantity of food and television sets as before, means income gains in both countries. United States consumers can trade some of their relatively cheap food for some of Japan's relatively cheap television sets. Consumption choices are no longer constrained by what is produced in each country. Since, in this example, both countries continue to produce the same bundle of goods as before, the income gain is entirely due to the onset of trade.

Larger income gains are available, however, because each country can increase its production of the good in which it has a comparative advantage. The relative price of food in the United States rises, and more food but fewer television sets are produced. In Japan the opposite occurs, the relative price of food falls, and less food but more television sets are produced. Thus additional producer and consumer benefits arise due to the ability of each country to specialize in producing the product for which it has a comparative advantage.

The third source of benefits results from improved competition that accompanies a reduction in trade barriers such as tariffs and import quotas. As a result of facing foreign competition in domestic markets, firms with market power find it more difficult to increase prices above marginal cost. The benefits of lower prices are passed along to consumers. This contributes to increased economic growth.

Many studies of trade liberalization have been undertaken; the work of Roningen and Dixit (1989) is representative. These writers show that with complete trade liberalization by the industrial market economies (IMEs), U.S. producers would lose $16.2 billion, consumers would lose $4.6 billion but taxpayers would gain $30.3 billion. Therefore, taxpayers could fully compensate producers and consumers for their losses and have $8.6 billion left over (Table 12.4). The figures for other industrialized countries and for the entire trading world tell a similar story, resulting in an estimated net economic gain of $35.3 billion.

The income losses for producers indicated above appear very large. However, for a number of reasons they are overstated. First, the proposed GATT agreement does not require complete trade liberalization and the changes are to be phased in over a number of years. Second, nontrade distorting income supports will be used to cushion some of the loss of more trade-distorting forms of protection. Finally, the trade liberalization analysis is static in nature and does not take into account the dynamic gains in efficiency and economic growth that are likely to be achieved.

The net benefits for the United States are even larger for unilateral liberalization. Because world price increases are smaller with unilateral liberalization, consumers gain whereas they are small losers in the multilateral liberalization scenario. However, for the same reason, the reduction in producers' incomes is significantly larger under unilateral liberalization. This causes governments to want to "share the pain" with other country's producers. It is much easier to sell policy reforms that are required of all members of a multilateral agreement than to get approval for unilateral support reductions. This is what the multilateral trade negotiations are trying to accomplish, i.e., provide an avenue through which all countries can make desirable policy changes at least cost.

Domestic price support programs and the border measures which accompany them do stabilize internal farm prices. However, these same measures destabilize world markets because protected countries do not fully adjust to changes in global supply and demand. Tyers and Anderson (1988) estimate that world wheat, coarse grains, and dairy product prices were 75, 13, and 136 percent more volatile in the 1980-82 time period than they would have been under free trade.

TABLE 12.4 Welfare Implications of Multilateral Trade Liberalization by Individual Industrial Market Economies, 1986-87.

	Multilateral Liberalization			*Net Economic Gain*	
Country	*Producer Losses ($ bil. U.S)*	*Consumer Gains ($ bil. U.S.)*	*Taxpayer Gains ($ bil. U.S.)*	*Total ($ bil.)*	*Per Capita (dollars)*
Australia	1.6	-1.5	1.1	1.1	71
Canada	1.3	0.2	3.8	2.6	101
EC-12	22.7	21.2	15.6	14.0	43
Japan	21.8	24.7	5.7	6.3	52
United States	16.2	-4.6	30.3	8.6	36
All IMEs	65.6	40.9	63.1	35.3	51

Source: Roningen and Dixit (1989).

Subsidies for agricultural production imply a tax on nonagricultural output, because resources (capital, labor, and land) are held in the agricultural sector. Research suggests that keeping one farm job in the United States costs $107,000 in nonfood output (Hertel, et al. 1989). Because of agricultural protection in the EEC, EEC manufacturing output has been reduced by 1.2 percent and the EEC has lost 2 to 4 million jobs (Stoeckel and Breckling 1988).

Advocates of trade liberalization acknowledge that some of the nontrade concerns raised by opponents are legitimate. However, they argue that these concerns are best addressed by domestic policies targeted at the problem, rather than by border controls which raise domestic commodity prices.

Finally, while intervention policies usually make farmers better off initially because their incomes rise, the value of the right to produce and sell a product is capitalized, over time, into the value of land and production quotas. It is this capitalization of benefits that makes it so difficult to remove support and protection, once provided, and it also explains why, under trade liberalization, producers in highly protectionist countries face the largest adjustment problems, even though their measured support may remain higher than in another country following liberalization.

Policy Alternatives and Trade-Offs

In the mid-1980s, changes in U.S. agricultural programs and trade policies helped stem the downward trend in U.S. competitiveness and exports. The reduction in U.S. loan rates and the initiation of export subsidy programs effectively reduced prices to importers.

Economists are divided on the benefits resulting from the most controversial decision—reintroduction of explicit export subsidies. Initially, these subsidies were targeted towards markets and commodities in which the United States was competing with the EEC. And they were paid in kind to reduce burdensome government stocks of grain. However, over time the export subsidies have been expanded to more countries and commodities, and are now being paid in cash. For the most part, the U.S. market share of heavily subsidized commodities has increased only marginally because competing exporters have matched the subsidy-inclusive export prices of the United States. While the benefits of the export subsidies to the United States can be debated, there is no doubt that they are unpopular with competing exporters, have harmed low-cost third world exporters, and have enabled importers of agricultural products to reap most of the benefits from the program (Anania et al. 1992). However, some would argue that the introduction and continued aggressive use of the Export Enhancement Program (EEP) played an important role in getting the EEC to unilaterally modify its Common Agricultural Policy (CAP).

Uruguay Round

The Uruguay Round of trade negotiations is an attempt to impose international discipline on the domestic and trade policies of member countries. However, agriculture has proven to be one of the, if not the, most difficult sector to negotiate. In December 1991, Arthur Dunkel, chairman of the GATT Secretariat's Agricultural Negotiating Group, tabled a draft agreement for agriculture. It contained four key provisions (1) comprehensive tariffication of all nontariff trade barriers; (2) reductions in export subsidies and domestic support programs; (3) a set of "green" or minimally trade-distorting policies that would not be subject to reduction commitments; and (4) an agreement to harmonize international sanitary and phytosanitary measures.

While most countries had some reservations about some aspect of the Dunkel text, it reflected 5 years of negotiating effort and a compromise among a host of competing positions. During most of 1992 the United States and the EEC engaged in a series of bilateral negotiations aimed at resolving their differences over agricultural trade matters. In November 1992, the United States and the EEC agreed to modify the Dunkel text in certain respects. These modifications are listed in Figure 12.3 that outlines the agricultural accord.

If the Uruguay Round is successful, agriculture will, for the first time, be subject to most of the international trade disciplines that now apply to nonagricultural products. The major exception is the continuation of export subsidies, albeit at lower levels. A successful round will result in the rethinking of the objectives and results of domestic agricultural policies and provide an impetus to move away from highly trade-distorting policies (export subsidies and import quotas) towards more transparent and less trade-restricting alternatives (tariffs and direct income aids) (Magiera et al. 1990). However, even if the Uruguay Round is not brought to a successful conclusion, budget and competitive pressures, in a more-integrated world trade environment, and a variety of other concerns will coerce changes in U.S. agricultural and trade policy. These changes will be resisted by many producers of protected commodities who favor the *status quo*. Nevertheless, it is implausible that domestic and related trade programs will all survive without change, no matter the outcome of the Uruguay Round.

Market Access Restrictions

- Convert all nontariff trade barriers to tariffs.
- Reduce tariffs, including those resulting from tariffication, by 36% using a simple average, with no less than 15% for each tariff line.
- Establish minimum access opportunities for those products where there are not significant imports, equal to at least 3% of domestic consumption in 1993, and increasing to 5% by 1999.

Export Subsidies

- Reduce budgetary outlays for export subsidies by 36% and the quantity of subsidized exports by 21%, using 1986-90 as the base period from which to measure reductions.

Domestic Support

- Reduce aggregate domestic support by 20%, with the exception of support which has minimal effects on trade or production, from a 1986-88 base period.

Sanitary and Phytosanitary Measures

- Use international standards to protect human, animal, or plant life and health.
- Introduce higher standards if they are scientifically justified.
- Accept different standards from other parties, if they are deemed to be equivalent.

Special and Differential Treatment for Developing Countries

- Allow developing countries up to an additional 10 years to implement the reduction commitments.
- Exempt least developed countries from any reductions.

Special Safeguards

- Restrict imports of a specific commodity if the price falls below an average 1986-88 reference price.
- Restrict imports of a specific commodity if the quantity imported exceeds either 125% of the average amount imported during the past 3 years or 125% of the established minimum access level.

FIGURE 12.3 Key Elements of the Proposed GATT Agreement on Agriculture. *Source:* Adopted from *Inside U.S. Trade—Special Report*. Washington, D.C., August 7, 1992.

Commercial Export Programs

U.S. programs to assist agricultural exports (Table 12.5) are both trade and budget issues (Ackerman and Smith 1993). Some question large federal expenditures to assist agricultural exports, especially when commodity supplies are tight. Others question the way funds are spent. However, commodity groups point to the large sums that foreign competitors spend to support their agricultural exports and to protect their own markets.

A GATT agreement would reduce export subsidies, but would not eliminate them. However, if no GATT agreement is reached, funding for commercial export programs could be increased, thus escalating the competitive subsidization currently underway. Higher Export Enhancement Program (EEP) funding may increase exports, but, depending on the U.S. supply situation, may also encourage imports of products similar to those sold under the EEP. This is particularly true for commodities produced within what is likely to become a North American Free Trade Area. However, the requirement that recipients be credit worthy could limit large increases in funding for credit guarantee programs.

High-Valued Product Exports

The importance of high-value products (HVPs) has risen sharply in both U.S. and world agricultural trade (MacDonald 1993). In the United States, although receiving the least amount of agricultural and trade assistance, HVPs are the category of agricultural exports growing most rapidly. In 1991, U.S. HVP exports exceeded bulk exports for the first time during a peacetime period.

Traditionally, U.S. agricultural trade policy has been focused on bulk commodities. This reflects not only the historical predominance of bulk exports, but also the focus of domestic support programs on grains, oilseeds, cotton, and tobacco. These programs give the U.S. government an interest and responsibility in the foreign marketing of bulk products that differs from its responsibility for marketing of HVPs. This difference in orientation shows in the proportions of bulk commodities and HVPs included in various U.S. export programs: only 10 percent of the EEP bonuses were for HVPs in recent years and just 24 to 35 percent of USDA's export credit guarantees were for HVP sales. Only the much smaller Market Promotion Program (MPP) was oriented towards HVPs. In contrast, 69 percent of the EEC's $10 billion average annual expenditures on export subsidies during 1986-90 was for the HVPs.

TABLE 12.5 U.S. Export Programs.

Program	*1988*	*1989*	*1990*	*1991*	*1992*[a]
			-------------$ Million-------------		
Total Food Aid:					
Value of all products	1,522	1,485	1,433	1,265	1,428
Total P.L. 480	1,157	1,191	1,099	1,070	1,071
Title I	693	722	648	424	374
Title II	463	469	451	465	456
Title III	NA	NA	NA	181	241
U.S. AID	86	187	118	na	na
Section 416 (B)	279	107	216	195	357
Total Export Subsidies:					
EEP (commodity value)[b]	3,301	2,802	2,355	1,908	3,017
EEP (bonus awarded)	339	312	917	968	1,068
Credit Guarantee Programs:					
Total guarantees approved	4,504	5,195	4,296	4,522	5,684
GSM-102	4,141	4,770	3,964	4,439	5,596
GSM-103	363	425	332	83	88
Market Development Programs:					
Total funds allocated	144	172	204	320	235
Foreign Market Development Program (FMDP)[c]	29	29	34	38	38
Targeted Export Assistance Program (TEAP)[d]	115	144	169	282	na
Market Promotion Program (MPP)[d]	na	na	na	na	197

na = not available.

[a] Current estimate.

[b] For 1985 through 1991, the value of the EEP commodities sold is based on either f.o.b. or c.i.f. sale prices. The 1992 sale value is based on f.o.b. prices.

[c] Expenditures for 1986 through 1989 and planned budgets for 1990 through 1992.

[d] Expenditures from 1986 through 1990. TEAP and MPP 1991 budgets are combined for 1991. The 1992 MPP entry is planned budget.

Sources: USDA, Foreign Agricultural Service. FMDP data extracted on September 25, 1992. TEAP and MPP data extracted on October 19, 1992.

Raising HVPs share of export program expenditures could help counter the EEC's subsidized exports. However, EEP bonuses for HVPs have been very large, sometimes accounting for 40 percent or more of the product price. Producers may benefit less than processors from subsidized high-value product exports.

Import Protection

Sugar, dairy products, and cotton textiles and apparel are major products for which the United States uses import restrictions to help maintain high domestic prices. Multilateral trade liberalization under a GATT agreement would increase the access of imports into the U.S. market. Proponents of maintaining quotas say the U.S. market would be overwhelmed by imports if the quotas were lifted and that U.S. producers would loose a significant share of their most important market. An alternative would be to reduce support prices and switch to direct income support. Such a program could provide income support for low-income producers while addressing the income, structural, and program cost concerns.

Summary

With or without a successful completion of the Uruguay Round of GATT negotiations, U.S. agricultural and agricultural trade policies are expected to change during the remainder of the decade. The fundamental issue underlying many agricultural policy debates will be the direction that U.S. agricultural policy should take: towards trade liberalization and competitiveness, or towards continued more protection, intervention and subsidization. How this fundamental issue is resolved will affect not just farmers, but U.S. consumers, taxpayers, and trade partners as well.

References

Ackerman, K. Z., and M. E. Smith. 1990. *Agricultural Export Programs: Background for 1990 Farm Legislation.* Staff Report No. 9033. Washington, D.C: Commodity Economics Division, Economic Research Service, USDA.

______. 1993. *Commercial Export Assistance.* Agriculture Information Bulletin 664-34. Washington, D.C: Economic Research Service, USDA.

Anania, G., M. Bohman and C. Carter. 1992. "U.S. Export Subsidies in Wheat: Strategic Trade Policy or Expensive Beggar-Thy-Neighbour Tactic." *American Journal of Agricultural Economics* 74(3):534-545.

Edmondson, W., and M. Robinson. 1991. "U.S. Agricultural Trade Boosts Overall Economy." *Foreign Agricultural Trade of the United States: Sept./Oct. 1991.* Washington, D.C: Commodity Economics Division, Economic Research Service, USDA.

Helmar, M. D., W. H. Meyers, D. J. Hayes. 1993. *World Agricultural Markets and Policy Outlook for the 1990s.* Unpublished paper, Dept. of Economics, Iowa State University.

Hertel, T., R. Thompson, and M. Tsigas. 1989. "Economywide Effects of Unilateral Trade and Policy Liberalization in U.S. Agriculture." eds. A.B. Stoeckel, D. Vincent and S. Cuthbertson, *Macroeconomic Consequences of Farm Support Policies.* Duke University Press, Durham.

Josling, T., S. Tangermann. 1992. *MacSharry or Dunkel: Which Plan Reforms the CAP?* Working Paper 92-10, International Agricultural Trade Research Consortium, July.

MacDonald, S. 1993. *Agricultural Export Assistance and High-Value product Exports.* Agriculture Information Bulletin. Washington, D.C: Economic Research Service, USDA.

Magiera, S. L., et al. 1990. *Reinstrumentation of Agricultural Policies.* International Agricultural Trade Research Consortium, University of Minnesota.

McClatchy, D., and T. K. Warley. 1991. "Agricultural and Trade Policy Reforms: Implications for Agricultural Trade." Paper for the XXI Congress of the International Association of Agricultural Economists. Tokyo, Japan.

Roningen, V. O., and P. M. Dixit. 1989. *Economic Implications of Agricultural Policy Reforms in Industrial Market Economics.* Staff Rpt. AGES 89-36. Washington, DC: Agriculture and Trade Analysis Division, Economic Research Service, USDA.

Stigler, G. J. and C. Friedland. 1989. *Chronicles of Economics Birthday Book.* Chicago: University of Chicago Press.

Stoeckel, A. B. and J. Breckling. 1988. "Some Economywide Effects of Agricultural Policies in the European Community: A General Equilibrium Study." eds. A. B. Stoeckel, D. Vincent and S. Cuthbertson, *Macroeconomic Consequences of Farm Support Policies.* Duke University Press, Durham.

Tyers, R., and K. Anderson. 1988. "Liberalizing OECD Agricultural Policies in the Uruguay Round: Effects on Trade and Welfare." *Journal of Agricultural Economics* 39(2):197-216.

U.S. Department of Agriculture. 1989. *Investigations of Changes in Farm Programs.* Washington, D.C.

______. 1991. *Economic Indicators of the Farm Sector: National Financial Summary, 1990.* ECIFS 10-1. Washington, DC: Agriculture and Rural Economy Division, Economic Research Service.

Commodity Policy Issues

13

Farm Prices, Income, Stability, and Distribution

Harold D. Guither, Harry S. Baumes, William H. Meyers

The Problem in Perspective

Low farm product prices, low farmer incomes, wide year-to-year price fluctuations, and demands for government assistance greatly influenced the development of farm price and income support policies in the 1920s and 1930s. Now, in the 1990s, prices, incomes, stability, and distribution of program benefits are still the issues being debated, but under a greatly changed agricultural production environment and amidst a call for less government intervention. Through these past 60 years, a process of revising and amending the basic legislation, primarily directed at commodities, has been followed. The Food, Agriculture, Conservation, and Trade Act of 1990, the most complex and detailed ever passed, represents the culmination of the process to date.

Working together through the years to develop food and agricultural legislation, Congress and the Executive branch have stated the objectives of national food and agricultural policy: "to provide an orderly, adequate, and balanced flow of agricultural commodities in interstate and foreign commerce" (1938); "to improve, maintain, and protect prices and income of farmers" (1961); and "to maintain farm income, stabilize prices, and assure adequate supplies of agricultural commodities" (1965). But goals and objectives are rather meaningless unless the current complement of policies is designed to achieve those goals and stated objectives. As one reviews the past 60 years, the goals and objectives of food and agricultural policy that provide for financial survival and prosperity of producers and a stable food supply at reasonable prices for consumers have remained unchanged, but the priority ranking has changed from one period to the next.

The key question when looking ahead to the Twenty-First Century is whether prices, incomes, stability, and distribution are still the critical and significant public issues that food and agricultural policy will be intended and designed to affect. If they are, do current programs respond adequately to these concerns? What program changes are needed? Will non-traditional areas (e.g., rural development, agriculture and the environment, food safety, and consumer issues) dominate the food and agricultural policy debate?

Farm Product Prices

Prices are the most apparent indicator of financial change in agriculture since they are known every day that the markets are in operation. In the absence of market failure or government intervention, changes in prices reflect the changes in supply and demand. Prices alone, however, do not reflect profit or loss for individual producers or for the entire agricultural industry.

Farmers' Incomes

Income generated by farming is determined by a combination of factors. Total cash receipts are determined by the price for each unit of product times the total product sold. Net income is that remaining after costs of production are paid. Consequently, prices received, amount of product placed on the market, and the costs involved in producing and marketing that product affect net farm income—all factors to consider in developing policies to increase or stabilize farm income.

If the concept of farm household income is included, then part-time farming and off-farm income must be considered part of the income issue. If policy is to stabilize income, level of income and equity elements must be addressed as well.

Stability

If one of the objectives of policy is to maintain farm income stability, then attention must be given to measures designed to achieve stability in production and prices. Production of individual commodities may fluctuate widely because of weather, disease, or other natural disaster. Should government assume the risk inherent in production due to weather? Also, if government needs to stabilize prices, commodity acquisition, dispersal, and storage policy become part of the equation. In a market system without government intervention, fluctuations in production affect market supply, and supply and demand interact to determine the price received by farmers and ultimately farm income.

Income stability is thus a complex and difficult issue. The private sector makes decisions to maximize profits, rather than to achieve stability. Public efforts to stabilize farm prices and supplies began with price and income support programs in the late 1920s. As we shall see later, these policies have not succeeded in stabilizing either prices or farm incomes.

Distribution

Distribution of government program benefits for the assistance of individual producers is a crucial issue in the light of federal budget deficits and the national debate about them. Since World War II, this issue has been complicated by the greatly changed structure of U.S. agriculture (e.g., increased specialization of production, regionalization of production, and mix of farm and off-farm income among farm families). It is further complicated by the mechanism of income transfers (to the sector) being tied to production of specific commodities. The result has been a highly skewed distribution of resource ownership and production. According to the 1987 Census of Agriculture, 699,000 farms, approximately one-third of the farms in the United States, received farm program payments. Other farms have received support in forms other than direct payments.

A critical issue is whether the largest producers should be eligible for federal agricultural support. Alternatively, should federal agricultural programs be targeted to medium- and small-volume producers? Should benefits be provided only to farms with incomes or production below certain specified levels? If supplies and demands are not in balance, efforts to control production will be ineffective if large producers are excluded from the program. Furthermore, there are wide variations in the distribution of payments among commodities and in different regions of the country (Reinsel 1990). Since prices, volume of production, and total incomes vary so widely, maintaining stability for individual producers or for agriculture as a whole continues to be a complex and difficult task for the U.S. government.

Current Policies Directed Toward Prices, Incomes, Stability, and Distribution

The Food, Agriculture, Conservation and Trade Act of 1990 continued some features of major farm price and income support policies that date back to the Great Depression. But these policies do not affect all producers in the same way.

Current programs deal with specific commodities. Support for the major program commodities—wheat, feed grains, cotton and rice—has been reduced to avoid government acquisition and accumulation of large quantities of commodity stocks. Since 1985, the "market-oriented" emphasis of agricultural policy with loan rates based on average prices of recent years, has resulted in market prices averaging above the loan rate. So prices have more opportunity to move up and down over a wider range, reflecting supply and demand interaction rather than governmental intervention. Current price support policies may do little, if anything, to maintain stability of prices and incomes, especially in years of sharp changes in production. They do, however, provide a safety net for farm prices and incomes.

For many years, the Commodity Credit Corporation (CCC) held surplus stocks that could be released on the market in short crop years. This resulted, at times, in government efforts to hold prices down by selling government-owned commodities. In response to farmers' apprehensions about government influence in the market and costs of such operations, the Farmer-Owned Reserve (FOR) was established. The FOR program permits farmers to maintain ownership of their commodity, while the commodity remains isolated from the market so as not to influence market price. The large crops of the early 1980s resulted in a substantial build-up of stocks in the FOR program as well as of stocks under direct CCC control.

The 1990 Act limited the quantities of grains eligible for placement in the FOR. Entry into the FOR is based on the market price in relation to the loan rate or stocks on hand in relation to expected use. A farmer may remove his commodity from the FOR at any time by repaying the loan, but not later than 27 months after the original loan expires unless the Secretary of Agriculture decides to extend the loan for an additional 6 months. The ease with which farmers can withdraw grain from the reserve allows for greater price responsiveness by farmers.

With the changes in numbers and size of farms, and in recent years the increase in deficiency payments for income stabilization, the distribution of program benefits becomes a major issue. Payment limitations have attempted to deal with this issue, but payment limitation restrictions have been relatively ineffective. The 1990 Act set annual payment limits of $50,000 per person for deficiency and diversion payments, $75,000 per person for gains from repaying a marketing loan or emergency compensation payments, $125,000 per person for wool, mohair, and honey payments, and a total of $250,000 per person. Disaster payments were limited to $100,000. The legislation enables persons to be partners in multiple-farm operations, though, and thus receive deficiency and diversion payments in excess of or in multiples of $50,000.

A General Accounting Office (1992) study of the 1990 payment distribution showed that 2,137 individuals, partnerships, entities and institutions received more than $100,000 in payments and 13,705 received more than $50,000. This is only one percent of all units receiving payments. Even though this distribution results from the nature of the market system and the current structure of U.S. agriculture, such a pattern of government payments raises concerns among some nonfarm groups and some farmers receiving smaller amounts.

Target prices and deficiency payments, intended to provide income support, have eroded with the federal budget crisis. Deficiency payments, although still sufficiently attractive to encourage many producers to participate in the programs, are paid on only part of the total acreage base of the farm as a result of the triple base and other flexibility provisions of the 1990 Act. The total direct payments reached an all-time record of $26 billion following the "farm crisis" of 1986, but the proportion of net cash farm income coming from direct payments has declined since 1987. Although deficiency payments provide income support for designated program crops, producers of many other commodities (e.g., milk, wool and mohair, soybeans, and other field crops) benefit from price support programs even though they are not eligible for deficiency payments.

Off-farm Income of Farm Operators and Their Families

Off-farm income has provided financial stability for many farm families in recent years. The aggregate amount of off-farm income has risen every year since 1960 except in 1977. The total off-farm income received by farm families nearly doubled from 1980 through 1990. Farm operators and their families earned about $35 billion from off-farm activities in 1980, and nearly $67 billion in 1990. Should government support be provided as a supplement when off-farm earnings are so large? Some would argue the need for continued support; but others question why support should continue in the face of this level of off-farm income.

Off-farm earnings per farm family are highest for those families with less than $20,000 in cash receipts from farming. Per-family earnings for those families in the less than $20,000 agricultural sales class averaged $35,206 in 1990. Those with sales of $100,000 to $249,999 averaged $18,096. Off-farm earnings were lowest for those who sold $100,000-249,999 in agricultural commodities (Figure 13.1). This information begs the following question. If support is desired for agriculture, should it be targeted for those in "need"? And, how should need be defined? What is the means test?

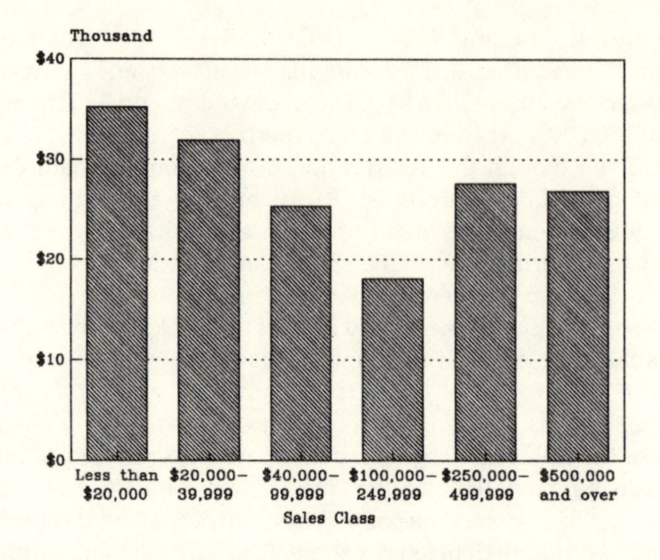

FIGURE 13.1 Off-farm Income Per Farm Family by Sales Class, 1990. *Source:* U.S. Department of Agriculture (1992).

Prices, Incomes, and Government Payments

Farm prices fluctuate substantially from year to year. During the 23 years from 1970 through 1992, the index of prices received by farmers dropped 7 times from the previous year, changed less than two percent in 5 years and increased 15 times from the previous year (Figure 13.2). Fluctuations for some individual commodities are greater than for all commodities combined.

Since prices are only one of the elements in the determination of farm income, a look at the longer-term trend in prices received along with that of prices paid provides a more complete picture of farm price and income relationships.

During the early 1970s, prices received by farmers rose more rapidly than did prices paid. For the decade, however, prices paid and prices received by farmers rose at comparable rates. During the 1980s, prices received stagnated and even declined during the middle years while prices paid moved higher (Figure 13.3).

In the absence of efficiency and technology gains and with a status quo in production practices, this trend suggests that profit margins and net incomes were less favorable for many producers in the late 1980s and early 1990s. Only through improved efficiency and greater volume of production per unit of input can an individual operation remain viable when costs increase more rapidly than do revenues.

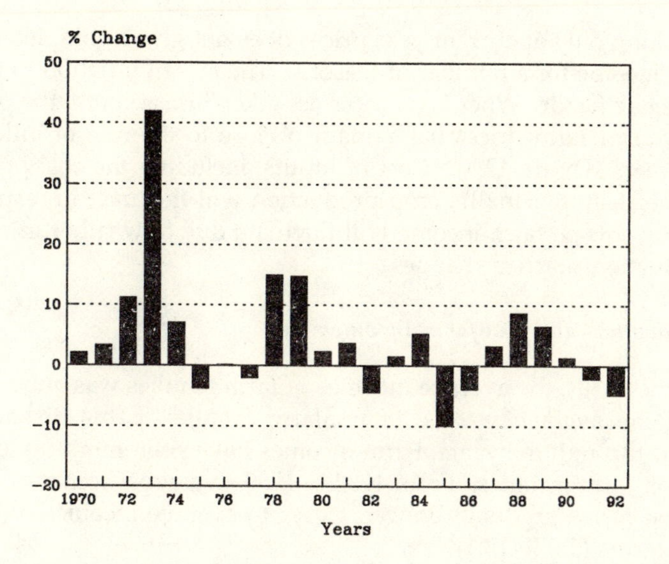

FIGURE 13.2 Variability in Index of Prices Received by Farmers, 1970-1992. *Source:* U.S. Department of Agriculture, National Agricultural Statistics Service.

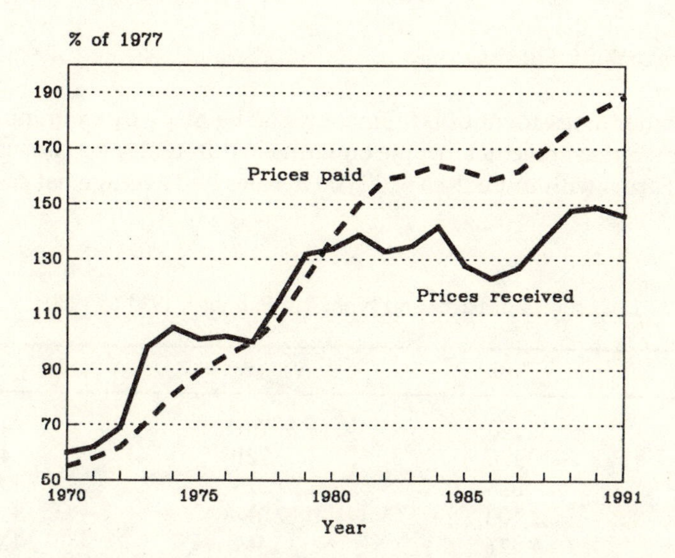

FIGURE 13.3 Indexes of Prices Received and Prices Paid by Farmers, 1970-91. *Source:* U.S. Department of Agriculture, National Agricultural Statistics Service.

Looking only at current year prices does not give a complete picture of farm income for a number of reasons. The rate of inflation in the economy is one factor. When real prices are taken into account, the purchasing power of farm prices has remained close to the rate of inflation in recent years (Figure 13.4). Cost of inputs, including the cost of capital, is another factor. Finally, crop production will fluctuate in response to natural events so farm income will fluctuate due to weather in addition to production practice changes.

Farm Incomes and Nonfarm Incomes

In the 1930s, the average incomes of farm families was about 80 percent of the average income of nonfarm families. This disparity has changed through the years. Farm incomes have risen more rapidly than have non-farm incomes. From 1930 to 1980, average incomes of persons living on farms gradually moved closer to average incomes of persons not on farms (Table 13.1).

Since 1980, average incomes have been compared on the basis of farm operator household income and U. S. household income. From 1980 to 1990, the average farm operator household income has moved above the average U.S. household income (Figure 13.5). In 1990 the average farm operator household income was $51,490 compared to $37,403 for the average U.S. household income.

Farm Income by Sales Classes

Another assessment of farm income can be made by examining farm numbers and average net income on farms in different sales classes (Table 13.2). Farms with more than $100,000 in sales had average net cash farm

TABLE 13.1 Incomes of Farm and Nonfarm Families, 1930 to 1980.

Year	*Farm*	*Nonfarm*	*Nonfarm/Farm*
1930	$ 170	$ 761	4.5
1940	176	721	4.1
1950	838	1,575	1.88
1960	1,100	2,017	1.83
1970	2,421	3,449	1.42
1980	6,471	8,074	1.25

Source: U.S. Department of Agriculture. *Agricultural Statistics* (various issues).

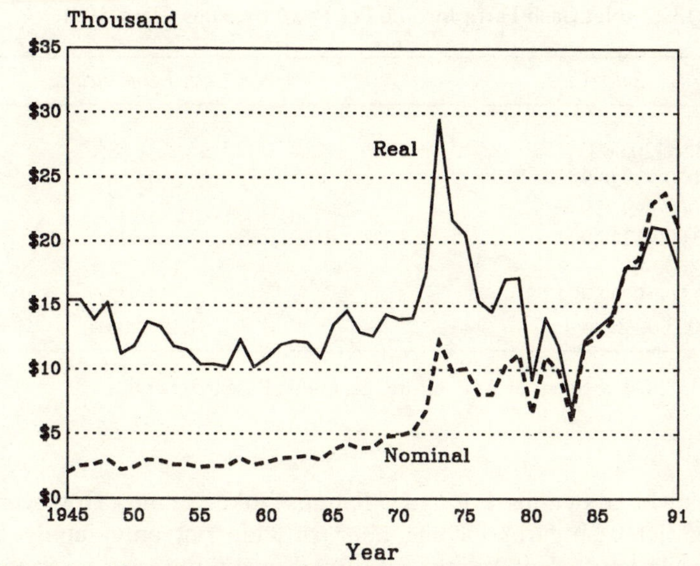

FIGURE 13.4 Nominal and Real Net Farm Income Per Farm, 1945-1991. *Source:* U.S. Department of Agriculture, Economic Research Service.

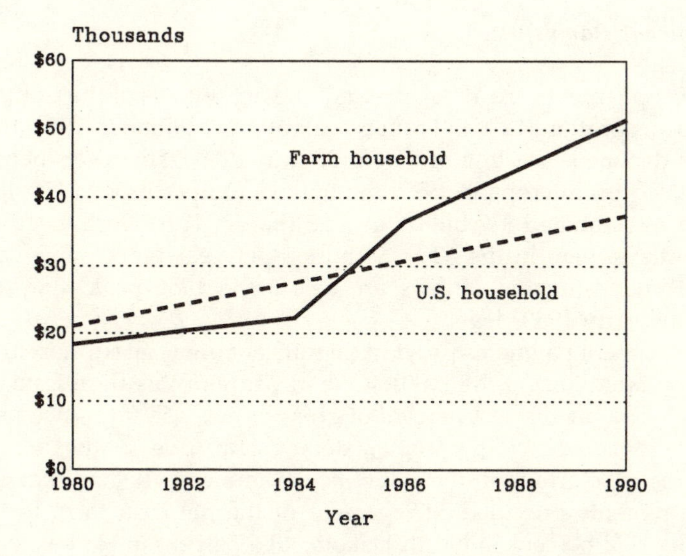

FIGURE 13.5 Farm Family and U.S. Household Income, 1980-1990. *Source:* U.S. Department of Agriculture (1992).

TABLE 13.2 Net Cash Farm Income Per Farm by Sales Class, 1991.

Sales Class	*Net Cash Farm Income*
Under $20,000	$ - 43
$20,000-$39,999	9,036
$40,000-$99,999	24,642
$100,000-$249,999	60,524
$250,000-$499,999	131,572
$500,000-$999,999	252,822
$1,000,000 & over	1,199,588

Source: U.S. Department of Agriculture, Economic Research Service.

incomes well above the average U.S. household income. However, this level of net cash farm income must provide not only family living expenses but sufficient investment capital to run the farm business—a major difference between the farm and nonfarm household. There is, however, a wide dispersion in net cash farm income across the different sales classes ranging to a low of $-43 for the smallest sized farms to a high of over $1 million for the largest sized farms.

Government Payments

Direct payments are a major part of the net outlays of the Commodity Credit Corporation (CCC). From a peak of $25.8 billion in 1986, the total outlays declined to a low of $6.4 billion in 1990. However, because of record feed grain crops in 1992, the outlays jumped from $9.7 billion in 1992 to an estimated $17 billion in 1993 (Figure 13.6). In the early 1980s CCC outlays were in the $2 to $4 billion-per-year range. Even though present and estimated outlays are below the 1986 peak, they remain above the early 1980s level.

Government payments provided significant financial support for many farm operators during the 1980s. One measure of this significance is the relation of payments as a percent of gross or net cash farm income. This percentage varies widely from state to state, due to the amount of program crops grown in the state. For example, in 1990, direct government payments provided 54.9 percent of the net cash farm income in Montana, 55.1 percent in North Dakota, 48.6 percent in Kansas, but only 3.3 percent in North Carolina. Overall, the percentage of payments comprising net cash farm income coming from government payments has declined nation-wide from a high of 29 percent in 1987 to 14 percent in

1991. Payments as a percent of cash receipts from total farm marketings peaked at 10 percent in 1987 and declined to 4.5 percent in 1991.

Since payments are based on acreage as well as yields, the major proportion of payments go to the larger farms. For example, in 1991, 2 percent of the farms (those with $500,000 or more in gross sales) received 15 percent of the total direct government payments. Farms with sales of $100,000 to $500,000, 13 percent of all farms, received 43 percent of the payments. Farms with $40,000 to $100,000 in sales, 15 percent of all farms, received 25 percent of all payments. More than half the farms had sales of less than $40,000 and received the smallest proportion of payments. The latter, however, depend largely on off-farm income to sustain the farm family. This illustrates the skewness in direct payments received by farm operators (Table 13.3).

Economics of Farm Price and Income Support

When price and income support programs were initiated, economists proposed supply control based on the principle that demand for farm output was, in the aggregate, quite unresponsive to price changes (i.e., was price inelastic). Thus a small percentage increase in supply would result in a greater percentage decrease in price, making the total value of the larger crop worth less than a smaller crop at higher prices.

Although this principle would apply to the total crop, individual producers are not affected in the same way, since the amounts produced on their farms may change more or less than the national average.

TABLE 13.3 Distribution of Direct Government Payments by Sales Class, 1991.

Sales Class	*No. of Farms*	*Direct Payments Received*
	------------Percent------------	
Under $20,000	58	10
$20,000-$39,999	11	7
$40,000-$99,999	15	25
$100,000-$249,999	10	28
$250,000-$499,999	3	15
$500,000-$999,999	1.2	10
$1,000,000 & over	.8	5

Source: U.S. Department of Agriculture. Economic Research Service.

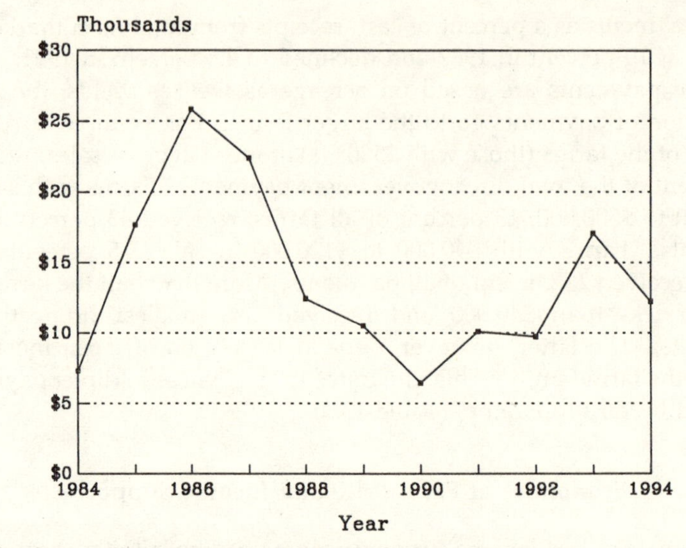

FIGURE 13.6 Total CCC Net Outlays, All Commodities.
Source: U.S. Department of Agriculture (1992a). (Estimated for 1993-94).

Farm price and income support programs for the major crops—feed grains, wheat, cotton and rice—cover only part of the total cash marketings from U.S. farms. Except for an early effort to reduce the supply of hogs coming to market, there have been no direct support programs for meat animals and poultry. Theoretically if grain supplies are reduced, livestock supplies will also be reduced and livestock and poultry prices will increase. However, due to the biological lags associated with livestock production, declines/increases are not nearly as rapid as in the crops sector.

Many of today's livestock and poultry producers buy their feed. This means that high grain prices translate into high feed prices, and profits from the livestock or poultry operation are likely to decline. Declining profits are a signal to producers to reduce production levels. Conversely, when grain prices are low, feed costs are low, and livestock returns may be much higher, prompting livestock and poultry producers to expand output.

Trade-Offs and Options for Price, Income, Stability, and Distribution Policies

Experience in the 1960s and 1970s proved that setting a support price higher than market price leads to an accumulation of substantial government stocks of program commodities. History has demonstrated that programs to restrict planting did not prevent the buildup of inventories. A shift to a more "market-oriented" policy occurred in 1985 with the setting of loan rates by formula based on past market prices.

Target prices and deficiency payments are intended to provide supplemental income to producers without distorting market prices. The deficiency payment rate, i.e., the difference between the target price and the loan rate or the market price (whichever is higher), is low or zero when market prices are high and high when market prices are low relative to the loan rate.

Although price and income supports may supplement income and provide some stability for crop producers, the results for livestock and poultry producers may be entirely different. Specialty crop producers may not get benefits at all from price and income support programs. The benefit to nonprogram producers is realized when program benefits are capitalized in land values and fixed assets. The entire agricultural sectors' asset base reflects capitalized program benefits.

If the goal is to provide payments to those who need it most, eligibility is a crucial issue for allocating direct payments among farmers. To adjust the basis for determining payments to some method that removes the linkage of payments to commodity program acreage would be objectionable to many producers, whereas others would argue that such a change would reduce and/or eliminate market distortions that result from acreage-reducing price and income support programs, and reduce the high cost of agricultural support.

Basing payments on financial need suggests a form of welfare and could lead to creative accounting practices. Currently payments are looked upon by many as a payment for participation in a program designed to stabilize prices and help conserve soil for future generations.

Payment limitations were established as one attempt to prevent very large payments from going to one person. However, operators engaged in more than one farming entity have received payments for more than a single farming operation and have received more than the single $50,000 limitation. Is this fair, equitable or desirable?

Payments for conservation practices shift program emphasis from commodity production to environmental and natural resource concerns. This approach directs programs away from farm price and income support. It offers farm operators a chance to earn government payments to support their income without a welfare connotation. But payment distribution through conservation programs could still favor the operator with the largest farming operations. The question becomes: enter the land in the Conservation Reserve Program (CRP), or continue to crop it? A prudent decision requires a solid cost-benefit analysis.

Reducing the amount of base crop acreage eligible for payment can reduce government outlays for direct payments. It also reduces income support for all program participants. As the payment base declines, the incentive for some producers to participate is also expected to decline. But nonetheless, a reduced incentive to participate in any given year is muted somewhat by the formula based on past plantings and considered plantings for determining base acreage. With a lower proportion of producers participating, the ability of the government programs to influence production, supplies, and price is reduced, and the farmer and consumer will need to rely on the market to stabilize prices and farm incomes.

References

Guither, H. D. 1969. "Support for Farm Prices and Income," in *Economic Questions for Illinois Agriculture*, EQ-1. Urbana, IL: Cooperative Extension Service, University of Illinois.

Lin, W. W., J. Johnson, L. Calvin. 1981. *Farm Commodity Programs, Who Participates and Who Benefits?*, Agricultural Economic Report 474, Washington, D.C.: Economic Research Service, USDA.

Lucier, G. A. Chesley, M. Ahearn. 1986. *Farm Income Data: A Historical Perspective.* Statistical Bulletin Number 740. Washington, D.C.: Economic Research Service, USDA.

Nelson, F. J. 1989. *Profile of Farms Benefiting from the 1982 Farm Commodity Programs.* Washington, D.C.: Agriculture and Trade Analysis Division, Economic Research Service, USDA.

Reinsel, Robert D. 1990. *The Distribution of Farm Program Payments,* 1987, U.S. Department of Agriculture, Economic Research Service, Agricultural Information Bulletin Number 607, June.

Stam, J. M., S. R. Koenig, S. E. Bentley, H. F. Gale, Jr. 1991. *Farm Financial Stress, Farm Exits, and Public Sector Assistance to the Farm Sector in the 1980s.* Agricultural Economic Report Number 645. Washington, D.C.: Economic Research Service, USDA.

U.S. Department of Agriculture. 1992. *Economic Indicators of the Farm Sector, National Financial Summary, 1991.* ECIFS 11-1. Washington, D.C.: Economic Research Service, USDA.

______. 1992a. *Agricultural Outlook,* Economic Research Service, July.

U.S. General Accounting Office. 1992. *Agriculture Payments,* Number of Individuals Receiving 1990 Deficiency Payments and Amounts, GAO/RCED-92-163FS, April.

14

Agricultural Supply Control

Fred C. White, James A. Langley, and Mark A. Edelman

Agricultural supply control programs (1) limit the quantity of a commodity that can be produced or sold or (2) limit the use of resources (such as land) in production of a commodity. The purposes of such programs may be to balance supply and demand, maintain a desired carry-over, maintain long-term productive capacity of resources such as land, and/or maintain desired prices. When farm programs support prices above open market-clearing levels, producers produce more increasing the quantity available, but consumers reduce consumption, resulting in excess supply (surpluses). Excess supply is the difference between production and consumption at prevailing prices when production exceeds consumption (Hallberg et al. 1989). To maintain prices at the support level, the government has to take action to address the problem of excess supply, either through taking surpluses off the market (buying on the open market) or by somehow limiting the quantity that can be produced. Supply control programs then become an integral part of any farm program aimed at supporting prices above open market-clearing levels.

Beginning with the Food Security Act of 1985, price-support levels have, for the most part, been reduced below prevailing market prices. However, supply control programs continue to be a significant aspect of U.S. commodity programs, with almost as many acres out of production in 1992 as in the 1960s (Figure 14.1). These programs receive constant scru-tinization because of the expense and their effect on efficient use of available resources. The United States historically has been virtually alone among agricultural exporters in assuming the burden of supply adjustment, but this role in shouldering the burden was altered during the 1980s, when Western Europe and other areas achieved self-sufficiency and begun producing surpluses under price supports.

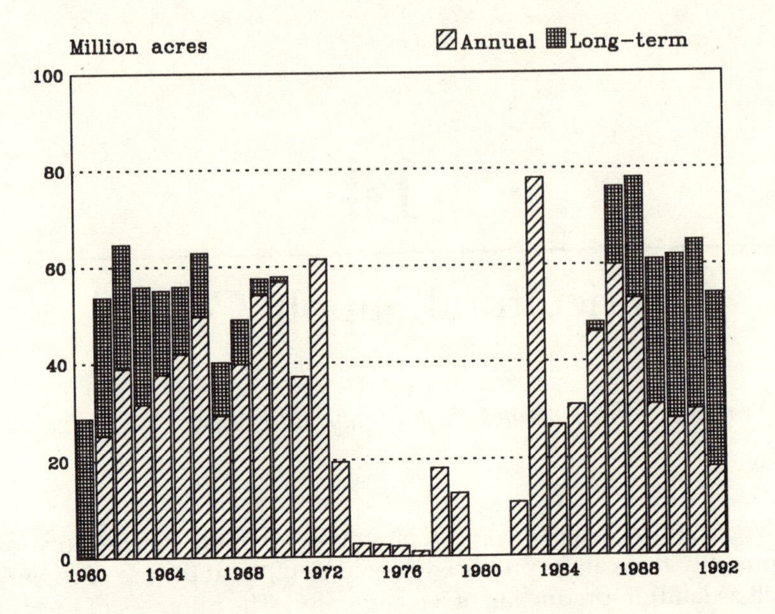

FIGURE 14.1 Reduced Acreage, Annual and Long-term in the United States, 1960 to 1992.

This paper presents an overview of supply control programs. Specific objectives are to: (1) describe the economic impacts of supply control programs, (2) discuss emerging issues related to supply control programs, (3) describe supply control programs presently in use, and (4) examine the consequences and trade-offs of alternative supply control programs.

Impacts of Supply Control Programs

Domestic Impacts

Supply control programs can be used to increase farm prices, stabilize prices, provide for orderly markets, and/or provide for food security. A reduction in production would tend to raise prices. Stabilizing production through production or marketing controls tends to stabilize prices. Commodity reserve programs can also stabilize prices through accumulating reserves when prices are low and releasing reserves when prices are high. Production controls, marketing agreements and orders, and reserves can improve orderly marketing. Reserves can help provide food security in the short run. In the long run, supply control programs that help maintain the productive capacity of the land and conserve water help provide food security.

Acreage controls and commodity reserves are more effective in raising farm prices and incomes for some commodities than for others (Gardner 1987). By reducing the quantity of a particular commodity being marketed, the supply control program would be expected to cause its market price to rise, as the smaller quantity of the product being marketed is more highly valued. Total revenue for this commodity will increase if the percentage increase in price more than offsets the percentage reduction in quantity. By producing a smaller quantity of the product, total production costs would be expected to decline. Hence profits for this commodity could increase from higher total receipts and lower total costs.

Supply control programs for grains can have significant impacts on livestock profitability in the short run. Lower quantities and higher prices of grains could increase the costs of producing livestock. However, the long-term impacts are difficult to determine because of industry adjustments to the initial impacts.

In the short run, consumers pay higher prices for those food products that include farm product ingredients that are limited through production controls. For example, when grains are limited in supply, consumers would pay higher prices for grain products, such as cereals and bakery products. Consumers would also have to pay higher prices for meat, eggs, and milk, because livestock feed would be more expensive. Rising food prices tend to be regressive, because low income earners spend a larger percentage of their income on food than do high income earners. Other supply control programs, such as reserves, may keep prices lower than would be the case in a free market, because existing reserves can be released during periods of shortage, such as caused by a drought year. In the long run, it is possible, that, with supply controls consumers might pay less than under the free market—if the supply controls act to conserve the productive capacity of natural resources, mainly land and water.

Conservation of natural resources can be addressed through supply control programs. The programs likely to be most successful in this regard are those targeted towards lands that are highly erodible or wetlands. Supply control programs can take out of production such lands already in production and ensure that they are planted to grass or trees. For such lands not yet in production, these programs can help restrict use to conservation-sensitive practices.

Supply can be limited to equal demand when prices are supported above equilibrium. Factors of production can be diverted to uses that society values more highly than production of commodities already in excess supply. Supply control programs can reduce government costs when direct price support programs are in effect. In many cases it is less expensive to fund supply control programs that limit production of a price-supported commodity than to pay the support price for unneeded production.

Supply control in an environment of prices elevated by government support programs is complicated by technological advances. Higher prices may enable producers to adopt new technologies that further expand production. With high price supports and rapid technological advance, surpluses continue to mount—making government efforts to control supply that much more difficult.

International Linkages

Supply control programs have different effects in a closed economy than in an economy open to international trade. It is generally easier to raise farm product prices through supply control programs if the product is not traded. Even then imports would have to be restricted with supply controls. For traded commodities, supply controls in the United States would limit quantities available for export and raise prices of these products. For many commodities the percentage increase in price would be greater than the percentage reduction in quantity exported in the short run. Therefore, in the short run, supply controls could increase export revenue for some commodities. However, the percentage increase in price would tend to decline through time, possibly reducing export revenue for some commodities in the long run.

The agricultural sector of the United States is an integral part of a global market and an interdependent world. Therefore, agricultural policies in the United States have an impact on other nations which may react by changing their policies, probably creating a reaction impact on our own interests. Likewise, the policies of other nations can have an impact on the interests of the United States agricultural establishment and cause the government to react through policy changes. These interactions emphasize the potential importance of international coordination of supply adjustment programs if international market stability and food security objectives are to be met.

Emerging Supply Control Issues

The prospects, in view of the constraints of a tighter budget, of achieving the major goals of farm legislation—stabilizing farm income, providing an abundant food supply, conserving resources, and enhancing the environment—is bringing a number of issues to the forefront.

International competitiveness, enhanced flexibility, and reserves are some of the major issues related to stabilizing income. Related considerations include annual farm income level, cost, program effectiveness, method of implementation, and potential uses of land removed from production. Numerous international developments, including actions of other countries, individually and in bilateral and multilateral agreements, impinge on supply control issues. Multilateral negotiations such as the General Agreement on Tariffs and Trade (GATT) and North American Free Trade Agreement (NAFTA) may impact supply management programs in the United States.

A major issue relates to establishing and maintaining the appropriate balance between short-term and long-term programs that idle land. After almost a decade of building a large long-term reserve of idled acreage, the United States must decide what to do as current Conservation Research Program (CRP) contracts begin to expire in 1996. There are currently 35.4 million acres enrolled in the long-term CRP, of which 22.6 million acres are from program-crop acreage bases. This level of acreage enrolled in the CRP has allowed the set-aside requirement under the annual acreage reduction program (ARP) to be set lower than would have been necessary to achieve the same degree of supply control without the CRP. What to do with acres coming out of the CRP is a topic of great interest, and identifying appropriate adjustments in annual acreage reduction and flexibility programs is an important emerging issue.

Establishing the appropriate level of short-term ARP acreage is dependent on the acreage level of the longer-term CRP as indicated above. An increase in the set-aside requirements under the ARP reduces government costs of price supports. However, as this requirement is increased participation in the program declines, because the differential in profitability between participation and non-participation in the program narrows. Introduction of flexibility options under the Food, Agriculture, Conservation, and Trade Act of 1990 has reduced the government's control over production of individual crops.

The United States has been active in establishing a free trade area in North America, first with the Canadian-U.S. Trade Agreement (CUSTA) and more recently under the North American Free Trade Agreement (NAFTA). Elimination of trade barriers with Canada and Mexico has important implications for supply control efforts in the United States. Imports from either Mexico or Canada could undermine some supply control programs. The problem could be further complicated by imports of "supply-control commodities" from other countries into Canada and Mexico thereby undermining U.S. supply control programs. Coordination of supply control programs in a free trade area may in some cases be advantageous.

In many nations, capitalism is evolving toward internal cooperation between government and business to implement stronger competition among nations. Such efforts may adversely affect the competitive position of U.S. agriculture. What role should the U.S. government play in supporting U.S. agriculture in the global market place? What international coalitions and policy frameworks should be developed to assure that the playing field in market access and supply adjustment is level?

Present Policy

Short-Term Acreage Reduction

The Food, Agriculture, Conservation, and Trade Act of 1990 authorized a acreage reduction program (ARP) and a paid land diversion program (PLDP). These programs apply to wheat, corn, grain sorghum, oats, barley, upland cotton, and rice. Acreage bases for each crop are established for all producers as the average acreage planted to that crop over the last three to five years. The percentage of the base acreage that is permitted to be planted each year is uniformly reduced through an ARP. Producers must comply with the announced ARP by idling the required percentage of their base acreage in order to be eligible for loans, purchases, and payments by the government. The magnitude of the ARP is dependent on the ratio of stocks-to-use. If the stocks-to-use ratio is relatively high, the ARP may be set at a higher level so as to idle more acreage and thus limit excess supplies. There are conceptual limits, as well as legislative limits, on how high the ARP should be set. If the ARP is set too high, many producers will not participate in the program and this would result in increased production as farmers plant as they please.

When supplies are excessive, diversion payments can be used to idle additional land under PLDPs. Producers are paid, at a specified amount per acre, to idle a percentage of their base acreage. The diversion payment is used as an incentive to reduce production—no other disincentives for nonparticipation are imposed on farmers. Higher payments would attract more producers to idle acreage.

Conservation Reserve

The Food Security Act of 1985 authorized 45 million acres of cropland and marginal pasture land to be taken out of production under voluntary long-term contracts. The Conservation Reserve Program (CRP) is aimed at highly erodible land, thereby combining supply control and soil conservation. Land has been enrolled in the program on the basis of sealed bids from the landowners. Contract periods are for 10 to 15 years. The government pays landowners an annual rental payment for taking the land out of production and planting it to grass or trees to establish cover and reduce erosion.

The Food, Agriculture, Conservation and Trade Act of 1990 established the Environmental Conservation Acreage Reserve Program (ECARP) which combined the CRP and other conservation programs. USDA is required to enroll 40 to 45 million acres in ECARP by the end of 1995. Over the period 1986-90, a total of 33.9 million acres had been enrolled in the CRP (U.S. Department of Agriculture 1991).

The 1990 Act established new programs to protect environmentally sensitive lands and to protect wetlands. Maintaining wetlands and other environmentally sensitive lands in reserve programs helps maintain environmental quality and prevents the conversion of these lands to crop production, thus serving a supply control function. Permanent or long-term easements are used in these programs.

Mandatory Control Programs

The peanut and tobacco programs are examples of mandatory control that have survived through evolutionary processes. Since 1982 the peanut program has included a two-price plan and poundage quotas. Peanuts produced within the poundage quota are marketed for domestic edible use. Peanuts in excess of the poundage quota are marketed for export. Quota peanuts receive the higher price. The lower price for nonquota peanuts is aimed at increasing exports. Government purchases of peanuts under the current program have been small. First, production of quota peanuts has been kept in line with the demand for domestic edible uses. Secondly, the loan rate for nonquota peanuts has been kept below the export market price.

Tobacco production is regulated through poundage quotas for burley and flue-cured tobacco. Acreage allotments are used for lesser-volume types. There is a price support program for tobacco, but production restrictions are used to help keep average prices above support prices. The tobacco price support program works through farmer-owned cooperatives that buy surplus tobacco with government loans. Producers contribute fees to a fund covering any loans the cooperative cannot repay. Hence, the government is responsible for only the administrative costs of operating the program. For a thorough discussion of the peanut and tobacco programs see Knutson and Smith (Chapter 15, this volume).

Commodity Storage

In order to take excess supplies of farm commodities off the market, the government can purchase and store the products or subsidize farmers or agribusinesses to store the products. Placing some of the current production in storage (i.e., keeping it off the market) can be used to bolster depressed prices when supplies are relatively abundant. Stocks can be released from storage when supplies are relatively scarce and prices rise. Conceptually, storing stocks can be used to dampen price fluctuations. This stabilization aspect of stocks is a major purpose of commodity storage programs.

Government storage subsidies for farmers or agribusinesses are more efficient than direct government purchases if the commodity will soon reenter commercial markets. The farmers or agribusinesses would bear some of the ownership costs while the government subsidizes the storage costs. With direct purchases, the government assumes both ownership and storage costs. Government ownership would be more effective than subsidies when supplies remain in abundance for such a long period that stocks have to be disposed of through noncommercial outlets, such as domestic or foreign food aid.

Excess government stocks have been used to pay farmers for nonuse of cropland; the stocks then move back into commercial markets. These programs are called payment-in-kind (PIK) programs because commodities, rather than cash, are used for payment. PIK programs simultaneously reduce production and stocks. When supplies from current production are abundant and stocks are large, the value of the stocks may be perceived as being little or nothing. Rather than continuing to pay storage costs on such stocks it might be more effective to use them in lieu of cash payments to farmers to idle cropland. Such payments are not reflected in the federal government's budget.

Commodity storage can be used as an integral part of price support programs. Rather than letting the price fall below the price support, the government would purchase excess supplies at the price support level and place them in storage. Conceptually, an efficient level of stocks would tend to be self liquidating over time. However, when prices are low, political pressures tend to cause excessive purchases for storage which in turn tends to discourage release from storage when prices rise.

Alternative Policy Approaches and Tradeoffs

Response to Scheduled Expiration of CRP Contracts

Long-term land retirement is clearly a policy issue to be resolved with by the next food and agricultural legislation. The ten-year contracts for land retirement under the Conservation Reserve Program (CRP) are scheduled to begin expiring in 1996.

Three basic alternatives to address the potential excess supply which could result from the expiration of these CRP contracts will be considered. First, the government could make no response, and the idled land would return to production. Secondly, the government could use additional short-run diversion programs to cope with any excess capacity resulting because of land now in 10-year contracts. Thirdly, another long-term land retirement program could be implemented.

First, without a new government initiative, landowners whose CRP contracts expire can return the cropland to production under current programs. Any program cropland returning to production must be farmed in accordance with a conservation plan. Producers will lose program crop acreage base if any returning program cropland is not planted or considered planted in the program. Another avenue for returning CRP land is in the 0-92 provision, where participating producers may leave the land idle and collect 92 percent of the eligible deficiency payments (providing, of course, that the 0-92 provision survives the latest budget round). The consequences of the first approach—i.e., no new government initiative—will vary depending on market prices at the time the contracts expire. If market prices are relatively high, it is expected that farmers would return most of their idled land to production. Such a scenario would be reminiscent of the situation with land idled under the Soil Bank in the 1950s and released in the 1970s during the "golden age" of agriculture when exports were rising rapidly. Under this scenario, the CRP would have taken land out of production when it was not needed and returned it to production when it was needed. If product prices are not relatively high at the time the contracts expire some of the land will not return to production, because it would not be economical to do so. However, some of the land would return to production, and could severely depress the already low market prices.

Secondly, short-term (i.e., annual) land retirement programs could be used to cope with the excess supply resulting from expired CRP contracts. Acreage reduction programs have been widely used to reduce production of specific crops. In some programs, farmers received direct payments from the government to idle a portion of their cropland. In others, farmers did not receive a direct payment for idling cropland, but they voluntarily idled some cropland in order to become eligible for other program benefits such as price guarantees or price supports. Payments for idling acreage have been used to encourage greater participation in the acreage reduction programs. With paid diversion, farm income may be increased in two ways—directly through government payments for diverted acres and indirectly through higher prices for the smaller aggregate quantity produced. Participation in short-term acreage reduction programs has been widespread throughout the nation, often taking some of the most productive land out of production. Idling productive land contributes to the effectiveness of short-term acreage reduction programs.

Farmers may attempt to circumvent short-term acreage reduction programs by idling their poorest land. Hence slippage occurs as production is not reduced in proportion to the idled acreage. Acreage reduction programs may limit farmers' ability to respond to changing relative prices. In order to get adequate participation in an acreage reduction program, the government may have to set the payment rate at such a high level that the overall cost of the program is prohibitive.

Thirdly, another long-term land retirement program could be implemented to remove excess capacity as CRP contracts expire. Entire or portions of farms are voluntarily taken out of crop production for multiple years under long-term land retirement programs. Cover crops such as grass or trees are established on the "retired" land to conserve the soil. Farmers receive an annual rental payment, determined by sealed bids, from the government as an incentive to participate in the program.

Long-term land retirement programs remove more production per dollar of government payment than do short-term paid diversion programs (Tweeten 1979). Conceptually, farmers should be willing to place land in this program if the rental payments equal or exceed the net returns from production. Hence the program would first remove from production that acreage with low net returns, (i.e. marginal land), improving farming efficiency. Land not in the program would be more productive than the land idled. The program can also be targeted at highly erodible or other categories of land which society desires to protect.

Participation in long-term land retirement programs tends to be concentrated in areas of marginal production, adversely affecting agribusinesses and local communities in those areas. Long-term contracts limit farmers' abilities to adjust to short-term fluctuations in crop prices. With the less productive land being idled, slippage occurs because production does not decline in proportion to the acreage idled.

Adjustments in Annual Acreage Reduction and Flexibility Program

Acreage reduction decreases the number of acres on which producers may earn deficiency payments. Reluctance to reduce target prices or payment yields, frozen at 1985 levels, tends to place budget savings as the guiding emphasis in controlling program acreage. Raising market prices through supply controls also reduces government payment exposure. Hence, higher ARP levels are a means of addressing budget pressures and low prices.

The Food, Agriculture, Conservation, and Trade Act of 1990 introduced numerous flexibility provisions which have an impact on the effectiveness of traditional supply control programs. A relatively high ARP for a program crop may be partially offset by planting that crop on the available flexible acres (15% required and 10% optional of crop acreage base) of other program crops. In addition, producers may voluntarily reduce plantings by *more* than a particular ARP would imply under the provisions of 0-92 and 50-92 programs. Under the 0-92 program, wheat and feed grain producers can devote permitted acreage to conserving uses and still receive deficiency payments on that acreage. Under the 50-92 program, cotton and rice growers who plant at least 50 percent of their base acreage can receive 92 percent of their deficiency payments.

Acreage reduction programs are currently implemented in terms of a crop acreage base determined separately for each program crop. Administering acreage programs in terms of either a total farm acreage base or normal crop acreage, instead of individual crop bases, has been proposed. Acreage reduction programs administered on a crop-by-crop basis are more effective in achieving the desired level of reduction. However, this method of implementation is more constrictive in terms of producers' responses to changing economic conditions and adoption of alternative crop rotations.

Acres removed from production under annual commodity programs must be placed in a conserving use that protects those acres from weeds and from wind and water erosion. Currently, the primary economic use of idled acres is to permit haying and grazing except during the five principal growing months of the year. The Food, Agriculture Conservation, and Trade Act of 1990 gives the Secretary of Agriculture discretionary authority to allow planting of designated crops (essentially any crop except program crops and fruits and vegetables) on up to 50 percent of the reduced acreage. Other provisions provide authority to permit planting of experimental or industrial crops on all reduced acres and on acres enrolled in the 0-92 or 50-92 programs. So far, none of the latter provisions have been implemented. However, interest continues to be expressed in allowing additional economic uses of reduced acres.

Economic incentives for producers to participate in annual supply control programs continue to be strong; but the difference in net returns per acre between participants and nonparticipants is narrowing. Reasons for this narrowing include implementation of normal flex acreage provisions (15% of base excluded from deficiency payments), higher market prices with correspondingly lower deficiency payment rates, and increased conservation program requirements. Due to budget concerns, there is a strong possibility that mandatory flexible acreage provisions may be expanded. If participant returns continue to decline, participation may begin to taper off, and the effectiveness of traditional supply control programs would be adversely affected. Participation rates are also dependent upon how ARPs are administered and the strength of the export market.

Given the budget deficit concerns and the reluctance of the Congress to fund voluntary incentives for pilot conservation programs (such as the wetland reserve), voluntary supply management incentives may continue to be curtailed in favor of a more regulatory approach that is less costly to the government. There may also be a growing possibility for producer-financed approaches to supply management and pursuit of desired conservation results.

The charge has long been made that acreage reduction programs adversely impact U.S. competitiveness in international markets, and provide incentives for foreign competitors to increase production. However, the European Economic Community (EEC) has implemented a 15 percent set-aside to control production and budget outlays. There may be increasing support, despite the accompanying increase in budget exposure, for expanding production in the United States so as to protect its world market share and maintain international competitiveness. Such an approach could reduce the trade deficit. However, maximization of a country's market share is not necessarily desirable because increasing exports to bolster market share would tend to reduce market price.

Supply Management in a Free Trade Area

Liberalizing trade with Canada and Mexico through free trade area agreements creates new challenges for the United States as related to supply management. Agreements on free trade areas—the Canadian-U.S. Trade Agreement (CUSTA) and the North American Free Trade Agreement (NAFTA)—are designed to remove the trade barriers among the participating countries. However, these agreements do not require common exterior barriers to third-party countries as would, for example, a custom union, such as the European Economic Community (EEC).

When fully operational, CUSTA and NAFTA will eliminate member-country trade barriers on agricultural products, and the United States will no longer be able to use supply control programs to raise prices of agricultural products significantly above those on the world market. High prices for selected commodities in the United States would encourage Canada and/or Mexico to expand production of these commodities or to import these commodities from other countries for export to the United States. Such a process would continue until the differential between the U.S. price and the world price is driven down to the point where it would not cover transfer costs from either Mexico or Canada.

Supply management programs aimed at conservation and environmental issues (i.e., by paying or requiring farmers to employ desired management practices) could still be effective in a single country in a free trade area. The primary goals of conservation and enhanced environmental quality could be achieved even though a free trade area exists. However, idling environmentally sensitive lands may not have a great impact on raising prices in a free trade area. The impact of any given reduction in production will be less on the larger free trade market than it will on the isolated market for an individual country.

A free trade area can make a unilateral reserve program less effective in stabilizing prices. For example, the reserves held by only one country might have to be used to help meet a production shortfall in more than one country. Hence the benefits of the reserves would accrue to more than one country. If only one country holds the reserves through a unilateral reserves program, that country might have to bear all the cost of the program. A coordinated reserves program is an approach that might align both benefits and costs. Other aspects that might be considered in establishing a coordinated reserves program are the relative instability of the various participants in the free trade area and the ability of different participants to support a reserves program. Finally, the optimal size (and cost) of a coordinated reserves program may be less than the sum of several individual national programs. Unexpectedly high yields in one country might help offset production shortfalls in others. Also, a given percentage shortfall in a single country would become a smaller percentage when spread over the entire free trade area.

Conclusions

The need for implementing new or modified supply control programs will increase as current CRP contracts begin to expire in 1996. However, the economic environment related to supply control programs is growing increasingly complex, particularly in the international arena. Involvement by the United States in a trading block may make unilateral supply control less effective, warranting international coordination of production limitations and/or reserves. Government policies in some foreign countries are aimed at increasing the competitiveness of their producers while policies in still other countries are aimed at restricting exporters' market access. Either approach affects supply control programs in the United States.

Finally, the political environment related to supply control programs is growing more complex. Conservation and environmental concerns are receiving more attention and have a direct bearing on supply control programs. Increased concern about federal deficits may mean reduced funding available for supply control programs in the future. The challenge is to develop efficient, effective, and less costly supply control programs that will achieve a broader spectrum of goals in a complex global setting.

References

Gardner, Bruce L. 1987. *The Economics of Agricultural Policies*. New York: Macmillan Publishing Co.

Hallberg, M. C., Jon Brandt, Robert House, James Langley, and William H. Meyers (eds.). 1989. *Surplus Capacity and Resource Adjustments in American Agriculture*. AERS 204. University Park, PA: Pennsylvania State University.

Tweeten, Luther G. 1979. *Foundations of Farm Policy*. Second Edition, Revised. Lincoln, NE: University of Nebraska Press.

U.S. Department of Agriculture. 1991. *Provisions of the Food, Agriculture, Conservation, and Trade Act of 1990*. Economic Research Service. Agriculture Information Bulletin No. 624, Washington, D.C.

15

Commodity Policy

Ronald D. Knutson and Edward G. Smith

Commodity organizations represent individual interests such as corn, wheat, cotton, peanuts, tobacco, or dairy. These organizations logically focus on the concerns of their member producers. The unique nature of these products has allowed commodity organizations to make specific program demands of Congress. Since only a single commodity is involved, the common economic focus makes development of a consensus policy position easier—although consensus is not always achieved. Accordingly, the central issue for any organization representing commodity interests is the circumstance (economic, political, or biological) utilized to justify the existence of a commodity program. Given this justification, Congress can then make the policy decisions.

Food and agricultural legislation has traditionally contained individual commodity titles upon which commodity organizations focused their attention. While for the major crops (feed grains, wheat, cotton, rice), these provisions have tended to be quite similar, the unique production conditions and end uses for each commodity have inevitably resulted in differences in key provisions for the different commodities. Moreover, producers of individual commodities find themselves at any moment facing unique problems and circumstances. As a result, commodity policy has been the historical focal point for farm programs.

As public policy educators and analysts, the authors of this chapter take no position regarding the worthiness of the case for individual commodity programs. Our emphasis, therefore, is placed on the major commodity issues, options, and their consequences from the perspective of producers, agribusiness, consumers, and taxpayers. Since broad farm program issues such as target price, nonrecourse loan, deficiency payment, acreage reduction, and conservation compliance are treated elsewhere, the unique problems, circumstances, programs, and policies facing the producers of individual commodities will be emphasized in this chapter.

Wheat Policy

Wheat is the nation's major food grain, and economic conditions impacting the wheat sector in the mid-1980s played a major role in shifting U.S. farm policy toward market orientation. Exports traditionally account for more than half of U.S. wheat utilization. Therefore, as the U.S. share of world wheat trade declined from the more than the 40 percent level of the early 80s to 27 percent in 1985, pressure peaked for policy changes that would restore U.S. competitiveness. Price support loan rates were lowered and a number of export expansion programs were introduced in the Food Security Act of 1985 and continued in the Food, Agriculture, Conservation, and Trade Act of 1990. After 7 years of this market-oriented experiment, questions remain as to its success. Continued budget pressure and environmental concerns will likely strengthen the position of those who feel loan rates should be increased and supplies managed in order to support farm prices.

Current Status

Wheat target prices have been frozen at $4.00 per bushel since 1990 and it is unlikely that they will be increased over the life of the 1990 act. Barring adverse weather, farm prices are likely to remain below the target price through 1995. Deficiency payments, therefore, should continue to be available each year resulting in farm revenue for participating wheat producers that is essentially frozen in nominal terms. Participating wheat producers, therefore, will likely continue to face a cost-price squeeze as input costs rise with inflation.

Despite lower loan rates and substantial expenditures on export subsidies, the U.S. share of the world wheat market was only slightly better in 1991/92 (32 percent) than it was in 1985 (27 percent). Wheat exports, however, continue to account for more than 50 percent of total U.S. utilization.

Annual acreage reduction requirements (ARP) have declined from the more than 20 percent of base acreage in 1986-88 to 0 percent for the 1993 crop. Wheat carryover stocks have declined, and features of the Normal Flexible Acreage (NFA) have reduced budget exposure. Carryover stocks to total utilization averaged 20 percent in 1991/92, the lowest level since 1973/74. The Conservation Reserve Program (CRP), however, has idled 12 percent (over 10.6 million acres) of the wheat base through 1992.

Issues

For many, the primary issues facing the wheat industry include:

- Given expected budget pressure, should and/or can current market-oriented policies be maintained?
- Do historically low stocks relative to use justify an acreage reduction policy?
- When the CRP contracts expire, what should be the policy regarding wheat base acreage coming back into production?

Policy Options

Status quo. The status quo would result in loan rates being maintained at their lowest possible level. Export expansion programs would be funded and pursued to the maximum. Target prices would remain frozen and the acreage reduction program would be implemented to allow the maximum production compatible with program goals. The marketing loan that begins in 1993 would be implemented and operated relative to the adjusted rather than the basic loan rate.

Wheat producers would face relatively stable nominal revenue, if the target price/deficiency program continues to operate as it has. Increasing costs of purchased inputs, however, would likely cause declines in farm income to the point that farm programs would be effectively decoupled. Government cost is likely to continue to be an issue due to deficiency payments and export subsidy incentives. If GATT is successful, government cost would fall but the industry would experience greater price volatility.

Increased Loan and ARP. This option would require no basic change in the current farm program language, only in its implementation. As opposed to the status quo, loan rates would not be adjusted below the basic rate determined by the current formula. Likewise, the ARP would be operated to enlist the highest proportion of set-aside acreage allowed, given projected stocks-to-use. The marketing loan could play a greater role under this scenario but, as is the case with cotton and rice, the non-recourse loan provision would assure that farm receipts (i.e., world price plus marketing loan benefit plus premium times quantity marketed) would remain at or above basic loan levels. Under this scenario, low stocks, coupled with poor weather, could result in wheat prices that approach or exceed the target price.

Initially, government program cost would decline in response to higher market prices. The extent of the decline would depend on how aggressively export subsidies were used to maintain exports. Volatility of producer prices and income would be greater than under the status quo. Program participation would decline due to the combination of increased ARP rates and higher prices. Exports, reacting to higher prices, would decline, increasing the need for increased ARP rates. Consumers would see higher wheat-product prices and agribusiness would suffer reduced sales volume. This option would run counter to the current U.S. trade negotiating posture in GATT.

Conservation Reserve Program. Approximately 30 percent of the land in the CRP has wheat base, so the wheat sector will be significantly impacted by decisions regarding the disposition of CRP land when contracts expire. Extending current contracts will cost taxpayers more money but the cost may be deemed socially and politically warranted if conservation and environmental objectives are achieved.

Some would argue that the economics and implementation of farm programs would force producers with base, in the absence of a contract extension, to return the land to production. This, however, does not have to be the case as long as the 0/92 program is in effect. If wheat producers do not feel the economic environment justifies bringing CRP back into production, they could utilize the 0/92 program and thus receive at least the guaranteed deficiency payment due on the wheat base. Of course, availability of this option depends upon Congress or the Secretary of Agriculture finding no reason why CRP wheat base would be ineligible for the 0/92 program.

Feed Grain Policy

U.S. feed grains are identified as corn, sorghum, barley, and oats. Together, they account for more than 54 percent of the total base acreage attributed to the major program crops. In the U.S., corn is easily the major feed grain with approximately 86 million base acres, representing 86 percent of the total feed grain production. Sorghum accounts for approximately 8 percent of production while barley and oats account for 4 and 2 percent, respectively. Exports represent approximately 20 percent of corn and barley utilization, while sorghum exports approach 40 percent of utilization. Oats are imported at a rate of approximately 15 to 20 percent of total utilization. Since corn is the dominant feed grain, the remainder of this section will focus primarily on the corn sector.

Current Status

Food, seed, alcohol, and industrial uses consumed 18 percent of corn utilization in 1991/92. This sector has grown consistently over the past 6 years, averaging an approximate 4 percent increase annually. Deviations from this steady growth will likely result from the economic and/or political environment for fuel additives. Corn feeding represents 62 percent of utilization and is impacted primarily by cycles in beef and pork production as well as by variations in poultry and dairy production. While exports represent only 20 percent of corn utilization, they play a significant role in U.S. world agricultural trade. The United States has lost market share in corn over the past 3 years, averaging approximately 64 percent of world corn exports in 1991/92. Problems in the former Soviet Union have contributed to much of the decline.

Like wheat, frozen target prices and rising input costs will continue to pressure incomes of participating feed grain producers for the foreseeable future. The corn ARP rate for 1993 is 10 percent, while the sorghum ARP rate is 5 percent, and the barley and oats ARP rate is set at 0 percent. The higher corn ARP rate reflects a stock buildup resulting from record-high corn yields in 1992. Average weather will likely result in ARP rates of less than 10 percent in corn and sorghum and a 0 percent ARP rate for barley and oats through 1995.

Collectively, the feed grains have placed roughly the same acreage base in the CRP program as wheat (10.6 million acres). As a percent of individual base, however, barley producers placed approximately 20 percent of the total into CRP through 1992. Approximately 15 percent of the sorghum and oat base and 5 percent of the corn base has been placed in the CRP program.

Issues

Major issues facing the feed grain sector include:

- Ability to maintain the current program given efforts to curtail the budget deficit. Annual CCC outlays to corn producers have resulted in the largest single commodity payment in 6 of the last 7 years.
- Domestic demand expansion will likely come from growth in the industrial and fuel utilization. It is unlikely that policy efforts in these areas will be confined to the 1995 food and agricultural legislation. Clean Air legislation may prove to be the major backdrop for feed grain policy. These will likely be efforts to target research on alternative uses of feed grains.
- Disposition of CRP contracts upon maturity will certainly impact the production of sorghum, barley, and oats. Feed grain producers also have access to the 0/92 program provision, therefore, the CRP issue will play out as discussed in the wheat section. U.S. corn is a significant factor in the world market, and a year or two of bad weather could make stocks policy an important issue.

Policy Options

The policy options for feed grains mirror those discussed for wheat, with the same basic impacts. However, since feed grains are major inputs into livestock and poultry production, efforts to enhance domestic prices through increased loans and/or increased ARP requirements will increase the cost of production in the livestock and poultry sectors. The longer-term impact would be an increase in livestock and poultry prices that would likely be felt by the U.S. consumer more than a comparable increase in wheat price.

Cotton Policy

Upland cotton accounts for approximately 98 percent of the cotton produced in the U.S.; extra long staple (ELS) pima cotton makes up the remainder. Farm program issues discussed in this section will relate to the upland cotton program; however data used in interpretation includes ELS production and utilization.

Current Status

The Food, Agriculture, Conservation, and Trade Act of 1990 extended market-oriented provisions that were introduced in the Food Security Act of 1985. Upland cotton is subject to a frozen target price, a minimum-bound formula-determined nonrecourse loan, and a marketing loan. The marketing loan allows producers to repay loans at less than face value if the adjusted world price for cotton falls below the loan level. If the adjusted world price is below the loan rate, the producers can repay their loans at the higher of either the adjusted world price or 70 percent of the loan rate. The marketing loan provisions were included in the 1985 act to maintain the competitive position of cotton in international markets.

As with other major program crops, budget pressure could cause policymakers to take a hard look at the cotton program. CCC outlays for cotton are expected to exceed $1 billion in FY 93. As a percentage of CCC outlays for all commodity programs, upland cotton is expected to receive from 10 to 15 percent of all payments. If the commodity program budget fails to grow, or even shrinks, look for other commodity groups to eye cotton's share as a potential source of available funds. This raises a question as to how cotton will fare within the reorganized House Agriculture Subcommittees. Cotton, and rice as well, have been placed under a general commodities subcommittee which includes wheat, feed grains, and soybeans.

How to maintain competitiveness in both domestic and foreign markets will certainly continue to be an issue. Cotton exports in 1991/92 were no higher than when the marketing loan program was first introduced in 1986, and are expected to decline in 1992/93. On the other hand, since 1986/87 domestic utilization has increased approximately 30 percent (it was 9.7 million bales in 1991/92), due to increased consumer preference for natural fibers. Season average farm prices have exceeded announced loan rates for 5 of the 6 marketing years the marketing loan has been in place. The exception was 1986 when farm prices fell below the loan rate by 260 points. That was also the initial year of operation and CCC stocks were being released through payment-in-kind (PIK) certificate payments. The marketing loan provisions will surely be addressed, especially since wheat and feed grains will have marketing loans in 1993.

The industry (i.e., input suppliers, processors, merchants, and exporters) favors a low ARP rate as a means of generating a large cotton supply. The 1990 act instructs the Secretary of Agriculture to set an ARP rate that will achieve a stocks-to-use ratio of about 30 percent. A 7.5 percent ARP rate has been announced for 1993; the industry claims this is too high. Producers are inclined to agree since they are faced with the 15 percent Normal Flexible Acreage provision which provides no payments on that 15 percent of their acreage. Given expected budget pressures, the ARP level and how it is set will be debated.

Approximately 9 percent of the upland cotton base has been idled in the CRP program. Release of this land upon contract maturity will be heavily debated due to perceived environmental impacts of cotton production.

Issues

Based on the above discussion, the following are likely to be the major cotton policy issues:

- Cutting the cost of the cotton program.
- Improving the effectiveness of the marketing loan in expanding exports.
- Using cotton land now in the CRP.

Policy Options

Status Quo. This option would maintain the basic cotton program that has existed since 1986. Target prices would remain frozen and the marketing loan operated as is currently the case. Assuming average weather, it does not appear that farm prices will exceed target prices for the immediate future. With revenues essentially frozen, cotton producers will see a cost-price squeeze and less incentive to participate. Taxpayer cost will remain a factor, and consumers will benefit from cheaper cotton. Agribusiness will likely benefit due to market-oriented provisions permitting cotton to be freely produced and sold.

Marketing Loan. As indicated in the issues discussion, the marketing loan program will be evaluated carefully against its stated objective and potential for program cost. Alternatives could include discontinuation or further fine tuning.

Discontinuation would likely increase CCC stocks as cotton becomes less competitive in domestic and foreign markets. Producer income would likely fall, and other agribusiness firms would feel results from reduced product movement. The accumulated stocks and reduced incomes would create producer pressure for increased loan levels and perhaps increased ARP levels. Higher farm prices resulting from policy implementation would reduce direct payments to producers and ease payment limit concerns. Consumers would pay marginally more for cotton-produced products and cotton-related agribusiness would be hurt. Total taxpayer cost would depend on the level of CCC stock accumulation and its ultimate disposition.

Cotton exports have not increased substantially over the last 6 years, so the value of the marketing loan in world trade is questioned. Do "state cotton traders" make it impossible for the United States to significantly expand exports because other countries will match any implicit U.S. subsidy? If this is the case, one would not expect the program to produce substantial increases in market share. If this is the case, is there a more efficient method of improving the competitive position of domestic mills (i.e., investment tax credits, commodity subsidy certificates, etc.)? Extension of the U.S. marketing loan program to wheat and feed grains is likely to lead to more research and debate into the operation of the marketing loan in a global market heavily influenced by state traders.

Conservation Reserve. Wheat and feed grains currently have the 0/92 program which could give producers of these products the incentive needed to leave land in the CRP at contract expiration. Cotton currently is subject to the 50/92 program and, as such, growers could feel compelled to bring cotton base back into production or lobby for paid contract extensions. As indicated, environmentalists will likely side with advocates for keeping environmentally sensitive land out of production. A movement to 0/92 may be enough of an incentive to extend the CRP for cotton. Agribusiness beyond the production sector will likely oppose such a move because of its volume-restricting impact. Policymakers, however, could decide to restrict 0/92 to only the cotton base previously enrolled in the CRP.

Rice Policy

U.S. rice base totals 4.1 million acres and is primarily grown in Arkansas, California, Louisiana, Texas, Mississippi, and Missouri. Long-grain production constitutes approximately 70 percent of total U.S. production. Medium-grain rice represents approximately 29 percent of production. Short-grain rice makes up the remaining 1 percent of the U.S. total. The industry's geographic concentration, in 6 states, makes it unique among the major program crops.

Current Status

Implementation of the rice program has cost approximately $600 to $800 million over the last 4 fiscal years. Rice program outlays represent from 6 to 10 percent of CCC commodity program outlays. Its geographic concentration and the large share of outlays relative to its acreage will focus the budget analysis onto rice. Geographic concentration, however, is a two-edged sword and strong congressional and administrative representation from the 6 states may swing in favor of the rice industry. In any event, the broader-based House Agriculture Subcommittee for General Commodities will force rice expenditures to be compared directly with those for corn and wheat. Previously, cotton, rice, and sugar had their own subcommittee, providing a degree of isolation from such comparisons.

As with cotton, the role of the marketing loan in maintaining competitiveness will be thoroughly debated. Since 1987, farm prices received for rice have been above announced loan rates, yet marketing loan gains have been realized. Questions are being raised as to the appropriateness of the marketing loan.

Although rice production is concentrated in 6 states, there are significant regional differences in production practices and in institutional marketing arrangements. The industry coalition has tended to reach compromises in the past but when budget cutting impacts are considered, regional pressure may increase the potential for fragmentation.

The 1990 act instructs the Secretary of Agriculture to implement an ARP program that will achieve a stocks-to-use ratio of between 16.5 and 20 percent. The program has been effective in achieving this goal for rough rice in aggregate. However, the 1991/92 long-grain rice stocks-to-use ratio (SU) was 11 percent, short- and medium-grain rice SU was 29 percent and, in aggregate, the SU for rice was 17 percent.

Farm program payment yields (FPPY) have been frozen since 1986 for the target/deficiency payment crops. Therefore, given the way the farm program yield was determined in the mid-80s, production technology from 1976 is reflected in producers' FPPY. The introduction of semi-dwarf, shorter stature rice varieties in the Gulf Coast areas during the 1980s have significantly enhanced yield potential, but have not been fully reflected in current FPPYs. Thus, there are regional and commodity differences in the effective program safety net that income supports create. Again, budget considerations may prevent any equitable adjustment resulting from differences in improved productivity.

Payment limits have brought about a substantial restructuring of U.S. rice farms. While rice is not alone in this farm restructuring phenomena, it is certainly the hardest hit. Annual deficiency payments ranging from $3.50 to $4.00 per cwt have resulted in producers with approximately 200 acres reaching the $50,000 payment limit. Since very few will dispute that a 200-acre rice operation is inefficient from a production standpoint, the rice industry has had to reorganize, change rental arrangements, and adjust certain farming activities just to maintain an operation of efficient size that can operate legally within the framework of the payment limit. While politically attractive and appealing to the general public, the payment limit provision has resulted in reduced efficiency, increased cost of producing food, and shown limited effectiveness in curtailing program spending.

Since rice is a major food grain in many countries around the world, trade issues related to rice abound. The successful completion of current GATT negotiations, therefore, could significantly impact international rice trade. Will the United States get greater access to the Japanese rice market? Can the Middle East market be regained?

Water quality, quantity, and related environmental issues will be important, given demands for water in rice production and its geographic location (i.e., coastal areas and along major waterways). Environmental issues are significant for all crops but rice will likely be watched particularly closely because of its water demands and releases into streams and estuaries.

Issues

Primary issues facing the rice industry include:

- Potential for reducing program costs.
- Effectiveness of the marketing loan.
- Rationalization of a policy in what appears to be a regionally diverse industry.
- Restoration of equity in treating different levels of yields among commodities and regions.
- Payment limits.
- Gaining access to the Japanese market for rice.

Policy Options

Status Quo and Marketing Loans. Impacts of maintaining the current policy for rice will be similar to those discussed for cotton. The same applies to the marketing loan discussion.

50/92. The 50/92 program has been used most extensively in Texas, and in the past few years in California and Mississippi. Questions have been raised about the appropriateness of the 50/92 program. Is it paying producers for idling land that would not be planted under any circumstances? Would its elimination, therefore, save money? If program spending has to be reduced, some might argue that 50/92 should be eliminated first. Texan, Californian, and other producers who enjoy the flexibility of this program may have different opinions. This is another regional issue that may divide rice interests.

ARP Determination. A case could be made for fine tuning the stocks-to-use formula for setting ARP levels that would include the type of rice grown. The industry will likely want to address this issue. A similar argument, however, can be made for wheat and cotton. This may be more fine tuning than Congress or the Administration has in mind. As an example, significant debate went into the decision in the Omnibus Budget Reconciliation Act of 1993 to require different ARP rates for corn, sorghum, and barley.

Update Farm Program Payment Yields. The issue will be discussed but the overriding constraint will be the size of the expenditure pie. Allowing yields to be brought current will transfer payments among commodities and regions. While this adjustment appears equitable, it will be a politically tough sell in view of budget constraints. Debate in this area will likely pit commodity groups against one another and raise regional differences as well.

Soybean Policy

Since the mid-1960s, U.S. soybean production has more than doubled. Some attribute this growth to the relative absence of government programs in soybeans, although the rapid expansion of broiler production, improved livestock nutrition, and the European Economic Community (EEC) agreement under the Dillon Round of GATT negotiations (i.e., not to impose import restrictions on soybeans) were certainly factors. In the 1980s, soybean production appeared to stagnate. Embargoes, increased foreign production (particularly in South America), and the enactment of the price support loan in 1977 might all have contributed.

Current Status

Important initiatives were taken in the Food, Agriculture, Conservation, and Trade Act of 1990 to bring soybeans into a program framework more nearly like that of the other major program crops. While a marketing loan had been authorized by the Food Security Act of 1985, it was mandated by the 1990 act for oilseeds including soybeans, sunflower seeds, canola, rapeseed, safflower, flax, mustard, and other oilseeds. However, it was directed to be implemented in a manner designed to minimize costs and loan forfeitures to the CCC, while encouraging exports. The loan rates remained at a relatively low $5.02 per bushel level with a 2 percent loan origination fee designed to defray program costs and presumably discourage participation in the loan program.

The flexibility provisions of the 1990 act were designed to encourage soybean production, thereby helping to recapture export markets. However, the main barriers to increased soybean exports were European Economic Community (EEC) soybean production subsidies which were found to be in violation of the Dillon Round GATT accord. These EEC subsidies became major stumbling blocks in the completion of an agreement in the Uruguay Round of GATT negotiations.

Issues

Major soybean policy issues include:

- Soybean production could be made more attractive by (1) raising the soybean price support, (2) lowering the corn price support, (3) a combination of the two, or (4) expanding the flexibility provisions of the 1990 act.
- When the 1990 act was enacted, some in Congress thought that a marketing loan with a recourse feature had been authorized—meaning that the loan had to be paid off and could not be forfeited to the CCC. To their surprise, a nonrecourse marketing loan had indeed been mandated. Therefore, questions of how to increase incentives for export expansion in the face of the nonrecourse loan program remain.
- The loan origination fee for soybeans is unique among the major program crops, resulting in inequitable treatment of soybeans relative to other crops. Its existence is attributable to efforts to control program costs, much like assessments in milk.

Policy Options

Status Quo. Expansion of the soybean program to a marketing loan program has been relatively uneventful and can be expected to remain so, as long as the basic loan rate remains low and assessment provisions to discourage its use continue.

Higher Loan Rate. An increase in the loan rate would enhance the attractiveness of growing soybeans. However, program costs would be increased through greater marketing loan payments, or soybeans could be forfeited to the CCC and thus priced out of the world market.

Recourse Marketing Loan. A recourse loan would eliminate the opportunity for forfeiture, and soybeans would thus not be priced out of the world market because of the requirement that they be marketed within the year.

Wool and Mohair Policy

The wool and mohair program has remained basically intact since direct incentive payments were introduced with the passage of the National Wool Act of 1954. The program's original objective was to assure supplies of wool for the military in times of war. Since then, the program has been continued to encourage production and protect the economic viability of the wool and mohair production sector.

The direct payment incentive program differs from the target price/deficiency payment programs offered for wheat, feed grains, cotton, and rice. Instead of receiving a calculated fixed payment rate on a predetermined farm program yield, wool and mohair producers' incomes have been supported based on the percentage of a national average price needed to bring average producer returns up to a parity-based formula target level. Therefore, since the program payment is determined as a percentage of the price received by the individual producer for the wool and/or mohair, he or she has an added incentive to produce a higher-valued product and to pursue the highest price available. The more the producer obtains from the market place, the higher the government payment. From a government cost standpoint, the incentive program operates to minimize cost. That is, as individual producers strive to market their wool and mohair for the highest possible price, they, in aggregate, reduce the percentage incentive payment necessary to reach the formula-determined target level.

Current Status

Despite the program's stated objective of encouraging production, shorn wool production has declined from 265 million pounds in 1960 to

84 million pounds in 1992. In 1992, wool producers received approximately $2.66 in government payments for each dollar received in the market place. This marked the 7th time in the last 10 years that the shorn wool incentive payment exceeded the market price. Prior to that, wool payments exceeded market prices only three times.

There are many reasons why the wool incentive has not been successful in enhancing production. The following are the most apparent:

- Wool receipts represents less than a third of the revenue (including the incentive payment) generated from sheep production. Production decisions, therefore, depend more on the price of meat than on the price of wool.
- Imports account for a majority of that used in textiles, thus market prices are impacted more by foreign supply-demand conditions than by those in the United States.
- Wool has faced significant competitive pressure from other fibers, both natural and synthetic.

Mohair production declined from 32.4 million pounds in 1965 to 8 million pounds in 1977. The support price over this period increased modestly from $0.72 per lb in 1965 to approximately $0.80 in 1976. Since then, however, the support price has increased 476 percent to its 1992 level of approximately $4.61 per lb, and mohair production has increased by 100 percent to 16 million pounds.

Market prices have declined as production increased. Mohair prices peaked in 1975 at $5.10 per lb but declined to less than $2.00 per lb by 1992. In the period 1988-1992, mohair producers received approximately $2.63 in mohair incentive payments for each $1 received in the market place.

Government payments to wool and mohair growers reached a record $191 million in 1992, having averaged $124 million annually during the life of the 1985 act. Although less than 1 percent of all commodity payments over the period, wool and mohair payments were generally increasing while payments to other programs were declining.

Issues

The major wool and mohair policy issues include:

- If wool is no longer considered a strategic fiber from a defense standpoint, is the wool program still justified? It has not been successful in encouraging production, although, the case can be made for mohair.

- Does the program enhance the economic viability of the wool and mohair industry? Again, the case can likely be made for mohair, but for wool it is less clear due to the relatively low contribution of the wool, relative to the meat, in terms of revenue from sheep production.
- Is a parity-based support formula derived from a 1958-60 base period relevant in determining appropriate support levels? The parity-based formula ignores any increase in productivity. This may not be a problem for wool production as yield per fleece has fallen from 8 lbs prior to 1980 to just under 8 lbs for the 1980s. Mohair yields, however, have increased from approximately 6 lbs per goat clipped in 1955 to 8 lbs in the late 1980s.

Policy Options

Status Quo. A continuation of the current program will likely lead to increasing program costs, especially if production is encouraged and market prices remain depressed. The parity-based support formula will likely escalate as prices paid increase relative to prices received.

Terminate Program. Termination of this program would adversely impact the economic viability of current wool and mohair producers. Sheep producers would likely feel the loss less than goat producers, due to the relatively higher contribution of mohair to the total revenue. Wool prices would increase only modestly, due to substitutes of foreign imports and synthetic fiber. Land prices would likely be less in the sheep and goat producing regions in the western states and Texas.

Limit Budget Exposure. This alternative could be accomplished through a number of methods, including:

- Capping government expenditures and incentive payments accordingly.
- Increasing the current 1 percent assessment placed on payments to producers.
- Change the parity-based formula to reflect productivity, and/or adopt a more-current base period.

These impacts would be the same as those expected from program termination, but less severe.

Dairy Policy

The two major components of dairy policy are the milk price support program and federal milk marketing orders. The latter use a Minnesota-Wisconsin basing-point system for establishing minimum prices handlers must pay for milk in the various regions. Price supports generally are the focal point of attention in food and agricultural legislation, whereas orders are authorized by enabling legislation not requiring regular reauthorization.

Current Status

Under the milk price support program, the CCC stands ready to buy butter, nonfat dry milk (NFDM), and cheese at prices intended to return to producers an average fluid milk price of $10.10 per cwt. The $10.10/cwt support level has existed since January 1, 1990, having gradually but persistently been reduced from its 1981 high of $13.49/cwt. The market price for manufacturing-grade milk, as measured by the Minnesota-Wisconsin (M-W) price series, has been above the $10.10/cwt level since the spring of 1991. A high proportion of CCC product purchases and uncommitted stocks in recent years have consisted of butter—reflecting the marked shift in consumer preferences for lowfat products.

With the reduced price support level, prices paid to producers for milk have become much more variable—with price swings of as much as $3.00 per cwt occurring between spring (high milk production) and fall (low milk production).

Issues

Dairy policy issues in 1995 will center around price instability, regional diversity, and butterfat surpluses:

- Price instability is a reflection of the interaction of highly inelastic supply and demand with nonuniform seasonal patterns of production and consumption. In the past, instability was tempered by a relatively high milk price support.
- Regional diversity reflects wide differences in production conditions, technology, and the size of dairy farms. The two largest dairy states in the United States are California and Wisconsin; in California the average herd size is 519 cows, in Wisconsin, it is 52 cows. Substantial economies of size are found in all regions, continuously putting income and survival pressure on the smaller and even the moderate-size producers.
- Butterfat has become the major dairy surplus problem. USDA has responded by reducing its purchase price for butter from $1.09 to $0.65 per pound and by raising the NFDM price support from $0.79 to $1.034 per pound. The intent is to stimulate butter

consumption, reduce incentives for butterfat production, and make U.S. butter more competitive in the world market.

Policy Options

Status Quo. Continuation of the current dairy policy signals price supports of less than the market price low in most months and years. Generally, only the price of butter would be supported, although continuing semiannual reductions[1] in the butter purchase price with corresponding increases in the NFDM purchase price can be anticipated. Butter could become competitive in the world market while NFDM would become less competitive as its price is raised in an effort to maintain the milk price support level at the current $10.10 per cwt.

Price Support Increase

Proposals to increase the milk price support level by up to $1.00 per cwt as a means of reducing milk price instability can be anticipated. As product prices to consumers increase, CCC purchase costs (primarily for butter and NFDM) would also rise. The increased CCC cost could be offset by self-financing assessments; these have been unpopular with producers.

Target Prices and Deficiency Payments

Target price/deficiency payment proposals for milk suggest elimination of the milk price support-purchase program and substitution of a target price. If the annual average M-W price falls below the target price, the difference would be paid, as a deficiency payment, directly to producers. Product prices would decline to market-clearing levels, that, under current conditions, would mean lower prices for butter and probably for NFDM. Deficiency payments could be limited to an historical base level of production—comparable to the concept of frozen farm program yields in crops.

Class IV Milk Pricing

Class IV milk pricing proposals are designed to discourage increases in production while making U.S. dairy products more competitive in the

---

[1]The 1990 act permits changes in the purchase price for butter and NFDM only twice a year.

world market. Producers would be given an historical production base. Milk produced in excess of the historical base would be priced at a level whereby the resulting products would be competitive in the world market; they may be sold only in the world market. This policy has similarities to the two-price plan described later for peanuts. Base production could be expected to take on an asset value, especially if the Class IV price were sufficiently low that it did not cover the out-of-pocket cost of producing an extra unit of production (marginal cost).

Sugar Policy

The U.S. sugar program is the key determinant of the domestic sugar market; the program uses a combination of price supports, import quotas, and (potentially) production controls to maintain the price of sugar at its support level. In addition to sugar producers, corn producers and manufacturers of high fructose corn syrup (HFCS), a commercial sweetener, have an interest in sugar programs. Sugar and HFCS share the caloric sweetener market approximately 50-50.

Current Status

Following a persistent decline since 1970, per capita sugar consumption stabilized in about 1986. Most of the substitution of HFCS for sugar took place in soft drinks. Concurrent with the decline in sugar consumption, the quota on sugar imports was reduced to the 1992/93 level of 1.362 million tons. This quantity of sugar enters the United States with a tariff of $0.0625 a pound. For quantities over the tariff rate quota of 1.357 million tons, the tariff rises to $0.10 a pound. This combination is designed to yield a U.S. raw sugar price of about $0.21 per pound.

Because of continuing downward pressure on the volume of sugar imports and the consequent reduced ability of import quotas to raise the price of domestic sugar to the support level, the Food, Agriculture, Conservation, and Trade Act of 1990 provided for the imposition of production controls in the form of marketing allotments on sugarcane, sugar beets, and crystalline fructose (a corn product) if estimated sugar consumption and estimated sugar production differs by less than 1.25 million tons. The 1.25 million ton differential is, in essence, a legislatively mandated minimum level of imports that was apparently established as part of U.S. foreign policy to maintain a raw sugar refining industry and generate foreign exchange for quota-holding countries.

Corn sweetener production in 1992/93 is expected to use 7 percent of the corn crop. HFCS prices follow sugar prices very closely.

Issues

Sugar policy issues include:

- Continuation of the sugar program. Most estimates suggest that, under free trade conditions, the world price of sugar would be approximately $0.125 per pound. Therefore, the U.S. sugar price is supported at approximately 70 percent above free market levels. Should this continue when the United States can buy sugar cheaper on the world market? This question likewise applies to HFCS and noncaloric sweetener substitutes.
- Additional challenges to U.S. sugar policy will occur when and if relations are normalized with Cuba, once a major supplier of sugar to the United States.
- The competitive position of HFCS is dependent on the sugar price support program, as is the survival of many sugar producers. Corn producers are thus also interested in sugar policy because of the importance of industrial uses of corn.

Policy Options

Status Quo. A continuation of current sugar policy will likely result in increasingly restrictive production controls on sugar as sugar production becomes more attractive relative to other commodities where income support levels have been persistently reduced. Pressure will likely develop to include HFCS in addition to crystalline fructose within the sugar production control program.

Target Price. Placing sugar policy in the same target price/deficiency payment framework as other crops would substantially increase government costs while reducing its market price. With a lower sugar price, the price of HFCS would likewise decline, dulling the incentive to substitute corn sweeteners for sugar. The existence of large sugar producers (predominantly in Florida and Hawaii) would create payment-limit problems in that payments could be denied to some of the most efficient segments of the sugar industry.

Free Market. Elimination of the sugar program would substantially reduce the sugar price and the HFCS price, and put many U.S. sugar producers and related agribusinesses out of business. Land values in major sugar-producing regions would fall.

Production would be particularly adversely affected in the higher-cost regions of Hawaii, the western United States, Texas, and the Great Plains.

Honey Policy

The honey program, mandated by the Agricultural Act of 1949, is among the most controversial of the commodity programs. One of the crucial early votes on the Food, Agriculture, Conservation, and Trade Act of 1990 was over continuation of the honey program. Honey producers won that vote and, in retrospect, paved the way for passage of more significant and less controversial commodity programs.

Current Status

Only 1 percent of the 212,000 U.S. beekeepers are commercial producers, who account for 60 percent of honey production. The value of the honeybee as a pollinator, however, has been estimated to far exceed its value as a producer of honey and beeswax.

Prior to the Food Security Act of 1985, the honey program provided an assured market for most of the honey produced through a price support loan. The support level progressively increased from $0.14 cents per pound in 1972 to a peak of $0.658 cents in 1984. CCC acquisitions escalated sharply in the 1980s as domestic users substituted imported honey for domestically produced honey. In 1985, honey imports approached the level of domestic production, meaning that most of the domestic production was acquired by the CCC. Honey stocks rose to record levels, as export subsidies and domestic food assistance programs were unable to utilize sufficient honey to prevent stock accumulation.

The 1985 act implemented a reduction in the support level, lower loan repayment rates (comparable to a marketing loan), and a $250,000 limit on loan forfeitures. The limit on forfeitures meant that if it came into effect, the support level would no longer be an effective floor.

The 1990 act continued the basic provisions of the 1985 act with a $0.538 support level throughout the 5-year life of the legislation. A 1 percent assessment was placed on commodities forfeited to the CCC. The result discouraged forfeitures and reduced government costs. "Per person" loan deficiency payments and loan forfeiture limitations were reduced to $125,000 by 1994.

Issues

Questions regarding the need for a honey program remain. If the value of honeybees in pollination far exceeds their value as honey producers, questions arise as to whether a honey program is really needed to support pollination. World honey prices are competitive with sugar, so does it make economic sense to attempt to maintain a profitable

U.S. honey industry? In addition, there are threats to the U.S. honey industry, e.g., Africanized honeybees, diseases, and pesticides.

Policy Options

Status Quo. Recent program changes have resulted in a drop in government costs from a high of $100 million in 1988 to $19 million in 1991. With lower market prices, imports have declined but remain a problem. Loan limits have caused the market price to fall below support levels.

Reduce Loan Limits. The $125,000 payment limit per person compares with $50,000 for other commodities, but is independent of the limit for other commodities. However, the $50,000 limit for other commodities does not apply to loan forfeitures. If the $125,000 limit was reduced to $50,000, the amount of price support provided by the program would be further eroded. The impact on pollination is unclear.

Eliminate the Honey Program. This alternative was advocated by the General Accounting Office (GAO) in a 1985 report. GAO contended that the honey program benefits only a few individuals, is costly, and is not needed—either for pollination or to assure a honey supply. If the honey program were eliminated, the price of honey would be expected to closely track world prices. As an alternative sweetener, there would be some substitution of honey for sugar, although this effect would be relatively minor within the United States.

Peanut Policy

The peanut program is a unique commodity program in that it is mandatory, utilizes a marketing quota, and establishes a higher minimum price for edible peanuts utilized in the domestic market than for peanuts exported or crushed for oil and meal. In this sense, it is a classic two-price plan, having some similarities to the EEC's Common Agricultural Policy. The EEC policy provides a higher domestic price and a world market competitive export price. Aside from its benefits to peanut producers and quota holders, many other producers and some policymakers view the peanut program as a highly successful model, wondering if it could be effectively applied to other crop and livestock products.

Current Status

The price support for edible quota peanuts, indexed to the cost of production, has increased continuously (from $0.197/lb in 1975 to $0.3375/lb in 1992). On the other hand, support for nonquota ("additional") peanuts has declined from a peak of $0.15/lb in 1979 to $0.066/lb in 1992; it has been relatively steady since 1985. As a result, the differential between the edible price and the "additional" price has increased markedly, meaning that U.S. consumers are consistently paying a larger share of the costs of producing peanuts.

With the requirement that quotas be adjusted in line with domestic use, stocks have fluctuated based largely on variation in yields. Yields are difficult to predict. About one-half the total U.S. production (approximately 5 billion pounds), is used for food (i.e., quota peanuts) 20 percent is crushed, 20 percent is exported, and the remaining 10 percent is used for seed or ends up in stocks. The U.S. has become, since it adopted the two-price plan in the 1970s, an important player in the export market. The U.S. share of world peanut exports is about 25 percent. Peanut imports are controlled, at a very low level, by Section 22 import quotas.

As a result of the high price for quota peanuts, quota rights have acquired considerable value, ranging widely among peanut producing states and regions, depending on the cost of production. These significant differences in quota values are sustained by limitations on the transfer of quota rights among counties and states. If the transfer was not restricted, peanut production would move to those counties and states with a comparative advantage, leaving adverse economic impacts on the areas vacated. Crops of lower value likely would then be produced in the former peanut production areas, profits would fall, and land values would decline. Nationally, these effects would be partially offset by increases in the new production areas.

Issues

The following peanut policy issues will likely be at the center of attention:

- The two-price policy for peanuts conflicts with U.S. policy positions favoring free trade and reduced barriers to trade. Comparability to the EEC Common Agricultural Policy undermines the U.S. trade negotiating position. The peanut program could not operate without the Section 22 provisions, (i.e., limitations on peanut imports).
- In contrast with other commodities where price and income supports have declined since 1985, support levels for edible quota peanuts continue to be adjusted upward, reflecting production cost changes. The Omnibus Budget Reconciliation Act of 1990, however, placed a 1 percent market assessment against applicable quota and "additional" peanuts, to be shared equally by the

grower and first purchaser. The consequences of lowering supports and/or increasing assessments will be debated?

- Should sales of quotas be allowed across county and state lines so that regional shifts in production in response to comparative advantages could occur?

Policy Options

Status Quo. The peanut program can probably exist as a unique program as long as Section 22 remains in effect. Consumers of peanuts have and must continue to be willing to bear the cost of the program in the form of higher prices for whole peanuts, peanut butter, and peanut candies.

Target Price. The peanut program could be converted to a target price program comparable to that of other major crops. This would transfer the cost of the program from consumer to taxpayers. With program costs being more apparent, changes in level of support could be adjusted on the same basis as for other crops. Domestic demand for peanuts would expand somewhat as the domestic prices of peanuts fall. On the other hand, the United States could initially become less competitive in export markets as more production is consumed by the higher-value domestic demand. Longer-term impacts in regard to export competitiveness would depend on the degree of expanded peanut production, especially in areas currently producing for the export market.

Program Elimination. The possibility that the U.S. would bargain away Section 22 during the GATT negotiations raised a specter of concerns that the peanut program might be done away with. The long history of mandatory regulations, extending back to the Great Depression, makes it very difficult to predict the consequences of program elimination. Surely, in current producing areas, profits and land values would decline. Would the United States produce enough peanuts even to satisfy its domestic needs? And is it important to anyone other than peanut producers, agribusiness, and the related rural communities? These are important policy issues for which data and analyses are, at best, sketchy.

Tobacco Policy

The tobacco program is unique in that it is continuing legislation and, therefore, does not have to be rewritten every 5 years. As a result, it generally is not considered as part of the food and agricultural legislation. It is also unique in that tobacco has maintained marketing quotas since 1938.

Current Status

During the 1980s, the tobacco program underwent several changes compelled by reduced consumption, increased imports, and the apparent incongruity between the U.S. subsidizing tobacco production while at the same time spending large amounts of money on research to find cures for diseases allegedly linked to tobacco consumption. Among the most important of these policy changes were producer assessments to pay for all program costs except administration (the no-net-cost feature of the program), setting the size of the marketing quota to domestic and foreign market needs, and reducing burdensome inventory stocks.

Of these, the no-net-cost feature of the program was perhaps the most profound and had the most far-reaching implications inasmuch as it has since been discussed for other commodities such as milk and honey. A total of 137,000 growers and 300,000 quota holders benefit from the continuation of the tobacco program which, in terms of the value of the crop produced, is the sixth largest in the United States. Typically, tobacco farmers have smaller, often low-income farms.

Issues

The issues facing tobacco policy include:

- The rationale for any type of government program designed to raise producer returns on a crop linked to health adversities.
- If a production quota program is to be maintained, what should be the level of the support price, and how should the quotas be allocated?
- Should the quotas be transferrable, resulting inevitably in a consolidation of tobacco production?
- In light of the adverse health effects, should production be further restricted, raising the price?
- What should be done with large stocks of low quality tobacco?

Policy Options

Status Quo. Continuing the current program maintains a domestic tobacco industry as well as the value of quotas held by more than 300,000 individuals, many of whom have retired. Tobacco is the lifeblood of many farm families, rural communities, and their residents in Kentucky, Maryland, North Carolina, South Carolina, Virginia, Georgia, and, to a lesser extent, a few other states.

Increase Quota and Lower the Support Price. Reducing the support price has the effect of making U.S. tobacco more competitive in international markets. It could also, unless compensated for by a tax increase, reduce the cost of tobacco products, encourage their use, and contribute further to adverse health impacts in the population. For quota holders, a lower support price means lower quota values, undermining asset values.

Lower Quota and Raise the Support Price. A more-restrictive tobacco policy might be advocated by those who want to discourage tobacco con-sumption while protecting quota values. However, a more restrictive tobacco policy could also encourage increased tobacco imports and there-by have few of the intended effects unless import quotas were established.

Eliminate Program. The competitiveness of the U.S. tobacco industry in the absence of a tobacco program is largely unknown. What is clear is that all asset values associated with the tobacco quota program would vanish. The economies of communities dependent on tobacco would likewise be adversely affected although, after a period of adjustment, it is quite possible that a tobacco economy would still exist although with fewer, but much larger, farms.

Future of Commodity Policy

Spending constraints have put considerable pressure on commodity policy. Signs of fragmentation are beginning to develop among commodity groups as each bargains for what it considers to be its share of the farm program pie. Yet, over the years, commodity policy has been amazingly durable, just as has the political influence of agriculture and the organizations representing farmers. Accordingly, it would be a mistake to bet that commodity policy is near an end.

All commodity programs are being impacted by the tendency to tie program benefits to environmental practices. Some feel that, over time, subsidies, to the extent they exist, will be made to cover a portion of the costs of conservation compliance. Moves in this direction will present substantial challenges to commodity groups.

Acknowledgements

The authors wish to acknowledge the ERS series of background papers published in 1990 for each of the major program commodities. Although not cited specifically, this series was used exclusively in developing this chapter. The same acknowledgement applies to other numerous USDA data publications relied on for numerical illustrations.

Marketing and Risk Management

16

Federal Marketing Orders

Walter J. Armbruster and Robert A. Cropp

Federal marketing orders for milk and for fruits, vegetables, and specialty crops implement policy established under the Agricultural Marketing Agreement Act of 1937 (AMAA). They allow producers, working together under government authority, to take collective action and achieve some control over the marketing of their commodity. The regulations affect behavior of marketing participants, altering the effective structure, conduct, and performance of the marketing system. Marketing orders are designed to achieve orderly marketing through coordinated activity at the producer and first-handler levels (U.S. Department of Agriculture 1975). The orders seek to do this by improving producer prices while protecting consumer interests and establishing orderly intraseasonal or interseasonal product marketing (Polopolous et al. 1986). Some states also authorize marketing orders for specific commodities produced in their states.

Provisions of current federal marketing orders are essentially the same as those originally included to deal with perceived problems and issues of the 1937 marketing system. Since 1937, many orders have been promulgated, terminated, or amended in attempts to deal with changing business practices or with market structure, conduct, and performance issues. But increasing competitive pressures and evolving societal issues require that orders be continually re-evaluated (Armbruster et al. 1993). Continuous modifications are necessary to assure that the regulatory programs do not hinder needed longer-term adjustments for the sake of short-term gains for existing producers. Eventually the underlying economics will bring about structural adjustment, and the regulatory programs must not delay and increase the cost of needed adjustments.

Order Provisions and Economic Rationale

Marketing orders for fruits, vegetables and specialty crops utilize minimum-quality standards and restrictions on volumes marketed in the primary market, i.e., generally, the domestic market. In recent years, more emphasis has been placed on market expansion through advertising, market promotion, and research programs.

Milk marketing orders use a classified pricing system in which use-classes are based on the final use made of the milk: generally, class I milk is milk used to produce fluid products, class II milk is used to produce soft manufactured products, and class III milk is used to produce hard manufactured products. The orders establish minimum prices that handlers must pay into an order pool for each class of milk. Producers receive a weighted average, or "blend," price that is based on these class prices and the amount of milk used in the different classes.

Marketing orders are based on economic underpinnings relating to the advantages that can be obtained through group action. All members of the relevant production and marketing industry must act in concert. Otherwise, some producers may obtain gains without bearing any of the cost of reduced marketings or of meeting quality standards (i.e., become free riders). Marketing orders provide a means of assuring that all producers participate and no one gets a free ride.

Economic theory suggests that diverting sales from the higher-priced, and less price-responsive, market to the more price-sensitive secondary market(s)—i.e., price discrimination—will increase total revenues for all sales. For some horticultural crops, limited quantities are marketed in the primary market and the remaining production is sold in the more price-responsive markets, stored, or sold for salvage. Fresh produce failing to meet minimum-quality requirements of the marketing order must be sold for livestock feed or diverted to other low-valued markets. Under the milk marketing order classified-pricing system, market allocation is accomplished by arbitrarily raising the price for milk used for fluid products (i.e., drinking milk) above the price for milk used to produce manufactured products (e.g., powdered milk). Pooling proceeds from classified pricing along with obligating plants to pay a common weighted average or blend price to farmers prevents free riding.

The scale of operations needed to obtain economic efficiency for handlers or processors may limit market outlets for producers facing highly concentrated buyers. Individual producers have little bargaining power. Collective action, among producers, to deal with the more-powerful marketing segment may create a more-competitive market outcome. Regulations to facilitate collective action alter the distribution of benefits and costs among market participants. As the marketing system evolves, the market power balances may change if producer cooperatives gain more membership and bargaining strength. Alternatively, if marketing firms segment the market to meet consumer demands, relative strength of buyers may increase based on their knowledge of the product attributes needed.

Marketing orders for horticultural crops and for milk have attempted to encompass relevant market areas to make the orders effective. The evolution of technologies in production, transportation, marketing, information, and management have challenged the ability of marketing order regulations to maintain relevance to marketing conditions. Marketing orders alter trade flows, and unless regularly reviewed and updated, might contribute to economic inefficiencies.

Regulations issued under marketing orders must also deal with potential imports if they are to effectively regulate the market. The efforts to reduce trade barriers through multilateral negotiations further challenges marketing orders.

Structural change in the global food marketing system with large multimarket firms interacting with producers may require different policy approaches than for those used with segments of agriculture where smaller, more-localized firms prevail. Marketing order regulations represent a regulatory burden on handlers and can involve consumer costs; either may restrict competitiveness of the order commodity (Heifner et al. 1981). Alternatively, they can facilitate economies of scale and market expansion.

Market control and price-discrimination marketing order provisions directly impact producers, marketing firms, and consumers. As competitive conditions change, marketing orders may impede or accelerate regional shifts in production. Technological changes involving biotechnology, information, and management may have implications for the marketing of agricultural and food products that necessitate adjusting marketing order provisions. For example, if instrumentation is perfected to instantaneously measure fruit and vegetable maturity, horticultural crop orders may need to adjust minimum maturity requirements. Research, promotion, and information collection facilitated through marketing orders may speed the adoption and enhance the effectiveness of new technologies.

Consumer issues and changing consumer demands may also require changes in marketing order provisions. For example, concerns about possible chemical residues in fluid milk, in other dairy products, and on fresh produce, along with efforts to segment markets and target products to consumer demands (e.g., convenience and nutritional concerns) may require changes in orders.

Horticultural Crops

Industry characteristics have led to completely different approaches to administering marketing orders for horticultural crops and for milk. Consequently, the policy issues in the two agricultural segments differ.

Horticultural marketing orders make use of three types of regulations affecting growers, though the regulations generally apply to the handlers. The most restrictive are volume controls that include shipping holidays prohibiting shipment during specified short periods, prorates regulating shipments to markets on a temporal basis, market allocations limiting amounts going to specified uses, reserve pools which require holding back a percentage of available production, and producer allotments that set a maximum quantity that individual producers can sell.

Another category of regulations sets minimum grade, size, and maturity standards for produce sold. These standards are intended to improve marketing efficiency by providing information about produce quality for buyers and sellers. On the other hand, they may be overly restrictive, preventing the sale of less-than-perfect produce that would find a ready market.

The third type of regulation involves market-support activities. These include advertising and promotion; production, marketing, and product research; and container and pack regulations.

Approximately one-third of horticultural marketing orders now authorize generic advertising and promotion programs and two-thirds have research and development programs. Production, marketing, and product research funded under marketing orders can improve marketing efficiency and competitiveness of growers by improving sales, yields, and quality and other characteristics desired by consumers. Container and packaging standards reduce aggregate marketing and handling costs.

Marketing orders have been evaluated for their impacts on producers, marketing firms, and consumers by a number of researchers. Task forces have examined price impacts (U.S. Department of Agriculture 1975) and economic efficiency and welfare implications (Heifner et al. 1981). The knowledge base about horticultural marketing orders may be summarized according to the types of regulations.

Impacts of the various types of volume control regulations differ, according to their restrictiveness. Shipping holidays prohibit shipments for brief periods such as a few days or weeks and are intended to prevent sharp price declines in the event of temporary oversupply. Historically, their temporary and infrequent use have made their impact on seasonal average price and sales relatively insignificant.

Prorates, which limit weekly shipments by handlers over longer marketing periods, are much more effective than shipping holidays in affecting annual average price and quantities marketed. Although authorized in seven marketing orders, prorates for California-Arizona navel oranges and lemons are the most publicized and controversial. The rationale for prorates is to stabilize intraseasonal prices by stabilizing shipments. Any cost savings from stabilizing shipments should reduce marketing margins and result in higher farm prices and, if marketers are competitive, reduce retail prices. Fresh-produce consumers will obviously pay more if prorates result in smaller aggregate shipments than would otherwise occur, or if they are so restrictive that fresh quality product is diverted to processing markets.

Continuous use of prorate regulations throughout a marketing season may reduce total marketings into the primary market, allowing producers to take advantage of price discrimination. In recent years, the California orange prorate programs have been controversial and were suspended on several occasions by the USDA. Late-season suspensions of handler prorates during the 1983-89 seasons increased quantities marketed fresh and reduced prices for fresh navel oranges (Powers 1991). In 1985, an 18-week suspension of the orange prorate appeared to decrease the FOB retail price spreads (Thompson and Lyon 1989). The policies implemented under the California-Arizona lemon prorate program have different impacts for producers, middlemen, and consumers. Consumers and producers would prefer constant prices with quantities adjusted in line with seasonal demand. Between 1961 and 1972, weekly shipments were quite variable and prices relatively stable. Shipments were much more stable and prices much more variable during the 1974-87 seasons (Carmen and Pick 1990).

There has also been considerable controversy over the years within the California orange industry, particularly driven by concerns of those wishing to expand their market share. Recently, some shippers in the market allegedly overshipped their prorates, violating the marketing order. This ongoing struggle to continue the program in the face of various violations and challenges is to be expected as industry structure changes and the regulatory institution based on earlier conditions slowly adjusts.

Market allocation volume controls do limit quantities that handlers may ship into regulated markets during a marketing season, thus regulating total seasonal marketing. They are predominantly used for semi-perishable dried fruits and nuts, and for the most part limit shipments to the domestic market. Market allocations may help stabilize intraseasonal prices and shipments. The success of market allocations in providing greater intraseasonal stability is debatable (Powers 1990b). The diversion of products to unregulated markets is a source of some concerns, particularly if it affects international trade flows or causes selling of products as animal feed.

Reserve-pool volume controls prohibit handlers from selling some portion of the current season's production. The withheld amounts may be placed in storage under the reserve pool and shipped into commercial markets or diverted to other uses if inventories build up significantly. Five marketing orders, primarily dried fruits and nuts, make use of reserve pools. Reserve pools can help stabilize quantities shipped between seasons to the extent that commodities stored during a large-crop year are marketed into the primary market at a later time. One recent study (French and Nuckton 1991) found that the public interest has generally been served by the raisin marketing order, in terms of reduced average consumer prices and larger quantities available and growers receiving some protection in years of unexpectedly large production relative to demand.

Producer-allotment volume controls require that handlers ship only the amount of produce for which a grower has an established marketing quota. Thus, allotments are potentially the most-powerful regulations because they indirectly control production, in that growers have little incentive to produce a commodity they cannot sell. Currently allotments are authorized only for cranberries, Florida celery, and spearmint oil, but have never been used for cranberries and are set at unrestrictive levels for celery.

The net effect of marketing-allotment programs depends upon how they are used. By preventing overplanting in the face of favorable price projections, they reduce losses from overinvestment and crop abandonment. The now-terminated marketing order for hops was found to help stabilize hop acreages and nominal hop prices, and to reduce cyclical variations in production, leading to more efficient resource allocation (Folwell et al. 1985). Allotments that are used to regularly restrict production benefit existing growers at consumers' expense. Any competing production, from outside the marketing order area, will restrict the potential gain from application of market allotments. One study (Taylor and Kilmer 1988) found that a limited price enhancement, above that in a perfectly competitive market, may have occurred for Florida celery. Orders authorizing marketing allotments provide some allotments for new and expanding growers each season, limiting somewhat the potential impact of the program.

Grade, size, and maturity regulations of marketing orders keep substandard produce off the market and improve the average quality of product marketed which in turn should strengthen consumer demand. However, setting standards for grade and size may indirectly control quantities marketed and thus raise consumer prices somewhat. A 1981 review of quality standards (Heifner et al. 1981) found little basis to conclude that they were used to significantly affect total supplies.

There has been increasing interest in using marketing orders to fund generic advertising and commodity promotion in an effort to increase demand. There are policy and efficacy questions about efforts to maintain or expand market share through generic commodity promotion programs. More information on these programs is available in Chapter 17 of this volume and in Armbruster and Lenz (1993).

Current Issues

A number of issues arising from today's global economy and environmental, consumer, and trade concerns pose challenges to continued success of horticultural marketing orders.

Structural Changes. As food marketing firms change ownership, grow, and expand their international operations in an increasingly global economy, marketing practices change. Will marketing orders enhance or restrict opportunities for U.S. producers and marketing firms to be competitive?

Quality control provisions of marketing orders provide short-term benefits but may have long-term benefits by increasing demand for high-quality products. Quantity-control marketing orders primarily focus on short-run issues, such as maintaining orderly marketing flows and stabilizing within-season price. Enforcement actions under marketing orders may negatively involve industry innovators seeking to take advantage of changing market conditions and new market opportunities. They come into conflict with existing regulations and an administrative structure that gives more power to dominant segments of the industry. Existing procedures for modifying handling regulations are in place and orders are periodically amended, primarily in response to industry requests. Nonetheless, marketing orders tend to support the status quo, at the possible expense of needed longer-run adjustments.

In recent years, marketing order regulations have, to some degree, deemphasized quantity-control provisions, as more attention has been paid to market-enhancing efforts (Buxton 1993). Commodity promotion programs are designed to help expand demand. There are questions about whether society gains from generic promotion programs and about the benefits obtained by the producers funding the programs. There are studies showing positive gains to producers from commodity promotion efforts. However, significant policy issues involve equity and efficiency implications of widespread use of these programs in the domestic market where aggregate demand growth is expected to be limited by the relatively slow population growth (Armbruster and Lenz 1993). For further discussion, see Chapter 17 in this volume.

The quantity control provisions of horticultural marketing orders have generally been viewed with disfavor by consumer groups, some political leaders, and free-trade proponents in recent years. They are also questioned by industry participants wishing to expand their operations. Noncompliance by orange shipping firms with the California-Arizona navel orange prorate programs draws into question whether industry members widely support the existing program. Secretary of Agriculture Espy has suspended the prorate provisions and initiated a thorough review of the order (U.S. Department of Agriculture 1993).

Continuing trade negotiations and U.S. pressures on other countries to not restrict production and marketing practices affecting international trade may also affect marketing orders using volume control measures. These orders treat export markets as nonregulated markets in which commodities diverted from regulated channels can be marketed.

Quality Standards and Chemical Use. Grade, size, and maturity standards may become less meaningful as marketing firms increasingly target market segments. Questions have been raised about the basis for the standards. Do they match consumer needs or do they merely provide producers a means of limiting quantities marketed without adequate attention to the different market segments? Changing consumer demand and marketing firms' interests in satisfying the convenience, health, and nutrition criteria for consumers may have implications for marketing orders. For example, it may not be appropriate to exclude smaller sizes from fresh market use if there are market niches for the smaller sizes and the consumers that buy them would not buy larger or more-expensive produce. Some handling regulations have been changed in recognition of these possibilities. Further research on consumer preferences could refine the use of minimum standards and keep the marketing orders abreast of market changes.

Technological developments may eventually reduce the need to use grade and size standards as indicators of quality. Based largely upon observable attributes, minimum standards for grade, size, and maturity have been used to convey information to buyers unable to determine quality prior to purchase. Some critics contend that excessive use of chemicals, and thus potentially harmful chemical residues, may be encouraged by the current sensory-based grade attributes (Office of Technology Assessment 1992). Growers may have incentives to use more chemicals in order to meet those standards and thus produce marketable products or to meet higher quality standards that will result in higher prices.

Trade Issues. Section 8e of the enabling legislation requires imports of produce items covered by a marketing order to meet the same minimum grade and size standards as applied to domestic shippers. The North American Free Trade Agreement (NAFTA) provisions specify that any measure regarding classification, grading, or marketing of a domestic agricultural product will also apply to like products imported from the other country for *processing*. There is increasing interest in utilizing Section 8e provisions and currently 25 of 39 active orders do so. Will it be necessary to modify import regulations related to Section 8e provisions if NAFTA is approved? For example, the domestic exemption for processing may also need to apply to commodities imported for processing.

Another issue involving imports and domestic regulation of grade and size relates to the desire of shippers in other countries that compliance with quality standards be established by inspection in their country. While some countries have an inspection system similar in rigor to that used in the United States, deterioration en route would be an issue. Current regulations require that products be inspected and certified at the point of importation, generally interpreted to mean the border crossing.

Regulatory Timeliness. The slowness of the regulatory system and it's potential for delays affecting perishable crops with particular market windows may make it increasingly difficult to use marketing orders in the changing market structures and relationships of the global economy. As competition from other countries and shippers continues, marketing orders will need to become more responsive. This may require reducing the number of regulations, changing the ways in which orders are administered, or finding ways to issue decisions that facilitate prompt response amid changing conditions.

The suspension of the California-Arizona orange marketing order by USDA, in reaction to increasing evidence of internal industry dissatisfaction, was cited earlier. The perceived lack of flexibility in marketing orders to accommodate changing market conditions and industry structure may be the root cause of the turmoil. Marketing orders essentially provide self-regulation by industry representatives through the auspices of the Secretary of Agriculture, who administers the orders and receives recommendations from industry committees. Several times in recent years the Secretary of Agriculture has rejected industry recommendations and suspended regulatory programs. When industry unity regarding continuation of certain marketing order provisions is not strong enough, the secretary may reject industry recommendations or suspend regulatory programs because they do not meet AMAA objectives.

In October 1992, PL102-553 specified that timely clearance of marketing order regulations within USDA would generally require a 45-

day response time. That requirement has been violated frequently, though less so in recent months. The Food, Agriculture, Conservation, and Trade Act of 1990 also requires that the Office of the U.S. Trade Representative (USTR) must be involved in approving Section 8e provisions. There is a 60-day deadline, but the USTR sometimes fails to meet the deadline and USDA is precluded from issuing domestic regulations until the import regulations are approved. Difficulty in approving regulations, in the increasingly challenging regulatory climate, raises further questions about marketing orders' ability to regulate markets as in the past.

Policy Alternatives

The current issues facing horticultural marketing orders are: (1) if they are really needed, and (2) what is the best way of accomplishing their underlying goals in today's marketing environment. Evaluation of continuing effectiveness is an ongoing process in the sense that regulations are implemented annually. However, the factors weighed in making annual decisions are short-run, affecting immediate market prospects. To deal with the issues identified above, a longer-term view of the issues and alternatives for adjusting marketing-order programs is necessary. Some alternatives for dealing with the major issues can be identified.

Structural Change. The marketing system and its structure, including relationships between handlers and producers, continue to evolve. It is appropriate to re-examine the strength of support among industry members for volume control regulations, in light of the changing structure and competitive environment, for the various commodities. Regulations must be sufficiently responsive and changes sufficiently rapid to take advantage of or counter changing international marketing opportunities and challenges. Thorough analysis of the current administrative procedures affecting volume control, quality, and market facilitating and development programs is needed to assure their effectiveness.

Re-examination of how well the marketing orders in each industry segment serve today's marketing needs could be accomplished by undertaking a thorough review process. Perhaps this process is underway in the case of the California-Arizona orange industry. Any thorough evaluation would examine the underlying structure of the industry, its programs, and the impacts of the marketing order provisions in the broadest possible context. The intent would be to examine where the marketing firms and relationships are heading and then restructuring of the marketing orders to facilitate progress toward desirable goals.

The review process should require input from various segments of the industry as well as those outside the industry who have an interest in the impacts or implications of the marketing orders. The procedure would go well beyond a formal hearing process recording the number of comments favoring or opposing a particular provision. This approach could minimize hindrances to needed adjustments and would require careful assessment of what provisions really serve both the short- and longer-run interests of existing growers, potential new entrants, and consumers. Such a comprehensive review would be costly, but could produce significant improvements that make the orders more relevant to today's evolving marketing system. A commitment to making changes derived from such a process is a necessary precursor. Resistance may be expected from those familiar and comfortable with the existing regulated marketing environment. To the extent needed changes are identified and implemented, net social welfare will be enhanced. Impacts among producers, marketing firms, and consumers can be expected to vary by industry segment.

Market Expansion Versus Quantity Control. With increasing interest in market expansion activities and reduced restrictions on trade flows among countries, marketing order policies designed to encourage successful commodity promotion may have merit. To the extent that they successfully create new demand, they could reduce the need for supply control provisions. However, market expansion may require longer-term efforts and be of limited use in dealing with temporary surpluses, so both tools may be needed. The potential problem of merely shifting domestic demand among commodities in the face of inelastic total demand must be considered.

In some industry segments that use quantity control provisions, quantities diverted from the primary U.S. market have been used to aggressively expand demand for new products or to develop foreign markets. A more-clearly defined policy related to encouraging such expansion efforts, perhaps allowing special allocations of added quantities, could be productive. Producers would benefit from any successes and consumers may gain from new product development, though they could face increased prices if alternative markets grew significantly.

To the extent that additional supply is triggered by price increases achieved through volume control, market expansion or new product development programs, the potential success of the efforts will be limited. This is particularly true if that expansion can take place outside the region covered by the marketing order. Increasingly this region must be viewed in the context of a global market as opposed to a U.S. or domestic market.

Quality Standards and Chemical Use. Current quality standards used under minimum grade and size regulations for horticultural marketing orders raise concerns about potential chemical residues, nutrition, convenience, and the availability of different quality characteristics for different end uses. Marketing order grade and size provisions could be evaluated in the context of these consumer concerns.

Technology is being developed to provide alternative methods of measuring quality characteristics; these will allow targeting quality more to particular uses than does establishment of minimum grade or size standards. If standard measurements for ripeness, sweetness, tenderness, and similar attributes become feasible, it may allow better tailoring of marketing efforts to consumer niches. Marketing order committees could be encouraged to explore opportunities and monitor their regulations for changes needed to accommodate technological progress. Exemptions from regulations for sales utilizing evolving technologies could encourage potentially beneficial progress in providing consumers with products having desired attributes.

Trade Issues. Section 8e provisions may need to be revamped if the NAFTA agreement is approved and a GATT agreement is reached. Use of minimum-quality regulations for other than assurance of product quality (such as to reduce imports) in order to enhance consumer demand, may be challenged. Careful monitoring to prevent challenges because of perceived trade policy violations due to the Section 8e provision may be important.

Trade agreements may open potential foreign markets, and marketing orders may be useful in assisting industry members to capitalize on such opportunities. Industry committees and USDA may need to review order provisions to identify revisions needed to accommodate trade developments.

Self-Regulation. As the concerns to be addressed by marketing order regulations increase, regulatory timeliness becomes more of an issue. This is particularly true for perishable or semiperishable horticultural crops. It may be appropriate to re-examine each of the marketing order programs to see if it really fits today's marketing system needs in the context of self-regulation, timing of action, and flexibility to meet evolving consumer demands.

A policy shift may be necessary. Input from a broader range of interested parties should increase the efficiency of the regulatory process for marketing orders. This could provide greater flexibility and ability to deal with emerging concerns and changing consumer demands.

Milk

A Federal Milk Marketing Order (FMMO) places certain requirements on the handling and pricing of Grade A milk in the market it covers. Unlike fruit and vegetable orders, FMMOs do not influence directly the quantity, volume, or grade of the product marketed. FMMOs regulate milk processors (called handlers) by requiring that they pay not less than order prices for Grade A milk, according to use. Grade B milk is not regulated by FMMOs. Dairy farmers (producers) receive a weighted average (or blend) price for their Grade A milk but are not otherwise regulated by FMMOs.

FMMOs are producer-initiated regulations. An FMMO is established on the basis of evidence obtained at a public hearing, and requires the approval of two-thirds of the affected dairy producers. FMMOs can be changed through regulatory hearings, by legislative action, and by court decisions.

About 80 percent of all Grade A milk and 70 percent of all milk was priced under one of 40 federal milk marketing orders in 1992. Grade A milk not priced under an FMMO is primarily located in California which has a state milk order.

The purposes of FMMOs are to: (1) establish and maintain orderly marketing, (2) establish reasonable prices to consumers and equitable returns to producers and milk handlers, and (3) assure an adequate supply of wholesome milk to consumers.

Recently, pricing provisions, and even the continued need, for FMMOs have come into question. Some critics charge that technological innovations in milk production, processing, transportation, and marketing make FMMOs out of place in a modern dairy industry. Upper Midwest dairy producers and some cooperatives and dairy interest groups claim that existing pricing provisions discriminate against them relative to producers in other regions. The U.S. Department of Justice questions the need for FMMOs and testified at the 1990 FMMO hearing in favor of their elimination. Its claim is that the structure of the dairy industry has changed to the point that dairy cooperatives themselves have sufficient market power to carry out most of the regulation and pricing functions provided by FMMOs.

A 1988 USDA Economic Research Service study concluded FMMOs could be modified to remove regional pricing differences among producers, provide savings to consumers, reduce government purchases of surplus dairy products, generate more efficient shipping patterns, and reduce interregional marketing costs. However, these changes would be accompanied by a reduction in overall producer revenues (McDowell et al. 1988). Similar conclusions were drawn by the U.S. Government Accounting Office (1988).

Current Issues

The relevance of FMMO provisions and related issues must be evaluated from the viewpoint of a modern dairy industry and the purpose of FMMOs. FMMO issues requiring resolution within the next few years include: (1) the continued need for and use of the classified pricing system; (2) the basic mover of class prices; (3) Class I price differentials and their bases; and (4) providing for multiple-component pricing. Consistency of FMMOs with the federal price support program and if indeed they are even needed must also be re-evaluated.

Classified Pricing. Classified pricing involves defining milk "use classes" and setting minimum prices in recognition of the differences in elasticities of demand for different milk products. The classes of milk are Class I milk, used for beverage purposes (i.e., whole milk, 2 percent, 1 percent, skim, and flavored milk); Class II milk, used for ice cream, cottage cheese, yoghurt, fluid cream, and bulk condensed milk for food manufacturing; and Class III milk, used for cheese, butter, milk powder, and evaporated milk. Class I milk commands the highest minimum price, while Class III milk commands the lowest minimum price.

Demand for fresh fluid milk products is more inelastic (less responsive to a change in price) than is the demand for manufactured products. Hence, total revenue from Grade A milk sales is enhanced through classified pricing. Producers benefit but, on the whole, consumers pay higher prices for dairy products than they would with one flat producer price for milk, regardless of use.

Class Price Mover. Since the early 1960s, the basic mover of minimum class prices has been the Minnesota-Wisconsin price series (M-W), the weighted average price paid for Grade B milk by a sample of butter, milk powder, and cheese plants located in Minnesota and Wisconsin. The M-W price for the current month becomes the minimum Class III price in all FMMOs and is also the mover for Class II and Class I prices.

The rationale for using the M-W price as the Class III price is that the markets for butter, milk powder, and cheese are national in scope and Grade A milk going, under FMMOs, into Class III uses competes with the same products made from Grade B milk. Using the M-W price as the basic formula price to move minimum Class I and II prices maintains a relationship between prices paid for milk used in manufactured products and milk used for fresh fluid milk.

The volume of Grade B milk, as well as the number of manufacturing plants buying Grade B milk in Minnesota and Wisconsin, have declined to a point where the M-W price may no longer be an appropriate reflection of the value of milk used for manufacturing. Further, market conditions in Minnesota and Wisconsin may be unique and thus fail to reflect national supply and demand conditions for manufacturing milk and dairy products. Use of the M-W price as the Class III price is currently (June, 1993) under critical review. The Food, Agriculture, Conservation, and Trade Act of 1990 called for a study of the M-W pricing rationale and possible alternatives, with the objective of replacement by June 1992. The Secretary's study was completed in 1991 and proposals, for alternative basic formula prices were considered at a public hearing held in June 1992. At this writing the final outcome is uncertain.

The policy issues to be considered here are: (1) should minimum class prices continue to be based on prices actually paid producers (competitive pay price), or should they be based on prices paid for products like cheese and nonfat milk powder (product price formula); (2) should the current M-W price survey be augmented by adding prices paid for Grade A milk or prices paid for Grade B milk in areas outside the Upper Midwest, or both; (3) should the same price series be used as the Class III price and as the mover for Class I and II prices; and (4) should the support price be used as the basic formula price.

Class I Differentials. Factors considered in establishing Class I differentials include: (1) additional costs of meeting Grade A sanitary standards; (2) costs of transporting milk from areas of production to areas of consumption; (3) cost of producing milk in the supply area; and 4) supply and demand conditions for milk, including the cost of alternative supplies (U.S. Department of Agriculture 1989, p 23). Class I price in any market cannot exceed, by more than the transportation cost, the Class I price in another market for very long before handlers will begin to purchase and transfer milk from the cheaper source.

The minimum Class I price for all FMMOs is the M-W price from two months earlier, plus a Class I differential that varies by market. Since the early 1960s "single basing-point pricing" has been used. For markets east of the Rocky Mountains, Class I differentials vary according to distance from Eau Claire, Wisconsin. The Class I price at Eau Claire is based upon the M-W price plus a Class I differential of $1.04 per hundredweight. For other FMMOs, the differential increases about $.21 per hundredweight for each 100 miles distance from Eau Claire. With this formula, Class I differentials range from $1.20 per hundredweight in the Upper Midwest order (Minnesota) to $4.18 per hundredweight in Southeast Florida (Miami).

The rationale for this geographic structuring of minimum Class I prices is based on the historical position of the Upper Midwest as the major U.S. milkshed. When handlers in other regions ran short of Grade A milk for fluid needs, they could draw from the vast reserve supplies of Grade A milk not needed for local fluid markets in the Upper Midwest.

The rationale for the factors determining Class I differentials may still be valid. The cost to produce Grade A milk still exceeds that of producing Grade B, but the difference is probably not more than 20 cents per hundredweight. The real issue is the continued use of single basing-point pricing.

The Upper Midwest is no longer the single source of reserve Grade A milk supplies. Only 17 of the 41 FMMOs in existence during 1991 had Class I utilizations of 55 percent or more (U.S. Department of Agriculture 1992, p 47). Thus, for many deficit fluid milk markets, there is often a source of Grade A closer than the Upper Midwest. Thus, continuation of single basing-point pricing may no longer be economically valid.

Producers do not receive the Class I price, instead they are paid a uniform or blend price. The Class I differential and the percent of milk utilized as Class I determine blend price. As Class I differentials and Class I utilizations increase with distance from the Upper Midwest, producer blend prices increase. For example, in 1991 the average blend price for the Texas order was $12.64 versus $11.24 in the Upper Midwest (U.S. Department of Agriculture 1992). The Class I differential was $2.08 per hundredweight higher for Texas and the Class I utilization was 53.9 percent for Texas versus 19.1 percent for the Upper Midwest.

Existing Class I differentials appear to have failed to establish prices based on the marginal cost of producing milk in the various markets. Until they do so, farmers will continue to produce grade A milk when they should be producing Grade B milk. Hence, Class I utilization and the resulting blend price will continue to fall. For example, the Class I utilization for the Texas FMMO declined from 67.7 percent for 1985 to just 53.9 percent for 1991 (U.S. Department of Agriculture 1987, 1992).

Single basing point-pricing and existing Class I differentials result in higher-than-necessary Class I prices to assure an adequate supply of beverage milk to consumers. Hence, consumers in markets distant from the Upper Midwest pay more than necessary for their beverage milk products. Excess Grade A milk is channeled into manufactured dairy products. Higher than necessary Class I differentials thus contribute to a misallocation of resources to the production and processing of Grade A milk in some FMMOs.

Until recently, pricing provisions in FMMOs have made the use of reconstituted milk uneconomical. This was to protect the classified pricing system and assure consumers a sufficient supply of fresh milk. However, technological advances have made it possible to produce a very acceptable fluid milk by reconstituting condensed or dried milk. Under a recent amendment, "concentrated milk that can be verified as having been used to make labeled reconstituted fluid milk products will be priced similar to bulk transfers of whole milk" (Cropp and Jesse 1993, p 1).

Multiple-Component Pricing. Most orders currently recognize only milkfat and skim milk values in pricing. Payments to producers are adjusted upward or downward based upon milkfat tests. However, changes in demand for dairy products and resulting changes in milk utilization have increased interest in multiple-component pricing in FMMOs.

Pricing milk on the basis of components in addition to milkfat (e.g., protein, solids-not-fat, or total solids) may be more economically equitable to producers and handlers alike. Additional protein or solids-not-fat in 100 pounds of milk increases the quantity of a manufactured dairy product (such as cheese) made from that milk.

Since the ratio between milkfat and protein or solids-not-fat from herds varies, the yield of manufactured dairy products from milk having identical milkfat tests can vary. Testing for and paying producers for multiple components in their milk would therefore more equitably reflect the market value of the milk. But the specific components (e,g., fat, protein, solids-not-fat, solids other than protein, etc.) to price and if multiple-component pricing should be applied to Class I (fluid) milk must be carefully evaluated.

"Industry-sponsored" multiple-component pricing plans outside of FMMO provisions are prevalent. Yet, apart from cooperatives that are exempt from paying the minimum order price to their producer-members, handlers are prohibited from reducing prices paid to producers by other than milkfat value, if the resultant price is less than the minimum FMMO producer blend prices.

However, five FMMOs currently utilize variations of multiple-component pricing, three others are under review to do so, and proposals to establish multiple-component pricing have been made by five other FMMOs. The dairy industry is moving rapidly to embrace this approach. However, multiple-component pricing is a very complex issue and the impacts of its adoption have not been well researched. Different component pricing plans in different markets will alter the economics of producing various milk products. In the long run, this will bring about some plant relocations. Understanding the various impacts is necessary in order to establish sound component pricing plans nationwide.

Consistency with Price Support Program. Under the AMAA and the Agricultural Act of 1949, which authorizes the dairy price support program, the Secretary of Agriculture is charged with establishing a structure of prices which will assure an adequate but not excessive supply of milk. At one time FMMOs were viewed as only a means of achieving adequate supplies of milk for fluid use. The prime focus of the price support program was on setting prices for manufacturing grade milk and butterfat. Today, the nation's milk supply is 93 percent Grade A (U.S. Department of Agriculture 1992) and only 44 percent of the Grade A milk under FMMOs is utilized for fluid purposes; the remainder is used for manufacturing (U.S. Department of Agriculture 1992). Hence, the FMMO and price support programs need to be considered jointly.

Using the M-W price as a mover of class prices in all FMMOs attempts to provide coordination between FMMOs and the dairy price support program. However, the price support program has carried the primary responsibility for adjusting price levels to encourage changes in milk supplies. But the level of Class I prices under FMMOs influences how much milk is produced and also influences consumer prices and consumption of fluid milk, both of which impact on the volume of milk for manufacturing. Excessive milk supplies have resulted in bringing down the support price for manufacturing grade milk from $13.10 per hundredweight in 1981 to its current level of $10.10 per hundredweight. At the same time, Class I differentials and single-basing point pricing provisions of FMMOs have not received serious consideration regarding their impact on milk supplies. In fact, the Food Security Act of 1985, in a period of large milk surpluses, raised (effective May 1, 1986) Class I differentials in 35 of the 44 FMMOs, while simultaneously including provisions for further reduction of the support price and for the voluntary supply management Dairy Termination Program.

Eliminate FMMOs. Some producers, dairy interest groups, and policymakers question the continued need for FMMOs. They argue that structural changes among dairy cooperatives have created the potential for adequate marketing expertise and market power to provide the orderly marketing and price stability functions of FMMOs.

The real question is if, in the absence of FMMOs, producers would see a greater need for cooperative representation in the marketplace. In the absence of a greater share of member producers and market share that may be anticipated without FMMOs, cooperatives could face difficulty in maintaining, let alone enhancing, prices to their members.

History shows that dairy cooperatives did establish "classified pricing" in the 1920s. But because of the free-rider problem, cooperatives had difficulty in maintaining classified pricing. Not until the passage of the Agricultural Adjustment Act of 1933, providing a program of "licenses" to assist producers, were cooperatives entirely successful in implementing classified pricing (U.S. Department of Agriculture 1989. p 8). The Act required all milk handlers in a given market to pay for milk according to use class and to pool the returns to producers on either a handler or marketwide basis.

One difficulty dairy cooperatives face is the free-rider problem. Benefits established by a cooperative or marketing agency-in-common for its members, and at a cost to the cooperative or agency, cannot be excluded from producers who are not in the cooperative. The smaller the market share of a cooperative or marketing agency-in-common, the greater the free-rider problem. Although all producers may benefit from a cooperative's marketwide services or programs, nonmembers may get higher net prices for their milk than do members because of costs incurred by the cooperatives and special deals offered by handlers. This weakens the ability of cooperatives to maintain membership and market power.

Policy Alternatives

There are some policy alternatives for dealing with these issues. Some have received considerable analysis and discussion within the dairy industry. Others have been subjected to much less scrutiny, and are not being considered for adoption in the near future.

Classified Pricing System. There are two basic alternative pricing systems; classified pricing, as exemplified by the current FMMO pricing system, and "flat" pricing in which all Grade A milk sold receives the same or "flat" price—regardless of use. Classified pricing raises prices of fluid milk products to consumers and returns a higher blend price to producers. Classified pricing tends to reduce seasonal price fluctuations, but may reduce the quantity of fluid milk purchased by consumers. Flat pricing, under certain circumstances, may reduce total costs to consumers and thus stimulate commercial sales. Flat pricing may produce significant price savings from seasonal supply-demand fluctuations.

To the extent that extra costs are involved in delivering milk for fluid use, as compared to milk for manufacturing, prices paid would need to cover these costs through payments to cooperatives for services rendered or to producers for production costs incurred. Flat prices do not imply uniformity of prices geographically, since transport costs would need to be covered. In an unregulated market, the prices to be paid will be determined over time. Changing to a regulated flat pricing system would require considerable analysis and careful policy design.

Class Price Mover. If minimum class prices continue to be established under FMMOs, the continual decline in Grade B milk production in Minnesota and Wisconsin means that an alternative to the M-W is needed. Alternatives to be considered include:

1. Producer pay prices from a more-inclusive Grade B milk supply in Minnesota and Wisconsin as well as considering supplies from other states.

2. An A-B competitive pay price in which prices manufacturing plants pay for both Grade B and Grade A would be included. Since most Grade A is under FMMOs pricing, some means of removing the FMMO price regulation, such as de-pooling these plants or subtracting out FMMOs pool draws, would be required for Grade A plants affected.

3. A product-price formula in which, unlike alternatives 1 and 2, the basic price mover would not be determined from competitive producer pay prices. The base price would be derived from a formula that would take into account the market prices of manufactured dairy products and possibly by-products, the appropriate yield of these products from 100 pounds of milk, and the plant cost (make-allowance) to manufacture these products from 100 pounds of milk. Hence, a product-price formula reflects what manufacturers, based upon product prices, can afford to pay for milk. Competitive factors, on the other hand, may force some manufacturing milk plants at times to pay more or less than what would be expected from a product-price formula.

4. The federal support price could be the basic price mover. Unless the support price is changed, the basic price would not change, nor would any of the minimum class prices. Since the current support price ($10.10 per hundredweight) is well below the full cost of producing 100 pounds of milk, established minimum class prices would be below effective market prices. Hence, market forces would determine all class prices most of the time.

5. The alternative to a single basic price for all classes would be a separate mover for Class I milk. Excess Grade A milk (i.e., surplus after Class I needs) is channeled into Class III manufactured products, so an appropriate price relationship between Class I and Class III needs to be maintained. But, because of differences in price elasticity of demand, there may be economic justification for not moving Class I and Class III prices the same absolute amounts when the supply and demand for milk change.

6. Some combination of the previous alternatives would also be possible. Analyzing potential impacts is challenging but this alternative may be better than any of the individual alternatives discussed.

Class Price Differentials. An alternative to single basing-point pricing would be the establishment of several basing points, each located in a reserve Grade A milk area. To reflect a more logical and economical movement of Grade A milk to meet fluid milk needs, Class I differentials would be determined from each basing point.

Another possibility for modifying Class I differentials would be to remove transportation cost as a factor in the differential. It has been charged that Class I differentials benefit milk producers in terms of higher prices, but are inadequate to move milk where needed. An alternative would be to establish a transportation pool, funded by all handlers, to reimburse part of the costs borne by those handlers that move milk for Class I needs. This would more closely match incentives and actions needed to maintain an orderly marketing system.

Multiple-Component Pricing. Most current FMMOs price milk by one component (fat) and total weight. Consumers now place a higher value on the skim portion of milk. However, additional skim solids in milk do not increase the yield or value of Class I beverage products. Unless consumers are willing to pay for high solids in their beverage milk products, Class I handlers will be unable to recover, from the market, the costs of paying producers for additional solids.

One alternative is to exempt Class I milk from multiple-component pricing, and applying it only to Class II and Class III milk, where product yields vary by milk composition. Producer incentives would thus be aligned with the value of their production without penalizing consumers. Another alternative is to raise the minimum solids standards for Class I beverage milk products. This puts the burden on the consumer to fund the incentives.

As stated earlier, USDA personnel report that eight FMMOs have changed to or are currently considering adopting some variation of multiple-component pricing. An additional five FMMOs have proposed such a change.

Eliminating FMMOs. The jury is still out as to whether dairy cooperatives will be able to find a viable alternative to the regulatory pricing provisions of FMMOs. If FMMOs are abruptly terminated, disorderly marketing will likely arise and create price instability, affecting producers and consumers.

But, in time, producers and dairy cooperatives may be able to organize, gain market power, and establish orderly marketing. In that event, producers, handlers, and consumers would benefit from an efficient and economical fluid milk distribution system. On the other hand, if cooperatives were to gain significant market power, inefficiencies and higher consumer prices could result. The net impact of no FMMOs at all remains unclear.

Summary

Marketing orders have been used for approximately 60 years. They are important elements of the marketing environment for milk and a number of horticultural crops. Their continued use deserves careful evaluation in light of the changes occurring in today's food marketing system.

There are always trade-offs in changing regulatory programs. Obviously there is a particular balance of producer benefits and consumer costs that is the result of marketing order regulation. In essence, government regulation enforces a consumer tax or transfer to the benefit of producers. The orders also provide some benefits to consumers. The extent to which reallocation of resources among producers and consumers is desirable is a matter for the political process to decide.

Interaction occurs between marketing order regulation, environmental issues, economic efficiency, competitive position in international economies, and equity of distribution of program benefits and costs. Trade-offs exist between producers of various size, between producers and consumers, and among marketing firms of various sizes. These are all factors to be considered in evaluating potential changes in marketing orders. Clearly, input from the various interests needs to be carefully evaluated, with the best economic and other relevant analyses incorporated into the discussion and decision process.

References

Armbruster, W. J., P. Christ, and E. V. Jesse. 1993. "Marketing Orders." Pp 183-191 in D.I. Padberg (ed). *Food and Agricultural Marketing Issues for the 21st Century.* Food and Agricultural Marketing Consortium, FAMC 93-1. College Station, TX: Texas A&M University.

Armbruster, W. J., and E. V. Jesse. 1983. "Fruit and Vegetable Marketing Orders." Pp 121-158 in W.J. Armbruster, D. R. Henderson and R. D. Knutson (eds.) *Federal Marketing Programs in Agriculture: Issues and Options.* Danville, IL: The Interstate Printers & Publishers, Inc.

Armbruster, W. J., and J. E. Lenz (eds.). 1993. *Commodity Promotion Policy in a Global Economy.* Proceedings of a Symposium, October 22-23, 1992, Arlington, VA. Oak Brook, IL: Farm Foundation.

Buxton, B. M. 1993. *Prorate Regulations in U.S. Citrus Industry*. Washington, D.C.: USDA Economic Research Service, Agric. Info. Bul. No. 664-56.

Carmen H., and D. Pick. 1990. "Orderly Marketing of Lemons." *American Journal of Agricultural Economics* 72:346-57.

Cropp, R., and E. Jesse. 1993. "Federal Order Class I Prices and Reconstituted Milk." In *Dairy Markets and Policy--Issues and Options*. Ithaca, NY: Cornell University.

Folwell, R. J., R. C. Mittelhammer, F. L. Hoff, and P. K. Hennessy. 1985. "The Federal Hop Marketing Order and Volume Control Behavior." *Agricultural Economics Research* 37:4(Fall).

French, B. C., and C. F. Nuckton. 1991. "An Empirical Analysis of Economic Performance Under the Marketing Order for Raisins." *American Journal of Agricultural Economics* 73:581-93.

Heifner, R., W. J. Armbruster, E. Jesse, G. Nelson, and Carl Shafer. 1981. *A Review of Federal Marketing Orders for Fruits, Vegetables, and Specialty Crops: Economic Efficiency and Welfare Implications.* USDA, Agricultural Marketing Service, Agricultural Economics Report No. 477. Washington, D.C.: USDA.

McDowell H., A. M. Fleming, and R. F. Fallert. 1988. *Federal Milk Marketing Orders: An Analysis of Alternative Polices*. USDA, Economic Research Service, Agricultural Economics Report No. 598. Washington, D.C.: USDA.

Office of Technology Assessment. 1992. *A New Technological Era for American Agriculture*. OTA-F-474, Washington, D.C: U.S. Government Printing Office.

Polopolus, L. C., H. F. Carman, E. V. Jesse, and J. D. Shaffer. 1986. *Criteria for Evaluating Federal Marketing Orders: Fruits, Vegetables, Nuts, and Specialty Commodities.* USDA, Economic Research Service. Washington, D.C.: USDA.

Powers, N. J. 1991. "Effects of Marketing Order Prorate Suspensions on California-Arizona Navel Oranges." *Agribusiness* 7:203-229 (No. 3).

______. 1990a. *Federal Marketing Orders for Fruits, Vegetables, Nuts, and Specialty Crops*. USDA, Economic Research Service, Agricultural Economics Report No. 629. Washington, D.C.: USDA.

______. 1990b. *Federal Marketing Orders for Horticultural Crops.* USDA, Economic Research Service, Agricultural Information Bulletin No. 590. Washington, D.C.: USDA.

Taylor, T. G., and R. L. Kilmer. 1988. "An Analysis of Market Structure and Pricing in the Florida Celery Industry." *Southern Journal of Agricultural Economics* 20(2):35-43.

Thompson G. D., and C. C. Lyon. 1989. "Marketing Order Impacts on Farm-Retail Price Spreads: The Suspension of Prorates on California-Arizona Navel Oranges." *American Journal of Agricultural Economics* 71:647-60.

U.S. Department of Agriculture. 1993. "Espy Moves to Amend Marketing Orders for Oranges: Suspend Volume Restrictions. *NEWS*. Release No. 0503.93.

______. 1975. *Price Impacts of Federal Marketing Order Programs.* Washington, D.C: Farmer Cooperative Service (Special) Report 12.

______. 1989. *The Federal Milk Marketing Order Program.* Agricultural Marketing Service, Marketing Bulletin No. 27. Washington, D.C.

______. 1992. *Federal Milk Order Market Statistics, 1991 Annual Summary.* Agricultural Marketing Service, Statistical Bulletin No. 839. Washington, D.C.

______. 1987. *Federal Milk Order Market Statistics, 1986 Annual Summary.* Agricultural Marketing Service, Statistical Bulletin No. 745. Washington, D.C.

______. 1992. *Costs of Producing Milk, 1989 and 1990.* Economic Research Service, Agricultural Information Bulletin No. 653. Washington, D.C.

U.S. Government Accounting Office. 1988. *"Milk Marketing Orders, Options For Change,"* Washington, D.C.

17

Commodity Promotion Programs

Olan D. Forker and John P. Nichols

Commodity promotion programs involve generic activities designed to enhance market demand for a commodity at the state, regional, national, or international level. Commodity promotion programs in the United States may be divided into two types: (1) those funded by producers of a commodity through a legislated checkoff program or marketing order and (2) those involving the use of public funds.

The first is by far the most important in dollar terms. Programs of this type are established under federal or state legislation where, by law, a certain amount of money (a checkoff) is collected from producers by the first handler and remitted to a Commodity Board to be used for generic research and promotion programs. These are considered self-help programs by some, since the funds come from the producers of the commodity. Others view the checkoff as a "tax" since every producer is required, by law, to pay the assessment. In either case it represents an increase in the cost of doing business for every firm involved in producing the commodity.

Congress has authorized "checkoff" legislation for selected commodities. The use of the funds is usually limited to generic advertising, promotion, research and market development activities (Table 17.1). Over the past decade, commodity checkoff legislation was passed for dairy in 1983, honey in 1984, beef, pork and watermelons in 1985, and soybeans, fresh mushrooms, limes, pecans, and fluid milk (processors) in 1990. The authorization has been translated into active research and promotion programs for dairy, honey, beef, pork, watermelons, soybeans, fresh mushrooms and limes (Forker and Ward 1993, Chapter 5). Prior federal checkoff programs exist for cotton, potatoes, eggs, and wool and mohair. Similar generic promotion activities are funded by producers under state and federal marketing order legislation existing since the 1930s. But

TABLE 17.1 Commodity Research and Promotion Programs Authorized by Federal Legislation.

Commodity	*Status (Date Implemented)*	*Authorizing Statute*
Wool	(1955)	National Wool Act of 1954 [Title VII of the Agricultural Act of 1954, 68 Stat. 910, Aug. 28, 1954]
Cotton	(1966-67)	Cotton Research and Promotion Act of 1966; amended in 1990 to terminate refund authority [Pub. L. 89-502, 80 Stat. 279, July 13, 1966]
Wheat	Inactive terminated in 1986	Wheat Research and Promotion Act of 1970 [Pub. L. 91-430, Stat. 885-886, Sept. 26, 1970]
Potato	(1972)	Potato Research and Promotion Act of 1971; amended in 1990 to terminate refund authority [7 U.S.C.A. 2611-2627, 7 C.F.R. 1207 (1988)]
Eggs	(1976)	Egg Research and Consumer Information Act of 1974; amended in 1988 to terminate refund authority [7 U.S.C.A. 2701-2718, 7 C.F.R. 1250 (1988)]
Flowers and plants	Rejected in 1983-84 referendum	Floral Promotion Act of 1981 [Title XVII, Pub. L. 97-98, 95 Stat. 1348-1358, Dec. 22, 1981]
Dairy	(1984)	Dairy Research and Promotion Act of 1983 (Subtitle B, Title 1, Pub. L. 98-180, 97 Stat. 1136, Nov. 29, 1983]
Honey	(1987)	Honey Research, Promotion, and Consumer Information Act of 1984; amended in 1990 to modify refund authority [Pub. L. 98-590, 98 Stat. 3115, Oct. 30, 1984]
Beef	(1986)	Beef Promotion and Research Act of 1985 [Title XVI, Subtitle A, Pub. L. 99-198, 99 Stat. 1597, Dec. 23, 1985]
Pork	(1986)	Pork Promotion, Research, and Consumer Information Act of 1985 [Title XVI, Subtitle B, Pub. L. 99-198, 99 Stat. 1606, Dec. 23, 1985]
Watermelon	(1990)	Watermelon Research and Promotion Act of 1985 [Title XVI, Subtitle C, Pub. L. 99-198, 99 Stat. 1622, Dec. 23, 1985]
Seafood	Law expired without vote	Fish and Seafood Promotion Act of 1986 [Pub. L. 99-659, 100 Stat. 3715, Nov. 14, 1986]
Soybeans	(1991)	Soybean Promotion, Research, and Consumer Information Act of 1990 [7 U.S.C. 6301-6311, 56 F.R. 31048-31068]
Fresh mushrooms	(Pending)	Mushroom Promotion, Research, and Consumer Information Act of 1990 [7 U.S.C. 6102-6112]
Pecans	(Pending)	Pecan Promotion and Research Act of 1990 [7 U.S.C. 6501-6513]
Limes	(1992)	Lime Research, Promotion, and Consumer Information Act of 1990 [7 U.S.C. 6201-6212, 57 F.R. 2988-2997]
Fluid milk (processors)	Pending request from processors	Fluid Milk Promotion Act of 1990 [7 U.S.C. 6401-6417]

Source: Forker and Ward (1993). Reprinted with the permission of Lexington Books, an imprint of Macmillan Publishing Company from *Commodity Advertising*: The Economics and Measurement of Generic Programs by Olan D. Forker and Ronald W. Ward. Copyright © 1993 by Lexington Books.

But during the past decade, the important change in the number of checkoff programs has come about through stand-alone federal legislation. The amount of money involved in all federal- and state-legislated commodity generic advertising and promotion programs is nearly $1 billion (Lenz et al. 1991).

The second type involves the use of public funds, primarily for export promotion activities. These programs, designed specifically for export promotion and foreign market development, are administered by the Foreign Agriculture Service of USDA. They include the Export Incentive Program (EIP), the Market Development Cooperator Program, and the Market Promotion Program (MPP).[1] Recipient organizations or companies are required to match the funds from the federal government, usually two for one. During the period 1955-1990, the federal government invested $151 million in foreign market development (the Market Development Cooperator and Export Incentive programs) and from 1986 to 1990, $495 million in the Targeted Export Assistance program (Henneberry et al. 1992).

Important Issues

By now (1993), the policy of authorizing commodity groups to operate checkoff programs is well established and not overly controversial, although acceptance by producers is not unanimous. The policy of using federal funds for export promotion is also well established; however, the level of appropriation and the use of such funds for brand advertising in export markets was the source of much debate in the 1992 appropriations process.

Issues that continue to be controversial and need to be considered in introducing new commodity promotion legislation in 1995 include the following:

Competing in Export Markets

- Is the use of public funds for export promotion in the public interest? And is the use of public funds for brand advertising appropriate?

---

[1] The MPP replaced the Targeted Export Assistance Program (TEA) in 1990. The TEA funds were used to support promotion activities in countries designated as having unfair trade policies. The MPP funds can be used independent of a retaliation objective.

Ensuring Program Efficiency

- What policy incentives can be used to make sure that the domestic commodity promotion programs, as operated, are in the public interest and operating efficiently? How can producers ensure an efficient and effective checkoff program?

Managing the Free-Rider Issue

- Is there a solution to the free-rider problem? If there is a solution, is it in the best interest of the industry and the public?

Facts Concerning the Issues

Research Shows Positive Benefit Potential

Export promotion programs are funded by Congress to encourage commodity groups, farmer cooperatives, and other marketing firms to be more aggressive in increasing export volume. The intent is to increase demand, and in turn total revenue to the commodity group receiving export promotion funds, thus increasing the economic well being of the agricultural sector. It is assumed that if the sector's income flow is increased, the overall economy will benefit. Studies of export promotion programs by Williams (1985) on soybeans, Lee et al. (1989) on citrus, and a study by the Australian Wool Corporation and Bureau of Agricultural Economics (1987) on wool exports to the U.S. all indicate positive returns and provide some evidence that the assumption of positive benefits from export promotion is reasonable.

For the domestic checkoff programs, the legislation is justified on the assumption that the program activities will result in benefits to the producers that provide the funds and to the economy. If the generic promotion activities increase demand by consumers and/or improve efficiency of the marketing and distribution system, then it is assumed that demand will be strengthened at the farm level. If the farm level demand is increased, this means a larger revenue flow to the producers of the commodity from higher prices and/or larger volume and thus a strengthened agricultural sector. Several studies of generic domestic promotion programs provide evidence of positive benefits to producers. Examples of positive benefits for several commodity groups are provided by Forker and Ward (1993, pp 196-250) from studies of generic domestic advertising and promotion programs for fluid milk, cheese, apples, catfish, orange juice, tomatoes, and butter. Returns from other commodity promotion programs might also be positive but economic analyses have not been conducted to determine the size or nature of the impact.

But a positive impact on sales does not necessarily mean that the program is economically effective or efficient or yielding the greatest possible benefit. The studies reported in Forker and Ward also demonstrate that the level of returns are dependent on the level of expenditures, the commodity being advertised, and the type of program effort. Other factors, such as price, market conditions, the extent of government intervention, and the degree of competition can also influence the extent to which the promotion effort provides direct producer benefits. Liu et al. (1989, p 46) show a 3.09 percent increase in the all-farm price of milk and a 0.34 percent increase in production as a result of the increased generic milk advertising following the establishment of the national dairy check off in 1984. Ward shows a 5 percent increase in beef demand from the national generic beef promotion effort (Ward 1992). Other studies show similar positive impact levels.

Use of Public Funds Expanding

Export promotion programs were expanded significantly by the Food Security Act of 1985. This Act increased promotion program expenditures from $35 million to an estimated $230 million in 1990. Since 1990 the MPP program alone has been funded at nearly $200 million a year. Expenditures under the traditional market development programs have been used for technical assistance, trade servicing, and consumer promotion. A relatively large portion of the budget has been used to cover administrative costs (Henneberry et al. 1992). TEA, and now MPP, funds have, however, been used almost entirely for consumer promotion, generic and branded. Branded promotion accounted for only 3 percent of Market Development Cooperator/EIP expenditures, but for more than 34 percent of TEA expenditures. While the administrative costs have been quite high for the Market Development Cooperator/EIP programs, the administrative costs of the TEA program represent a small percent of the total.

Domestic Checkoff Programs

Federally legislated checkoff promotion and research programs are funded by producers of the commodity. Funds generated from assessments, refund levels, and the extent of producer approval are presented in Table 17.2. For dairy, beef, and pork the funds indicated are only that portion of the assessment forwarded to or retained by the national board. The balance is kept by qualified state or regional organizations. For example, the 15 cent per hundredweight mandatory assessment of the national dairy program generates over $200 million annually, but only

TABLE 17.2 Income, Refund Levels, and Vote Record of Federal Commodity Promotion and Research Programs.

National Board	Year Implemented	Assessment Income National[a] (thousands)	1990 Refunds (thousands)	Latest Vote[b] Year	Latest Vote[b] Approval (percent)
National Dairy Research and Promotion Board	1984	$76,670		1985	90
Beef Promotion and Research Board	1986	43,149[c]		1988	79
United Soybean Board	1991	40,000[d]			e/
Cotton Board	1966-67	31,780	$10,659[f]	1991	
National Pork Board	1986	24,762[c]		1988	78
American Egg Board	1976	7,348[f]	g/		
National Potato Promotion Board	1972	6,073	1,116[h]	1991	81
Honey Board	1987	2,663[c]	345[i]	1991	91
National Watermelon Promotion Board	1989	954	101	1989	52[j]

a Only the income received or retained by the national board. Source is AMS, USDA.
b Latest referendum.
c Includes assessments on imports.
d Estimated annual income, assessments began September 1, 1991.
e Referendum to be conducted in 1993-94.
f Refund provision terminated in 1991.
g Refund amount not available.
h Plan was amended to assess imports and terminate refunds in August 1991.
i Producers and importers voted to terminate assessment refund provision by 72 percent in August 1991.
j 72 percent of volume.

one-third is forwarded to the national board. For most commodities, the various organizations attempt to coordinate the use of the funds to obtain maximum impact.

Refund Policy

Until 1990, several programs required a refund provision. Although each qualified producer was required to pay the assessment, producers of cotton, eggs, potatoes, honey, and watermelons could ask that their assessment be refunded. When a substantial number of producers receive the benefits without incurring the cost, the commodity group has a free-rider problem. In the late 1980s, the portion refunded was about 35 percent for cotton and 18 percent for potatoes, with smaller percentages for honey and watermelons. In 1991, the refund provision was eliminated for cotton, potatoes, and eggs. Assessment rates and the portion of the commodity covered by the assessment are presented in Table 17.3. For the commodities where import volumes are substantial, imports are usually also assessed. For some commodities, small-volume producers are exempt from the assessment.

For several years it was the policy of the federal government that individual producers should have the option of obtaining a refund if they were in disagreement with the philosophy or purpose of the checkoff effort. With generic programs, the benefits are not transparent. Therefore, there is a natural economic incentive to withdraw funds, giving rise to the free-rider problem as experienced by the cotton, egg, and potato programs. The free-rider problem also manifests itself in the form of imports that are not assessed. And, in the case of export promotion, generic programs that do not differentiate effectively on the basis of the country of origin also have a free-rider problem.

Oversight Responsibility

Program planning and implementation for the effective use of the available funds are the responsibility of the commodity Board of Directors and its staff. USDA is responsible for oversight. Each commodity promotion organization has its own unique organizational structure. The make up of the Board is specified in the enabling legislation and the Boards are appointed by the Secretary of Agriculture. Some Boards utilize mostly in-house staff to plan and manage the program activities. Others contract with other commodity or trade organizations. Most of the commodity groups utilize market research to guide them in planning and implementating programs. Although some of the major organizations use economic analysis to measure the impact of sales and estimate an overall return on the advertising investment, few use economic analysis to guide them in funding and allocation decisions (Lenz et al. 1991).

Public Concerns

Export Promotion

The public appears to be concerned about the use of public funds for export promotion activities, especially the portion that is provided large

TABLE 17.3 Federal Commodity Promotion Programs: Assessment Rate and Coverage.

Commodity	*Authorized Rate*	*Coverage* [a]
Beef	$1 per head	All cattle producers and importers
Cotton	$1 per bale plus up to 1% of bale value	Producers and importers
Dairy	15¢ per cwt	Dairy farmers
Eggs	Up to 10¢ per 30-dozen case (current rate is 5¢)	Producers with 30,000 or more laying hens
Fluid milk[b]	20¢ per cwt of all fluid milk products marketed	Processors who market consumer-type packages
Honey	1¢ per lb	Producers and importers of more than 6,000 lb per year, all 50 states, Puerto Rico, and District of Columbia
Limes[b]	1¢ per lb	Producers, producer-handlers, importers of more than 35,000 lb yearly, all 50 states, Puerto Rico, and District of Columbia
Mushrooms[b]	1st year, up to 1/4¢ per lb; 2nd year, up to 1/3¢; 3rd year, up to 1/2¢; subsequent years, up to 1¢	Producers, importers, of more than 500,000 lb yearly, all 50 states, Puerto Rico, and District of Columbia
Pecans[b]	Prior to referendum, 1/2¢ per lb for in-shell; afterward, up to 2¢ per lb; twice the rate for shelled	Growers, grower-shellers, importers, all 50 states, Puerto-Rico, and District of Columbia
Pork[c]	0.25% of 1% of market value; may increase by 0.1 annually, not to exceed 0.50%	All producers of porcine animals and importers
Potatoes	2¢ per cwt or up to .5 of 1% of 10-year average price	Producers growing 5 or more acres; importers, Irish potatoes, all 50 states
Soybeans	1/2 of 1% of net market value of soybeans sold	Producers
Watermelons	Fixed by Secretary, not to exceed 2¢ per cwt for producers and handlers	Handlers and producers growing 5 or more acres

[a] All producers in the 48 contiguous United States unless otherwise noted.
[b] Programs not implemented yet in early 1992.
[c] Rate in 1992 is 0.35 percent.

Source: Agricultural Marketing Service, U.S. Department of Agriculture, April 1991.

corporations to match, and therefore subsidize, their foreign market development activities, i.e., brand promotion. Public concerns include management of programs to ensure maximum benefit to U.S. producers, limiting the benefits which flow to large multinational corporations, and placing limits on the number of years a single organization can receive support to focus on a particular market, particularly when its market share may be relatively high. These important concerns could be addressed through changes in policy and program regulations guided by specific analysis of program impacts.

It is possible that for some commodities brand promotion is more effective than generic promotion in increasing overall demand for the commodity. And it is also possible that cost-sharing export promotion is an effective way to increase export demand and is more efficient than some of the export enhancement or export subsidy programs.

Program Efficiency and Public Interest

Concerning domestic checkoff programs, producers are concerned, and justifiable so, that the programs that they fund translate into a good investment. This means that the programs should result in increased revenue flows to the members of the commodity group—and the increases be large enough to more than offset the assessment and thus provide a reasonable return on the investment. The measurement of the increase should be in comparison to what the income flow would have been with-out the promotion program.

The public is concerned that the programs might lead to higher prices for the products that they buy, without any offsetting benefits to them. This issue has two dimensions. First, if the program does increase demand, then it follows that the total expenditure on the commodity will be higher than it would have been otherwise. This probably means that consumers will be paying more on average and in total than they would have been without the program. It is probable that the producers will realize some benefits. However, the second and more controversial issue is whether this also benefits consumers. One might argue that if well informed and sovereign consumers purchase more as a result of the information provided by advertising then they are better off. The associated increase in expenditures means that in aggregate consumers perceive a higher value relative to cost or they would not have made the additional purchases. So the question has to do with whether the promotion programs produce information or technology that has value to consumers in excess of the assessment cost.

The Free-Rider Problem

The major issue with respect to the free-rider problem is one of equity. When free-riders exist, the return on investment, if any, to the producers who fund the programs is eroded in proportion to the volume of the commodity for which refunds are obtained. If the producers who fund the programs had an effective way to differentiate their portion of the sales from those who asked for refunds, then this would not be an issue. But by the very nature of generic efforts, if they are effective, everyone benefits regardless of the extent to which they share in the cost of the effort. For generic programs this is an equity issue, not an economic issue.

Theoretical Foundations

The theoretical basis for commodity promotion programs comes from advertising theory, information theory, demand theory, and the theory of competition. Theory treats advertising and promotion activities as conveying information. If the information is of value to the consumers receiving the information, they will alter their consumption patterns and thus their purchase behavior. The purpose of commodity advertising is to provide information that will encourage consumers to alter their behavior to increase the demand for the commodity being advertised. If consumers are willing to purchase more of the product categories of the commodity at the same price or the same quantity at a higher price then the producer demand is likely to be increased as well. The magnitude of the shift will depend on the size and effectiveness of the advertising and promotion effort. By estimating the magnitude of the shift, questions about program benefits and public interest can be answered.

Research to date provides some evidence about the shape of the advertising response function (Figure 17.1). Some sales occur without advertising (level d). There is likely some threshold level that has to be reached before a positive impact will occur (a_0a_1). Above that level, one can hypothesize an S shaped function that will first increase at an increasing rate (from a_1 to a_2), then at a decreasing rate (a_2 to a_3) to some maximum level. By developing measures of the functional relationship for the various commodity programs, some answers can be obtained as to whether or not the funds are being used in an efficient manner. Studies of past programs provide insight about effectiveness and provide some evidence that can be used to predict future impacts (see previous section). Of course new programs, new creative advertising approaches, and reallocation of funds among program activities can provide different results in the future.

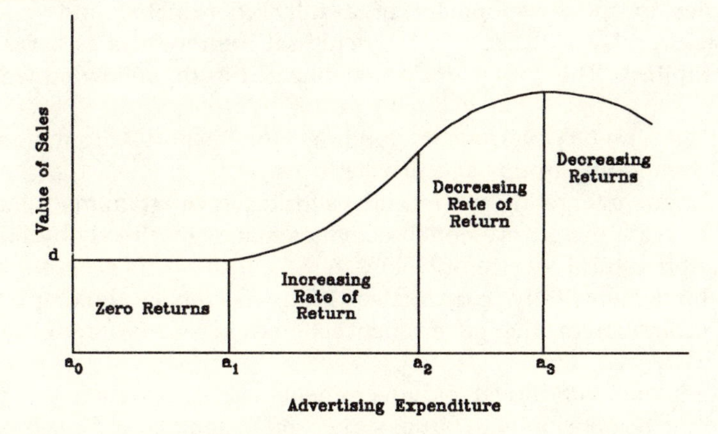

FIGURE 17.1 Theoretical Generic Advertising Response Function.

The theory of competition is relevant as well. Some argue that the legal authority to raise checkoff funds and use them to influence the market enables the commodity group to change the balance of market power more to its advantage than would be the case without such authority.

In addition, the information generated by the promotion and research effort can provide technical and market information to producers and processors that will enable them to create new products and in response to changes in consumer preferences change product attributes, e.g., producing leaner pork or leaner beef. It is assumed that benefits generated at the consumer level will filter down to the producers of the commodity. The portion of the benefits accruing to producers depends significantly on the nature of competition among the processors and other participants and on the extent and kind of government intervention in the marketing and distribution system.

Policy Alternatives and Tradeoffs

Several policy alternatives exist for each of the three important issues. In this section we present policy alternatives and discuss the tradeoffs involved.

Competing in Export Markets

Policy alternatives range from complete government withdrawal (apart from minimal facilitating functions) from export promotion subsidies to the development of a fully coordinated and enforced industrial policy with food and agricultural commodities as key export opportunities. The options may be delineated in the following way:

- Withdraw government funding for export promotion by commodity groups and private firms.
- Improve targeting and incentives for better program management.
- Increase export promotion activities managed directly by Foreign Agricultural Service of USDA.
- Implement fully coordinated industrial policy linking export promotion to other government actions and sectors of the economy.

Complete withdrawal of government assistance shifts the entire promotion responsibility to farmers, checkoff programs, and agribusiness firms. Benefits from this option are reduced government costs and increased reliance on market signals to direct export investments. Adopting a policy of withdrawal may be inconsistent with government objectives of increasing international competitiveness, maintaining appropriate balance of payment levels, or supporting other foreign policy initiatives. The social value of enhancing food and agricultural exports may exceed the returns that would accrue to private domestic exporters and farmers, thus leading to an under-investment in export promotion activities. With continued promotion by other major exporting countries, U.S. market share could drop significantly.

Contrasts in policy options may be best illustrated by considering an industrial policy approach to export promotion. In many recent trade policy debates it has been argued that the United States is the only major industrial country without a well-developed industrial policy. It is argued that government needs to take the lead in defining long term strategy and creating the incentives to get investment in the "best" industrial sectors. Agricultural commodities and food could benefit from such a policy if it can be shown that our inherent productive capacity and future growth of markets provide a high-priority opportunity.

Concerns expressed in regard to the industrial policy approach are likewise substantial. It requires extensive information and analysis to support decisions that are likely to be influenced by other important political factors. The bureaucratic nature of planning and implementing such a program may delay prompt response to major changes in dynamic world markets. There is no guarantee that governments can pick winners in the competitive global marketplace.

Other options represent an effort to compromise between the two extremes. When government provides direct support for export promotion, substantial public interest follows. Congressional oversight and agency management of these programs can evolve to clarify objectives, increase targeting, and tighten effectiveness of monitoring. Assisting smaller firms or emphasizing value-added products can be pursued. This option represents a continuation of current policy as it responds to demands for increased accountability in an environment of tighter fiscal resources.

An additional option may be a stronger direct government role in carrying out coordinated export promotion programs. This would involve integration of all government export promotion programs (including both agriculture and commerce) into a single agency, perhaps with more emphasis on agency involvement in direct implementation of export promotion projects and activities. Benefits would include a more-coordinated government effort and more-informed choices regarding the selection of products and commodities to be promoted. Problems include the implied increase in agency responsibilities, including staff requirements, government cost, and bureaucracy. Also, agricultural interests may be concerned because current export promotion expenditures made by government are much higher for agricultural and food products than for most other export sectors.

On balance, the choices will be constrained by the resources government has to invest and the public's willingness to support an active government role in influencing export market competitiveness. The actions of other exporting countries will also condition these choices.

Ensuring Program Efficiency

Commodity promotion programs, funded solely through checkoffs or in combination with public funding, are held to a test of management and economic efficiency by producers and responsible government agencies. The issue is determining the best policies for encouraging or enhancing efficiency and responsiveness. Several options are possible:

- Initiate mandatory periodic referenda.
- Require increased analysis and reporting of effectiveness.
- Increase agency and congressional oversight.

One of the ways of focusing attention on performance of commodity checkoff programs is to require that producers vote on program existence at regularly scheduled intervals. The obvious benefits of a mandatory scheduled referendum is that it increases the pressure on management to both provide and communicate effective programs. Producers would not have to go through a petitioning process to have an opportunity to vote. The drawbacks are that it may distract management from planning and implementing programs and cause an increase in efforts on self-serving activities that can be easily communicated to producers. If referenda were scheduled too frequently it could cause greater emphasis on programs with short-run payoffs to the detriment of long term investments in building markets. Another concern is the cost of conducting the referenda; the cost is not insignificant.

Commodity checkoff programs should have internal incentives to continually evaluate programs as an on-going management responsibility. Many programs conduct evaluations and assessments but the objectives and methods vary widely.

One policy option is to require checkoff boards and managers to conduct or contract for specific analysis of program efficiency and effectiveness. Benefits include improved information for managerial decision-making and better information for producers to use in judging the value of the checkoff within the context of their business and industry needs. In many cases data are the limiting factor. Through required evaluation, checkoff boards and managers could be encouraged to collect and maintain data necessary for useful program evaluation. However, it is not clear that mandates are necessary to achieve these benefits as many groups conduct assessments as a routine business practice. For small checkoff groups with limited resources, mandates to conduct extensive evaluations can be ineffective and counterproductive.

Some form of collective or cooperative sharing of evaluation data, methods, and results might be useful. This would permit the sharing of overhead costs inherent in data collection and analysis. Common issues and problems in checkoff program management could benefit from coordinated efforts.

Another option is for the USDA to take an increased role in conducting analyses of effectiveness. Alternatively, the government and the promotion organizations could jointly fund a research and evaluation center to continually evaluate the impact of various programs and the approaches used. This arrangement would permit careful third-party assessment insulated from other pressures that may exist within a checkoff program. It could also enhance the comparison of impacts across programs and help identify interactions of effects among commodities. This could be a useful part of public policy analysis. Such analyses, however, would require a substantial increase in research activity by the responsible agencies. Also, a significant investment of time and effort would be required to coordinate the collection and reporting of data from checkoff programs.

Regardless of the approach, tight budgets and producer involvement should provide incentives for efficient program planning and management. It is important that these or other incentives continue to exist.

Managing the Free-Rider Issue

Commodity promotion programs will always be challenged to find ways to deal with the free-rider issue. Programs for several commodities have evolved from voluntary checkoff activities to mandatory national assessments precisely for the purpose of dealing with free-rider problems. Several policy options remain for many commodity groups:

- Institute mandatory, nonrefundable assessment.
- Collect checkoff assessments on imports.
- Increase focus in export markets on differentiable products.

Non-refundable mandatory nationwide assessments provide the most direct way of reducing the free-rider problem. In recent years commodity checkoff legislation has moved in this direction. Such an option eliminates, by definition, the free-rider problem. Such a policy increases the funding base for promotion and reduces the need to continue to "back-sell" the program to producers just to keep the refund rate under control. The board leadership and management can focus on longer-term market development activities with less fear of losing financial support in the short-run.

One difficulty of this option is that refunds are often put forward as a way to ensure board and management accountability. When refunds are not possible there is the potential for alienation among disaffected producers or producer groups that could do long-term damage to the organization. Refunds provide producer recourse to perceived or actual lack of managerial performance. Balancing these interests is a continuing challenge for the board and management.

In domestic markets imports can be a free-rider on efforts to expand the domestic market. Extending the collection of a promotion assessment to imports is the primary policy option. Most recently this was done in the cotton industry and provides a model for other groups to examine.

The benefits of collecting on imports are an increase in resources and coverage for the promotion programs and a greater sense of fairness on the part of domestic producers, who previously had carried the entire burden. Such efforts might also lead to greater cooperation or less resistance to the removal of other types of trade barriers.

Difficulties arise, however, in administering import checkoff programs, including the definition of product forms covered. Also, domestic commodity promotion boards will be required to expand representations to include the interest of importers. This could be viewed as detrimental to interests of domestic producers.

In export markets there will generally be a free-rider problem when generic commodity-oriented promotion programs are implemented.

Commodities with similar characteristics can be obtained from other supplying countries. A policy option for addressing this issue might be to encourage a greater focus on more easily differentiated U.S. products. This might result in greater emphasis on further-processed and even branded products. The benefit comes from ensuring that increases in export market demand will be supplied by U.S. products which meet program criteria. The tradeoffs are the difficulty of defining eligible value-added products and the concern that benefits may flow more to processors and exporters than to producers of the raw commodity in the United States.

There are no clear-cut ways to resolve the free-rider problem of collective commodity promotion programs. These policy options have been tried in various forms. The challenge is to refine them and extend their use to a broader array of commodity situations.

Present Policy Approaches Relevant to the Issues

Export Promotion

The U.S. government provides information, trade leads, and other assistance in exporting through its system of agricultural trade offices. Direct financial support in excess of $200 million per year has also been provided to match industry checkoff funds for export promotion. Brand advertising has been encouraged as an efficient way to capitalize on existing U.S. marketing strengths. The Foreign Agriculture Service is charged with developing and managing the program to achieve the greatest public benefit. Periodic reviews by Congress and the General Accounting Office (GAO) have identified a number of issues regarding the balance of benefits. USDA has responded through periodic changing and updating of regulations.

Program Efficiency

Currently the USDA has modest oversight over commodity promotion programs. In addition, most programs are permitted to continue indefinitely unless the Secretary of Agriculture is petitioned to hold a referendum by 10 percent of the producers affected. The enabling legislation of the Dairy Promotion Program requires that economic analysis be conducted annually by a third party and the results reported to Congress. Beyond this requirement it is expected that the boards of directors, as producer representatives, be responsible for making sure the funds are used effectively. Most commodity promotion programs have adequate control over their funds to invest in analysis that will yield information needed to make sure the funds are operated effectively.

The Free-Rider Problem

Legislation during recent years has moved toward mandatory national checkoffs with severely limited refund possibilities. Also, for most major commodity groups the enabling legislation authorizes the same assessment level on imported as on the domestically produced volume. The present trend is for more of the commodity promotion programs to have the authority to assess imports.

Summary and Conclusions

Commodity promotion programs have evolved from small state or regional organizations funded through marketing orders or voluntary checkoffs to large national mandatory assessments. Success of these programs depends on federal enabling legislation and, in export markets, public funding. In the face of reduced federal fiscal resources, price supports and related deficiency payments will become even more limited. Commodity promotion programs, appropriately supported by federal legislation, provide a unique approach to assisting producers to integrate further into the marketing channel. But a system of analysis and accountability is necessary to make sure that the programs are in the public interest and in the best interest of the producers that fund the programs.

References

Armbruster, W. J., and L. H. Myers, Eds. 1985. "Research on Effectiveness of Agricultural Commodity Promotion." Proceedings from Seminar, Arlington, VA. April 9-10, 1985. Oak Brook, IL: Farm Foundation.

Armbruster, W. J. and R. L. Wills. 1988. "Generic Agricultural Commodity Advertising and Promotion." A.E. Ext. 88-3, Northeast Regional Committee on Commodity Promotion Programs (NEC-63), Published by the Dept. of Agricultural Economics, Cornell Univ., Ithaca, NY.

Australian Wool Corporation and the Bureau of Agricultural Economics. 1987. *Returns from Wool Promotion in the United States.* Occasional Paper 100, Australian Government Publishing Service, Canberra.

Forker, Olan, and Ronald Ward. 1993. *Commodity Advertising: The Economics and Measurement of Commodity Programs.* New York, NY: Lexington Books, A Macmillan Imprint.

Henneberry, S. R., K. Z. Ackerman, and T. Eshelman. 1992. "U.S. Overseas Market Promotion: An Overview of Non-Price Programs and Expenditures." *Agribusiness* 8(1):57-58.

Hurst, S. and O. Forker. 1991. "Annotated Bibliography of Generic Commodity Promotion Research (Revised)." A.E. Res. 91-7, Dept. of Agricultural Economics, Cornell Univ., Ithaca, NY.

Kinnucan, Henry, Stanley R. Thompson, and Hui-Shung Chang. 1992. *Commodity Advertising and Promotion.* Ames, IA: Iowa State Press.

Lee, J. Y., L. H. Myers, and F. Forsee. 1979. *Economic Effectiveness of Brand Advertising Programs of Florida Orange Juice in European Markets.* Gainesville, FL: Florida Department of Citrus ERD Report 79-1.

Lenz, John, Olan Forker, and Susan Hurst. 1991. "U.S. Commodity Promotion Organizations: Objectives, Activities, and Evaluation Methods." A.E. Res. 91-4, Department of Agricultural Economics, Cornell Univ., Ithaca, NY.

Liu, D. J., H. M. Kaiser, O. D. Forker, and T. D. Mount. 1989. "The Economic Implications of the U.S. Generic Dairy Advertising Program: An Industry Model Approach." *Northeastern Journal of Agricultural and Resource Economics* 19(1):37-48.

Nichols, J. P., H. W. Kinnucan and K. Z. Ackerman, Eds. 1991. *Economic Effects of Generic Promotion Programs for Agricultural Exports.* Agricultural and Food Policy Center, Dept. of Agricultural Economics, Texas A&M Univ., College Station.

Ward, R. W. 1992. *The Beef Checkoff: Its Economic Impact and Producer Benefits.* Food and Resource Economics Department, Institute of Food and Agricultural Sciences, Gainesville, FL.

Williams, G. W. 1985. "Returns to U.S. Soybean Export Market Development." *Agribusiness* 1(3):243-263.

18

Policy Alternatives to Protect Against Crop Failure in U.S. Agriculture

Jerry R. Skees and Daniel B. Smith

Risk in agriculture has been identified as one justification for government intervention for well over half a century. It is argued that agricultural production is uniquely exposed to natural elements. The United States has tried a variety of policies designed to mitigate hardships created by natural disasters. Though the issues are not new, the debate in recent years has intensified. The history of disaster assistance dates to the 1930s. More recently, Congress has provided considerable support for alternatives that require farmer contributions for risk management (e.g., Federal Crop Insurance), while also providing ad hoc assistance for each crop year since 1988 (Skees 1992).

Ad hoc disaster assistance has since 1988 averaged $1 billion per year. Simultaneously, the annual cost of the Federal Crop Insurance (FCI) program has averaged in recent years, three-quarters of a billion dollars (U.S. General Accounting Office 1992). The concerns over federal cost have caused many policy-makers to question whether these two types of assistance programs can continue side-by-side. An early version of the President's fiscal year 1994 budget recommended replacing the current crop insurance program with an area-yield plan. The design would be similar to the Soybean Group Risk Plan Pilot Project.

The budget deficit has forced Congress to consider new alternatives to deliver crop failure assistance. The search for a proper mix of disaster assistance or crop insurance or modified crop insurance will continue to challenge policy makers. This chapter examines four alternatives: (1) disaster assistance, (2) the current crop insurance program, (3) the Group Risk Plan, and (4) revenue insurance. The intent is to assess the consequences of each alternative designed to protect against crop failure in the United States.

Space constraints prevent a review of other alternatives. Of the other disaster assistance programs offered by the federal government, the most costly is the low-interest loan program offered through Farmers Home Administration. Farmers have typically been eligible for such loans only when their area is declared a disaster. In addition to low-interest loans, there have been various emergency feed and conservation assistance programs in place over the years. Each of these types of assistance programs operate much as does direct disaster assistance. Therefore, there is less need to review these.

Farm-Level Risk Management

In the 1990s, successful farmers are likely to be those who learn how to manage risk. The 1990s will very likely be more risky. Among factors to consider are: (1) recent changes in the commodity programs, (2) weather patterns, and (3) changing policy risk.

Because of the way that U.S. commodity program payments have been tied to a fixed Agricultural Stabilization and Conservation (ASCS) yield, the move to flexible base has increased farm level risk. For program crops, farmers now receive deficiency payments on 15 percent fewer acres. Thus, this amount of stable income is no longer in place. Future reductions in the level of deficiency payments are likely. Such reductions will increase risk. In addition, the general direction to less government and more market should mean more price risk.

Further, many studies are now showing that bad weather in the 1980s was not atypical (Strommen 1992, Gonzales et al. 1991). Though claims were made that 1988 was a once in 50 years occurrence in the Midwest, several years preceding 1956 had weather events even more severe than the 1988 drought. There is increasing evidence that the 1960s and 1970s were atypical weather decades. The 1980s were more typical of the previous 100 years of weather. This suggests that major droughts should be an expected part of farm planning.

Farmers are also faced with increased policy risk from a variety of sources. Emerging environmental concerns will increase the burden on farmers to change their practices. Many new practices will very likely increase risk. In addition, lessons from the late 1970s and 1980s have shown that U.S. agriculture is more vulnerable than many industries to changing macroeconomics policies. Low real interest rates in the 1970s encouraged expansion and resulted in higher land prices as well as increased exports. The shift to a tight money supply to control inflation resulted in a high real interest rate and an unfavorable exchange rate.

The ability to read and react to change—to manage risk—will, in large part, decide which farmers survive the 1990s. Events of the 1980s have left farmers, lenders, and the government far more cautious and more aware of the need to consider risk management strategies.

Farmers use a variety of strategies to manage risk. Many of those strategies have little to do with government programs. Diversification of crops and livestock is important. Farmers have also learned that spatial diversification, whereby they rent or purchase farms on a north-south pattern, can reduce the risk from hail or increase the chances that some of their acreage will receive rain. Technological advances such as improved irrigation systems, drought-resistant crop varieties, and timely applications of pesticides have each contributed to reduced risk for modern farms.

In addition to the market mechanisms that farmers use, many farmers are now using tools that government has helped put in place for risk management (crop insurance and removal of the ban on trading options in the futures markets). Moreover, educational programs provided by the Extension Service are now showing new ways to combine forward-pricing strategies with crop insurance to provide even more risk protection. These programs emphasize the pricing opportunities available to farmers early in the growing season; a strategy that may be used to counter effects of both price and yield risk. Early in the season, the market is uncertain about the size of the crop and price premiums may be available. Crop insurance can allow farmers to take advantage of these opportunities because farmers may be more comfortable forward pricing an unknown yield when they have crop insurance. For those in major production regions, using a put option with crop insurance provides added and complementary protection.

Just as crop insurance protects a farmer when the yield drops below a specified level, put options protect the farmer when price drops below a specified level. The attractive feature is that while farmers are protected on the downside, they have the opportunity to enhance their own price by receiving any price above the level at which they have chosen to protect themselves.

Multiple Peril Crop Insurance (MPCI) works the same way. MPCI is a federal product that is administered by the Federal Crop Insurance Corporation (FCIC) and sold by private companies. MPCI contracts are available for coverage of 35, 50, 65, or 75 percent of the farmer's average yield. In a major production region, the likelihood of low yields and low prices at the same time is small. Again, a forward pricing strategy and crop insurance are complementary. By coupling a put option with crop insurance, crop insurance (if there is a crop loss) can protect farmers against crop shortfalls so they can take advantage of higher prices. If there is a bumper crop, the put option can protect them from a low price while they take advantage of high yields.

Alternate Performance Criteria for Crop Failure Policies

Before assessing the consequences of the alternatives presented here, it is essential to identify certain performance criteria. Many performance goals that can be identified for crop failure policies are conflicting, and in the final assessment, policymakers must decide which merit the most emphasis. Four major categories of performance will be evaluated: (1) budget costs, (2) equity, (3) efficiency, and (4) consistency with other policies. Figure 18.1 is provided so the reader can make his or her own assessment of each alternative discussed in this chapter.

Since budget concerns have dominated the agricultural policy debate, it is critical to identify alternatives that will provide assistance at lower cost. To some extent, it is also important that these alternatives have a predictable annual cost, and that they have a cost efficient delivery system.

Equity considerations may also be important to policymakers. One measure of performance may be an alternative that provides adequate financial assistance to help most farmers withstand and recover from the effects of natural disasters. It may be desirable to design alternatives that do not favor one set of farmers over another because of commodity or region. It may also be important to provide similar amounts of assistance to farmers suffering similar losses. Most would also agree that disaster assistance programs should guard against fraud and abuse and minimize the type of micro-level problems that are inherent in most insurance alternatives.

Efficient allocation of resources is also important. It must be understood that programs with more public assistance will be bid into land prices and cause resources to be allocated less efficiently. To provide effective programs that meet the test of time, disaster assistance programs should not create incentives to adopt farming practices that increase the likelihood and size of losses. Also, programs should not increase production on highly erodible soils. Farmers may also like assistance programs that are consistently available over time to allow for long-range planning. Finally, most would agree that disaster assistance programs should protect against crop failure as a result of weather events and not due to poor farm management practices. Therefore, disaster assistance programs should not protect inefficient farmers and poor managers from business failure.

To the extent possible, disaster assistance programs should complement other agricultural programs. It may also be desirable that disaster assistance programs be consistent with multilateral trade agreements such as the General Agreement on Tariffs and Trade.

	#1	#2	#3	#4
Budget Exposure - Overall Performance				
Losses Paid				
Delivery Costs				
Subsidy				
Predictability				
Equity Considerations - Overall				
Aid to large numbers				
Fairness by region & commodity				
Fairness based on losses				
Minimum abuse				
Efficiency Considerations - Overall				
Improved market efficiency				
No incentives for risky practices				
Conservation of soils				
Consistency for farm planning				
Paid on weather/ not management				
Ease and ability for administration				
Compatibility w/ other govt programs				
Commodity programs				
Trade agreements				

#1 = Free Disaster Programs

#2 = The Multiple Peril Crop Insurance Program

#3 = The Group Risk Plan

#4 = Revenue Insurance Programs

FIGURE 18.1 Ranking the Performance for Four Alternatives.

Disaster Assistance

Disaster assistance programs are designed to be responsive to the immediate needs of farmers suffering from crop losses due to a natural disaster. When there are serious crop failures, experience has shown that government is likely to respond with ad-hoc disaster programs. Such actions are justified by claims that without disaster assistance, many farmers may go out of business. It is also argued that participation in the current crop insurance program has been inadequate. Depending on the structure of the aid, disaster assistance can be more costly than crop insurance. In recent years, this has been true. Delivery cost for disaster assistance is lower than that for crop insurance. The ad hoc nature of disaster assistance has been troublesome for budget planners. These costs are not "recorded" by the Office of Management and Budget and therefore are not considered in budget planning.

Direct disaster assistance is generally accessible to anyone having a crop loss. Consequently, it may be considered more equitable than crop insurance, if the major objective is to provide income transfers. However, disaster assistance may be considered inequitable because farmers in certain regions receive more benefits than do farmers in other regions. Ad hoc disaster assistance is even more troublesome than a standing disaster program. Regions with sufficient political power must suffer a disaster before assistance is provided. Eligibility requirements have been arbitrary for many crops that were not in the traditional ASCS programs. The Congress has continued to expand the number of eligible crops when enacting ad hoc disaster assistance. For most of these crops, the ASCS offices simply have no history on yields or acreages for farmers. Opportunities for abuse have been significant. With the inclusion of non-program crops, recent disaster assistance has not been compatible with commodity programs.

In the 1991 disaster assistance legislation, Congress allowed farmers to claim losses for both the 1991 and 1990 crop years. It was impossible to assess crop losses from the 1990 crop year. In addition, availability of federal funds limited payments to less than 50 cents on the dollar.

Frequent aid heightens expectations that Congress will respond in case of natural disasters. If this encourages farmers to plant more risky crops in certain regions, then disaster assistance in itself will ultimately increase crop losses. Experience of the 1970s showed that farmers in high-risk regions do increase production of crops less well suited to the region (e.g., cotton in the High Plains of Texas and wheat in parts of Colorado). Since the disaster assistance costs the farmer nothing, areas with high risk stand to gain the most. In these regions, income from frequent disaster payments will likely cause farmers to bid up rents and land prices. Such a combination can increase the financial risk for farmers as the new risk becomes the policy risk that may encourage the government to eliminate the entitlement. Thus, policy risks can replace production risks. Many have argued that ad hoc disaster assistance is a cause of reduced participation in the current FCI program. By all counts, society's resources are allocated inefficiently when disaster assistance is free.

Disaster assistance may also encourage production in high-risk regions with marginal soils. Regions that have high production risk are frequently areas that are highly erodible with thin soils. Therefore, any program that encourages production of row crops in high-risk regions will also lead to more soil erosion.

During the debate prior to passage of the Food, Agriculture, Conservation, and Trade Act of 1990, the Office of Management and Budget (OMB) forced the issue of federal cost. The OMB placed a recommended $500-million cap on all agricultural disaster assistance spending. In response, then Secretary of Agriculture Clayton Yeutter proposed elimination of the FCI program in favor of a permanent disaster assistance program that could be met within the $500-million limit. That program would pay farmers only when average yields on their farm dropped to 65 percent of the county average. Among the most serious problems with this proposal was that the frequency of such events is very low for the major grain producing regions in the United States (Hourigan and Skees 1991). No one believed it would prevent Congressional bailouts. Again, the budget constraint problem played a major role in the decision process. The Congress did not seriously consider Secretary Yeutter's proposal. Instead several provisions in the 1990 act were designed to "fix" problems with the FCI.

Performance of Crop Insurance Alternatives

Alternatives that require farmers to share in risk protection have the potential to fit most performance criteria. The two primary performance goals that can be met with insurance alternatives are efficiency and equity.

There are two major reasons why government involvement is needed for Multiple Peril Crop Insurance (MPCI). First, crop losses are generally correlated. Thus, the exposure is extensive. The private sector is unable and unwilling to take on the type of risk associated with a major widespread drought like that of 1988. Even the international reinsurance industry is not interested in this risk. Government intervention is needed to compensate such widespread losses. With the current FCI program, a major role of the government is that of reinsurer. The other justification for government intervention in MPCI is that subsidies are needed to get acceptable participation levels. Without premium and administrative subsidies from the government, premiums would be too high for many farmers (Barnett et al. 1990).

The subsidy level for MPCI is an issue. If the goal is to enhance farm incomes, insurance alternatives will not perform well unless heavily subsidized. Unsubsidized insurance will only stabilize income. However, most alternatives designed to enhance income have serious shortcomings in that benefits are bid into assets such as land prices. Over time, when the cost of land is considered, incomes adjust to previous levels. The consequence can be price supports that are higher than a market alternative causing the United States to lose markets—both domestic and international. To the extent that insurance alternatives are priced according to the risk, they will not be bid into land prices and they do not jeopardize markets.

Properly working insurance alternatives should improve the ability of efficient farmers to survive major shocks. Survival of efficient farmers is an increasingly important performance criterion. In addition, crop insurance is appealing because, as economic theory supports, it provides the opportunity for more efficient allocation of society's resources than does disaster assistance. The price farmers pay for risk protection is related to the risk they face. If the subsidy is not too high, farmers in high-risk areas will be reluctant to risk putting out a crop.

Since farmers pay based on the risk they face, crop insurance also has the potential to be more equitable. Depending on the structure of the subsidy, the program should be less costly than a disaster-assistance program. Finally, crop insurance can provide both disaster-assistance and risk-management assistance. The individual farmer with crop insurance, has received protection even if the county or region has not been declared a disaster area.

Changes in Federal Crop Insurance

The debate of the 1970s brought about passage of the Federal Crop Insurance Act of 1980. This legislation made significant changes in what had been an experimental program for over 40 years. Significant changes included a premium subsidy of up to 30 percent and a transition to the private sector for delivery of federally subsidized crop insurance. Congress envisioned that these changes would attract participation for more than 50 percent of the eligible acres. Under these circumstances, it was believed that farmers would be protected from natural disaster and there would be no need to provide direct disaster assistance.

During the 1980s, the United States experienced several years of crop disasters. Participation in Federal Crop Insurance did not reach the anticipated level, and by 1988, only 25 percent of the eligible crop acreage was insured. After passing a substantial aid bill in 1988, Congress enacted legislation enabling The Commission for the Improvement of the Federal Crop Insurance Program to examine the issues. In 1989, the Commission presented administrative and legislative recommendations.

Even with these attempts to "fix" FCI, Congress provided ad hoc disaster-assistance for crop years 1988-1993, despite rhetoric from both the Congress and the Administration that those who could purchase crop insurance would not receive disaster-assistance.

Currently, crop insurance participation covers between 35 and 40 percent of eligible acreage. Still, problems with the current FCI program are significant. It suffers from adverse selection and moral hazard (Skees and Reed 1986, Glauber et al. 1993). Moral hazard is the term used to explain behavior changes that occur after insurance is purchased. These changes increase the risk of the contract. They can range from very subtle changes in farming practices, such as reduced pesticide or fertilizer applications, to blatant abuses that may in fact be fraudulent. Adverse selection occurs when farmers know more than the insurance agency about the risk they face and choose to participate based on that knowledge (i.e., farmers with risk greater than those represented in the contract will choose to participate, while farmers with less risk will not participate).

The premium rate increases required in the 1990 act will not solve the problems of adverse selection and moral hazard. Raising rates addresses only the actuarial problem that is a symptom of the real problems of adverse selection and moral hazard. Raising rates does *not* solve participation problems or make crop insurance the type of program that might provide adequate risk protection to many farmers. Effective underwriting (including establishment of good measures of yield) is much more important than raising rates. A proper focus on this dimension would reduce loss overruns and enhance the attractiveness of the program.

During recent policy debates about crop insurance, few people were willing to discuss some important problems. The moral hazard and adverse selection problems will *not* be solved with most of the proposals offered by policymakers. Farmers know more about their yields than the government will ever know. An incentive system has not been in place to put appropriate accountability into the yield-determination process. Sometimes, insurance agents have incentives to increase yields beyond a reasonable level to make a sale. Under these conditions, farmers whose potential yields are below the insurance offer will always be attracted to the program. This is particularly true in markets where management explains more of the variation in yields among farmers than does weather or other factors (e.g., soil type). In these markets, participation is low because adverse selection has destroyed the market through higher and higher rate adjustments. Crop insurance is working best in markets where most risk is weather related. In North Dakota, Montana, and Washington participation levels exceed 70 percent of the planted acreage. It appears these farmers are receiving fair and adequate risk protection.

In other areas, where adverse selection and moral hazard have dominated, participation is low and excess losses (indemnities greater than premiums) have been high. Southern soybeans represent the most dramatic case. For the initial years of the individualized yield program, Southern soybean farmers were offered yields that were too high for many farmers. Since farming practices dominate yields and yield risk, adverse selection occurred. Soybeans accounted for roughly one quarter of all excess losses (indemnities greater than premiums) during the 1980s. This bad experience has been loaded in premium rates for Southern soybeans to the point that some farmers would have to pay $20 or more per acre for crop insurance. As a result, participation is low and those who do participate are now among the highest risk farmers.

The cost of the current program is also a major consideration. Although the government subsidizes premiums for the current product at approximately 25 percent of total unsubsidized premium cost, premium subsidies represented only about 25 percent of all FCI costs during the 1980s. Excess losses represented 35 percent of all federal costs. Delivery and administrative costs account for the remaining 40 percent. Reducing excess losses and administrative costs is thus important to the future of any FCI program.

Although the FCI program has been accompanied by large excess losses (payouts from 1981-90 exceeded premiums by $2.5 billion), these losses have not been uniform. Soybean and wheat losses comprise more than half the total excess losses, and four states accounted for 72 percent of all excess losses for soybeans (Arkansas, Georgia, Louisiana, and Mississippi). Montana and North Dakota accounted for 60 percent of the wheat losses (Glauber, et al. 1993).

Using the private sector to deliver FCI has met with mixed success. On the positive side, there has been a substantial increase in acreages protected by FCI. On the negative side, it is unclear that the previous reinsurance agreements between the U.S. government and the private companies provide proper incentives for improved underwriting, with the low level of risk required of companies being a major question. In addition, private companies have been allowed to pass producers on to the government for nearly all of the risk in a high-risk pool. Major changes are underway to restructure the reinsurance agreement to encourage the private sector to improve underwriting. More of the risk is shifted to companies by increasing the share of losses that companies must absorb when large payouts are made. Still, the question of an appropriate reinsurance agreement is not a simple one, since Federal Crop Insurance is to be made widely available. Companies will likely abandon regions with bad-loss experience if the government is not willing to absorb a large share of FCI losses.

Another reform has been the use of a nonstandard classification system by the Federal Crop Insurance Corporation (FCIC). This system identifies farmers who have been the most serious abusers of the current program and surcharges their rates. Certain provisions of the 1990 act provided more authority to track such farmers.

Despite some significant changes, voluntary crop insurance, in its current form, is not likely to forestall ad hoc disaster-assistance. Further, mandatory crop insurance has serious problems. Though the aggregate farm loss ratio of the current program has been around 2 during the 1980s, not all farmers share equally in the distribution of benefits. Some farmers receive more than 2 dollars for each dollar of premium they pay. Still, the moral hazard and adverse selection problems for some crops have resulted in increased rates so that many lower-risk and high yield farmers could expect to receive far less than a dollar in indemnity payments for each dollar paid in premiums. Under such conditions, requiring crop insurance for farmers with high debt loads can reduce the odds of survival (Skees and Nutt 1988). Congress made further reforms in the FCI program with passage of the Omnibus Budget Reconciliation Act of 1993. These reforms mandate that the FCIC achieve a loss ratio of 1.1. In order to accomplish this, the rules for establishing farm coverage yields have been changed. These changes penalize farmers who have fewer than four years of data with which to establish their yield coverage.

Most alternatives considered so far represent variations of the current mix of disaster assistance and MPCI. Many emphasize restructuring the reinsurance agreement between the FCIC and private insurance companies. This would pass more of the risk on to the private sector and place more of the burden of fixing the microlevel problem with the companies. Although this has a certain appeal, it does not settle the question of whom will be involved with rating and oversight. Since there is a subsidy, the government will want control over these functions. This requirement will very likely limit what can be done—even with a new reinsurance agreement. Other proposals would have a low level of MPCI given to individual farmers in case of major disaster. These proposals have many of the limitations mentioned above because they do not solve the problems of adverse selection and moral hazard. Further, they are generally rated as quite expensive (cost estimates of a free 50 percent MPCI generally exceeds a billion dollars per year). The political advantage of these alternatives is that they would establish a standing disaster program at a low level that would be coupled with crop insurance. It is believed that this would reduce the urge to implement ad hoc disaster programs.

The Group Risk Plan (GRP)

Seeking an alternative that will be cost effective and provide for risk-management needs has been difficult. One alternative that has features of both disaster assistance and crop insurance is currently being pilot tested by FCIC for soybeans, wheat, and forage. This alterative is called the Group Risk Plan, or GRP (Halcrow 1949, Miranda 1991, Barnaby and Skees 1990), and is now working in the Canadian province of Quebec. Congress provided FCIC the authority to expand GRP "to the extent practicable" by the 1993 budget act. Thus it is now likely that 8 major crops (corn, soybeans, wheat, grain sorghum, forage, cotton, peanuts, and barley) will have GRPs by the 1994 crop year. The single most attractive feature of the GRP is that it solves most of the adverse selection and moral hazard problems. Farmers have incentives to produce their crops because they are paid based on area yields—not their individual yields.

As the current pilots are designed, farmers receive payments if the county yield drops below a trigger yield selected by the individual farmer. The trigger yield is established at the coverage levels of 90, 85, 80, 75, 70, or 65 percent of the expected county yield. For example, if the expected county yield is 30 bushels and a farmer chooses a coverage level of 90 percent, then the trigger yield for that farm is 27 bushels. Whenever the farmer's yield drops below 27 bushels, the farmer receives a payment. Farmers can select any protection level per acre up to the maximum established, as 50 percent more than the expected county yield times the expected price. This is done to attract farmers with yields above the county average. Using a 30-bushel yield and a $5/bushel expected price for soybeans would calculate out as 30 x 1.5 x $5 = $225. If the county yield drops to 20 bushels, that represents a 26 percent shortfall (27-20/27). The percentage shortfall is multiplied by the protection to determine the payment (e.g., .26 x $225 = $58).

In research that compares the current GRP design with the current MPCI program, more than 60 percent of nearly 3,000 farms would have received superior risk protection from GRP during the 1980s. These data were taken from 10 years of FCIC records for soybean farms (Hourigan 1992). Farmers who have never purchased crop insurance should also find GRP attractive. Many of these low-risk farmers have not purchased MPCI because its price exceeds their risk. Current mandates to improve the actuarial performance of MPCI will only exacerbate this problem. Farmers concerned about widespread catastrophic risk will also be attracted to GRP. It is relatively inexpensive and provides protection against events such as drought and hurricanes. Another attractive feature of the GRP is that its paperwork is considerably less than that required in the current MPCI program.

Despite its attractive features, GRP has two fundamental problems: (1) some farmers will not be paid when they have a loss, and (2) some farmers will be paid when they don't have a loss.

The most serious shortcoming occurs when the farm yield is low and the GRP does not pay. A precondition for GRP is that the farm yield must be correlated to the county yield. For those farmers who farm in a part of the county where soils are different or are exposed to flooding, GRP could be less effective. Therefore, private sector initiative will be important. One alternative would have the private sector offer a private MPCI product that pays only when GRP does not pay. This alternative is attractive for several reasons: (1) private companies would be protected from catastrophic losses due to the GRP (i.e., GRP would replace the need for government reinsurance); (2) private companies would have improved incentives to underwrite and eliminate losses; (3) farmers could receive improved risk protection; and (4) companies could make an early assessment of losses and pay promptly.

The other problem with GRP is that some farmers will receive benefits when they don't have a loss. There are several things to focus on here: First, it occurs rarely. In a study of 3,000 case farms, this occurred less than 5 percent of the time on farms with the 90 percent coverage level. Second, farmers have paid a premium based on the county yield—they are entitled to collect. The simple fact that farmers can collect when they don't have a loss is fundamental in providing incentives for farmers to continue to try to grow a crop during bad conditions—moral hazard is eliminated. Since GRP is an option on an index, the problem of payout when there is no loss is not as serious as it may first appear. GRP works just as a put option. Put options are indexed with the futures prices to protect against low prices. However, it is possible for local cash prices to be quite different from the futures market price. This "basis risk" is also present with the GRP, as farm yield can be quite different from county yield. With put options, it is possible for local cash prices to be high even when futures market prices are low. Under these conditions farmers receive two payments—one in the cash market and one from the put option. Therefore, there is precedent for contingent markets like GRP. Paying the farmer based on county losses even if he has no losses is not a serious shortcoming since the subsidy level is not too high.

The GRP should be considerably less costly. Administrative costs will be reduced because there is no need for farm-level loss assessments, and compliance needs are greatly reduced. Subsidy cost can also be less because premiums will be lower and a 30 percent subsidy may not be required. Excess losses should no longer be an issue, because adverse selection and moral hazard will be significantly less. In the 96 U.S. counties where GRP will be tested on soybeans, there were $170 million in excess losses during the 1980s. Had the same acreage been insured under GRP at the maximum levels, these losses would have been less than $10 million.

Revenue Insurance

Another alternative that has been discussed is revenue insurance. The Canadians are now combining their stabilization and crop insurance programs into a Gross Revenue Insurance Program (GRIP). This policy change has been touted as the most significant change in Canadian agricultural policy in over 50 years. Although it is a move to market-oriented alternatives, there are two important features that make the Canadian GRIP something less than market oriented: (1) the level of the subsidy is significant (farmers' share of the total cost of the program is generally less than 50 percent); and (2) the degree of price protection offered is higher than expected market prices. The GRIP has come under considerable attack. The opportunities for moral hazard and adverse selection problems are serious. Further, there are concerns that this policy has been more favorable to some crops than others.

Still, the Canadian experiment with GRIP merits watching. Each province has a great deal of leeway in how the GRIP is implemented, and this provides a very fascinating opportunity to try alternative designs.

The keys to a successful individual revenue insurance program are proper design and implementation. Individual revenue insurance programs are susceptible to considerable moral hazard and adverse selection problems. Thus, an alternative that may offer needed protection for many farmers would be group revenue insurance. Two types of programs can be considered: (1) area revenue insurance for an individual crop or (2) area revenue insurance based on the portfolio of crops grown on the farm.

In the case of area revenue insurance for an individual crop, a homogeneous production zone would be identified. The combined expected zone price and yield would trigger the payment. Farmers would purchase protection from revenue shortfalls in the zone. Such a policy would be less expensive in major production regions where price and yields are negatively correlated (i.e., move in opposite directions).

Area revenue insurance based on the portfolio of crops would be more complex. Sweden's crop insurance program before 1988 provided such insurance. In that program, the Swedish government collected farm-level statistics on the acres planted to various crops. It also established production zones with expected yields and prices. When the yields dropped below a threshold level, calculations were made to decide if the farm should receive a payment. By using portfolio theory, the Swedes established revenue thresholds based on the farm's individual crop mix and the correlation among crops within the zone. Thus, areas with high-risk and little crop diversification received less in benefits from this program.

Such a program could be developed in the United States. However, several factors are critical: (1) some successful experience with a zone crop insurance program (i.e., the GRP); (2) a commitment to development of information systems that would facilitate development of homogeneous zones (this would move across county boundaries); and (3) willingness of farmers to report all planted acres.

Conclusions

The debate regarding crop insurance versus disaster assistance will continue. Policymakers need to keep in mind that as the United States reduces the level of income and price supports, risk-management will become more important for U.S. farmers. Farmers are learning to combine crop insurance with marketing strategies that use futures markets. Still, in recent years, the integrity of both programs has been called into question. Efforts are needed to fix both important risk-management instruments (i.e., crop insurance and futures markets) for U.S. farmers. Appropriate incentives are needed to improve the underwriting for the current crop insurance program.

The GRP experiment should be watched closely. This alternative has the potential to solve many problems associated with the current FCI program. The challenge will be in developing timely yield data for homogeneous production regions of adequate size. In addition, there is an educational challenge in helping farmers understand the conditions needed for them to receive risk protection from this alternative.

Ongoing efforts to fix the current FCI program are significant. However, it is unlikely that these efforts will solve the problems of low participation and high cost within the time frame demanded by the political process. Therefore, the issue of crop insurance versus disaster assistance is likely to be present in the 1995 legislation. Alternatives that will satisfy likely political constraints will be needed during the debate. Federal cost is the dominant constraint. Thus, any alternative should focus heavily on reducing federal cost. At the same time, alternatives should be sensitive to the risk-management needs of U.S. agriculture.

The GRP, group, or zone revenue insurance alternatives may provide the next frontier for U.S. agricultural policy if the pilot proves successful. Insurance alternatives have the potential to play a major role in future U.S. agricultural policy for commercial farmers. Mixing private and public efforts to structure appropriate insurance alternatives that can replace costly disaster assistance presents a major challenge.

References

Barnaby, G. A., and Jerry R. Skees. 1990. Public Policy for Catastrophic Yield Risk: An Alternative Crop Insurance Program. *Choices*. Second Quarter, pp 7-9.

Barnett, Barry J., Jerry R. Skees, and James D. Hourigan. 1990. Explaining Participation in Federal Crop Insurance. Presented Paper at the American Agricultural Economics Meetings, Vancouver, Canada, August 3-7.

Glauber, Joseph, W., Joy L. Harwood, and Jerry R. Skees. 1993. An Alternative for Reducing Federal Crop Insurance Program Losses. Agricultural Economic Report Number 668, ERS, USDA. Washington, D.C.

Gonzales, Clemen, Haiping Luo, Jerry Skees, and Mary Marchant. 1991. Combining Longer Series of Weather and Climate Data With Short Series of Yield Data to Enhance Information About Yield Risk. Presented Paper at the American Agricultural Economics Association, Manhattan, Kansas, August.

Halcrow, Harold G. 1949. Actuarial Structures for Crop Insurance. *Journal of Farm Economics* 21:418-43.

Hourigan, James D. 1992. A Farm-Level Analysis of Alternative Crop Insurance Designs: Multiple Peril Versus Area-Yield, Unpublished M.S. Thesis, University of Kentucky.

Hourigan, James D., and Jerry R. Skees. 1991. An Evaluation of the Administration's Disaster Assistance Program. Presented Paper at the Southern Agricultural Economics Association, Fort Worth Texas, February.

Miranda, Mario J. 1991. "Area-Yield Crop Insurance Reconsidered." *American Journal of Agricultural Economics* 73:233-42.

Strommen, Norton, D. 1992. "Climate and Crop Yield," *National Geographic Research & Exploration* 8:10-21.

Skees, Jerry R. 1992. "Background Research and Educational Material for a Pilot Test of the Group Risk Plan for Soybeans." Presented to the Board for FCIC, September 17.

Skees, Jerry R., and Perry J. Nutt. 1988. "The Cost of Purchasing Crop Insurance: Examining the Sensitivity of Farm Financial Risk." *Agricultural Finance Review* 48:37-48.

Skees, Jerry R., and Michael R. Reed. 1988. "Rate-Making for Farm-Level Crop Insurance: Implications for Adverse Selection." *American Journal of Agricultural Economics* 68:653-59.

U.S. General Accounting Office. 1992. Crop Insurance: Program Has Not Fostered Significant Risk Sharing by Insurance Companies. RCED-92-25. Washington, D.C.

Rural Programs, Research and Education

19

Rural Development and Human Resource Adjustment

Brady J. Deaton, Glen C. Pulver, and Thomas R. Harris

Public concern for the development of small towns and rural areas is imbedded in conceptions of fairness, efficiency, and equal opportunity. Consequently, rural development is only one component of what must be a national development issue. Rural development does require special considerations of space, density, multiple decisionmakers, cultural and historical diversity, and the biological and ecological nature of agriculture.

From the perspective of national economic development, rural and urban policies cannot be separated. Rural and urban communities are inherently intertwined within an evolutionary social structure. Understanding this interdependence will be a critical foundation for developing a more enlightened public policy.

As guardians of the vast spaces that produce the nation's and a share of the world's food and fiber, as well as principal users and protectors of the natural environment, rural residents are critical participants in effective agricultural and rural development policy. The continuing technological revolution in food production, combined with the diseconomies of small scale in public infrastructure and private enterprise, create a dual squeeze on the economies of small towns and rural areas. Both sets of forces tend to displace labor, thereby contributing to sparse populations, a weaker property tax base, more-limited purchasing power, and less economic diversity.

A series of socioeconomic indicators reflect growing disparity between rural and metropolitan regions of the United States. Per capita income, rates of poverty, school dropouts, functional illiteracy, school achievement scores, unemployment and underemployment, all reflect disparities of alarming proportions. An understanding of the changing economic base of rural America, and the factors affecting the potential for rural development, is a crucial first step in building a comprehensive rural economic development policy.

Statement of Problem

Unfortunately, both academic opinion and prominent public policy participants seem to hold very superficial and relatively uninformed views of rural and urban America. Consider, for example, Robert Reich's observations in 1988 that America is becoming a "bi-coastal economy" in spite of "enormous sums being spent on farm support programs (over $50 billion last year alone), bail-outs of rural banks (an estimated $20 billion over the next few years), and various public works programs, vast stretches of rural America are now occupied by people who are unemployed or underemployed. Much of rural America is being vacated" (Reich 1988, p 3).

The presumption that farm program supports should somehow reverse the fortunes of rural communities reflects an unrealistic view of the rural economic base. Only one of every three jobs in the nonmetropolitan counties of the United States is agriculturally related—14.7 percent on the farm, 1.2 percent in input industries, 6 percent are in the processing industries, and 8.5 percent are in food and fiber wholesaling and retailing. Agricultural industries are also important employers in urban communities. The broader agricultural sector employs 21.2 percent of the total U.S. work force. The majority of these workers live in urban areas. Farmers purchase trucks, farm machinery, fuel, fertilizers, and other inputs, that are often produced in or near large cities. Farm products are also processed and distributed there.

The personal income generated by farming is of declining relative importance, even in the most farm-dependent states. A healthy farm economy will influence the immediate employment prospects of less than half the rural residents. On the other hand, farming is an industry that provides important economic stability. With few exceptions, farming does not suddenly leave a county and move to another state or country, as happens with some manufacturing plants particularly. Structural changes that alter the importance of the agricultural industry to a county will generally take place over several years, even decades, and thereby provide a reasonable planning period for local leaders to seek new job opportunities.

Rural America is a complex economy with diverse sources of employment and income. The income of farm families is tied in a major way to nonfarm sources and has been for half a century. Over 40 percent of farmers work off the farm over 100 days during the year. The 1.5 million farm families with annual gross farm incomes of less than $40,000 are almost totally dependent upon nonfarm income. Nonfarm earnings are of critical importance to many farm families with gross incomes above $40,000 from farming. The majority of that nonfarm employment is not directly connected with agriculture.

Thus, the future of rural America, including its farms, is heavily dependent on nonfarm rural development policy as well as on agricultural policy. Historic commitments to agricultural policy are not sufficient to change the economic face of rural regions, even in the most farm-dependent counties. Likewise, urban America has a high stake in the future of U.S. agricultural policy.

Rural Conditions

The quality of life of many rural communities is deteriorating while it flourishes in other communities. A broad public concern has emerged, i.e., to gain a better understanding of the diversity of rural communities in order to identify effective public policies that will improve the quality of rural life. Improved employment opportunities and sound educational and health services are among the highest priorities.

Urban Perspectives

Ken Deavers (1992) identified three characteristics that make rural areas different from urban areas: (1) small scale, low density settlement patterns, (2) distance from large urban centers, and (3) specialization of rural economies. Deavers incorporates the costs of distance, the relative importance of density, and variation in economic specialization.

Emery Castle (1991) argues that the "rural-urban" distinction lies on an arbitrary line of demarcation. Because human density is the distinguishing characteristic between that which is urban and that which is rural, a continuum rather than a distinct line of demarcation is involved." As income and population have grown within the United States and in other countries of the world, the relative value of space is inversely related to agricultural technology, inversely related to costs of transportation and directly related to society's demand for the environmental and aesthetic qualities of the rural environment. This last condition holds important implications for future demands for rural space in the United States.

Less-dense rural settlement patterns place significant demands on health care, education, and other critical public services. On the other hand, space drives innovation in program delivery and technology adaptation to serve the needs of more-remote communities. Satellite communications, for example, will play a major communication role in future service planning and public education in rural areas. In assessing these issues, it is vital that the diversity of economic sectors and rural conditions be recognized.

Regional Differences in Income Sources

In 1982, the U.S. Department of Agriculture (USDA) classified the 2,443 nonmetropolitan counties in the United States by economic dependence. This classification system continues to be instructive for understanding the effects of a given economic policy or program.

Farming Dependent. Residents of 29 percent of the nonmetropolitan counties received more than 20 percent of their labor and proprietor income from farming between 1975 and 1979. The largest number of these counties are in the Plains states, with smatterings in the Southwest and Southeast. In 1979, the year of peak farm earnings in the United States, the people in these farming-dependent counties received 19.8 percent of their total personal income from farming. Dividends, interest, and rent accounted for 17.6 percent and transfer payments for 14.5 percent. Other nonfarm sources accounted for 48.1 percent of personal income. A major portion of the nonfarm earnings in these counties is farm connected (e.g., nonoperator farm rents, input suppliers, grain warehouses, farm product processors).

Manufacturing Dependent. In a slightly smaller number of nonmetropolitan counties, residents received 30 percent or more of total labor and proprietor income from manufacturing in 1979. These counties comprise 28 percent of all nonmetropolitan counties. The rural Southeast, Great Lakes, and Northeast regions are heavily dependent on manufacturing.

Retirement Counties. Another 21 percent of the nonmetropolitan counties are identified as retirement counties, i.e., their economies are dependent on dividends, interest, rent, and social security payments. These counties are widely dispersed, with concentrations in the Ozarks, the Southwest, Southeast, and Upper Great Lakes.

Others. Mining-dependent counties are concentrated in Appalachia, the Western Plains, and in Louisiana, Oklahoma, and Texas. Counties with at least one-third of their area in federal ownership are in the West. They are generally very rural. The specialized government counties, i.e., those with 25 percent or more of their labor and proprietor income from government payrolls, are dispersed throughout the country. The persistently impoverished counties are located primarily in the South.

Resource Diversity

The regions most severely affected by income and/or employment reductions in the goods-producing sector (i.e., farming, forestry, manufacturing, mining) are those that are relatively remote from urban areas—such regions as the farming communities of the Plains, the mining towns of the West, and the small mill-towns of the Southeast. People who live in regions outside easy commuting distance to cities with populations of 15,000 or more, have difficulty finding alternative employment. Cities of this size and larger appear to offer economies sufficient to provide a basis for location and expansion of high-technology manufacturing and service-producing businesses—the high-growth industries of the 1980s. The great majority of people who live in nonmetropolitan America live within one hour's commuting time of a city of 15,000 or more people. Many rural communities are closely connected to interstate highways, are near commercial airports and are tied directly to excellent telecommunications systems. Others are on poorly maintained secondary roads, hours from any regularly scheduled airline, and have antiquated multiparty telephone systems.

Community development options and prospects are diverse. Each rural community is a special combination of fundamental resources (soil, topography, water); characteristics over which the community may have little control (distance to urban centers, interstate highways, regional airports); local infrastructure built in the past (schools, streets, businesses); and human capital created by previous decisions (educational level, social structure, human health). As a consequence, opportunities for economic development differ.

Improving production efficiency in farming may offer a prime economic development opportunity in one region. The attraction of tourism may represent the highest probability of generating more jobs in another. A region with a large number of small manufacturers may generate substantial economic benefits through retention and expansion of existing firms. The key to survival in other rural regions may be in stimulating business start-ups or capturing a greater share of consumer purchases. Since opportunities vary among communities, action strategies must be tailored to specific local situations.

The diversity of rural economies across the country is further illustrated by the variation in poverty rates and relative responsiveness of different rural regions to conditions of the national economy. Deavers (1989) points out that rural economies in the Northeast region outperformed metropolitan economies in job creation during the 1980s. That was the only region of the United States to do so effectively. Rural poverty in the 1980s generally grew significantly. Rural poverty is heavily concentrated in the South where 54 percent of the rural poor reside, and the nonmetropolitan poverty rate exceeds 21 percent (Deavers 1989, p 5).

Deavers further points out that "farm poverty is now a relatively small share of total rural poverty, and few farming-dependent counties are among the nation's poorest. In fact, in terms of average per capita income, the farming-dependent counties have been above the average of all rural counties for the past 20 years" (Deavers 1989, p 5). Some rural areas grew very rapidly during the 1980s, particularly those with high amenity values, retirement communities, tourist areas, and other communities favored by national economic trends.

Clearly, the industrial composition of rural communities will shape their responsiveness to national fiscal and monetary policy. Favorable trends in the national and international economies for some sectors may be particularly advantageous to some rural communities while harmful in others. "Rural areas have a significant stake in stable macro policies that achieve the highest possible rates of overall economic growth consistent with reasonable price stability. Such policies may reduce the pace of adjustment required in rural communities, although they will not eliminate structural change" (Deavers 1989, p 7).

In assessing rural economic policy and recent changes, it should be noted that the U.S. economy has been in a relatively stagnant state for the last 18 years. Peterson (1992, p 340) points out that "the annual average rate of growth for median family income has dropped from a healthy 2.7 percent between 1948 and 1973 to a minuscule 0.37 percent since then; and, most important of all, productivity growth has fallen from 2.5 percent annually in the 1948 to 1973 period to a mere 0.82 percent since then.

Bernat's (1992) recent analysis yields some implications of these national conditions for rural areas. He found that 17 percent of all nonmetropolitan wage and salary employment is in manufacturing. He found that nonmetropolitan employment has shifted into lower paying durable goods manufacturing since 1970. Also, the manufacturing and employment generated in rural communities has "constituted a disproportionate share of year-to-year fluctuations in total nonmetro employment."

Moreover, manufacturing jobs accounted for approximately 50 percent of the total annual variation in employment over the 19 year period from 1970 to 1988. On the other hand, manufacturing jobs accounted for only 17 to 21 percent of total nonmetropolitan jobs during that period. In other words, the shifts in industry mix have contributed to lower pay per job and to greater annual variability in manufacturing employment

(Bernat 1992, p 33). More-stable economic sectors suffered a relative decline over this time period, while the more cyclical industries increased their share of rural employment.

Projections through 2000 indicate that goods-producing industries will maintain only a 19 percent share of the total national employment market. Employment in retail and wholesale trade is projected to grow 8 percent faster than the national employment growth rate (Harris 1988, p 2). The reduction in the relative importance of the goods-producing industries as a source of employment in the United States is due to: (1) the increased productive efficiency of goods over time through technological advances; and (2) the movement of some industries into cheaper labor markets in other countries. The movement of such industries has had a disproportionately negative impact on small and rural communities of the United States.

Demographic Changes

After a decade of rural population and economic growth in the 1970s, the 1980s revealed continuing deterioration of rural conditions with migration losses of a magnitude matching the 1950s and 1960s. Driven, in part, by high unemployment in rural areas, the out-migration was further fed by hospital closures and deteriorating social infrastructure. The 1980s also saw a widening of the rural-urban income gap for the first time since the Great Depression. Rural incomes as a percentage of urban incomes fell from 77 percent to 72 percent between 1979 and 1989. Also, infrastructure investment—a critical factor in the migration of financial capital, the consequent job-creating ability of local businesses, and the attractiveness of localities for private sector investment that is increasingly influenced by the international market—has been greatly reduced.

The consequences of continued demographic shifts are complex and require a careful assessment of economic and sociological factors. The continued out-migration of the young from rural communities, particularly in Midwestern rural communities, has resulted in a growing concentration in those areas of elderly people who can then no longer depend on extended family and relatives for support. This accentuates the problems of elderly care and health care.

The loss of population tends to depress real estate values, and leads to a deteriorating tax base with consequent negative implications for financing education and other public services. At the same time, recipient areas are facing their own problems of adjusting to growing populations; thus urban crime and deteriorating social conditions may be severely amplified by the immigration from rural areas. This may lead to other problems in education and welfare dependency. For example, of the 43 Standard Metropolitan Statistical Areas (SMSA) experiencing population increases of 20 percent or more from 1980 to 1987, 37 were in the states of California, Florida, and Texas where school dropout rates were among the highest in the nation. By sharp contrast, high school graduates in the upper Midwest have among the highest standardized test scores in the nation, and their school dropout rates are among the lowest in the nation.

Changing Capital Markets

Flows of human capital and financial capital are highly correlated. People must have jobs to maintain the health and vitality of their communities, so job creation becomes an important factor in the decision-making of each locality, as well as an important policy consideration for state and federal government. The deregulation of the banking system has led to concerns that investment capital will not be available for local job creation as funds tend to be drained off into more-profitable ventures around the world. There may be some truth to this concern, but the evidence is quite mixed (Markley 1984, U.S. Department of Agriculture 1990).

One aspect that has clearly hurt many rural communities is the less-personal, more-bureaucratic approach of the financial system to local community needs. Investments in local job-creating opportunities is based on impersonal, objective criteria determined by a credit committee often removed from the local environment. Families who have established their reputations and dependability over generations can no longer count on easy acceptance at their bank for support of their financial ventures. Considerations of family background and local cultures are no longer sufficient reasons for obtaining investment capital.

Theoretical Foundations for Analyzing and Understanding Rural Development and Human Resource Adjustment Issues

To add focus to our discussion of theory, policy alternatives and trade-offs, and present policy approaches, we adopt the definition of Deaton and Nelson (1992, p 87) for rural development:

> The allocation of physical, social, and human capital in a spatial pattern that provides the possibilities for: (1) adequate income for all families; (2) education for leadership, entrepreneurship, workers, and citizenship; (3) access to health services; (4) regional and community economic development that is nationally generative of new economic opportunity; (5)

leadership and organizational structure appropriate to ensure economic and social health; and (6) a healthy and inviting natural and built environment.

Given this definition, a conceptual framework for rural development must encompass numerous subject areas, incorporate multiple units of analysis, and recognize that the process has a time dimension (Deaton and Nelson 1992, p 87). Regions of the United States (as well as of other nations) are also influenced by international market forces which interact with technological change, spatial patterns of development, *agglomeration economies*, and unique combinations of factors that may lead to unanticipated shifts in the local economy. *Agglomeration economies* are economies of size and concentration that result in reduced per unit cost of producing products and services. Richardson (1973) argues that this concept can be separated into social, household, and business agglomerations which result in the strong attraction of population concentrations to capital and labor.

No single theory of economic and social change fully addresses all of these issues. A combination of theories, however, provide a useful perspective which can guide empirical analysis and policy approaches.

Neoclassical and Cyclical Theories

Consistent with neoclassical economics, market forces have tended to narrow, economic disparities among regions of the United States over time even though the income distribution within regions has remained as diverse as ever. Movements of capital and labor combined with economies of scale, though, appear to create patterns of cumulative growth that generate new divergencies within regions even as regional disparities decline.

At the regional level, sharp changes occur from time to time. An example is the decline in the Great Lakes region between 1979 and 1986, when the region fell below the national average income for the first time on record (Hansen 1988, p 109). Regions of the country appear to go through cycles of change just as industrial sectors go through unique cycles of change. The interaction of technological change and regional change produces a complex set of cyclical phenomenon.

Supply Side Theories

Deaton and Nelson (1993, p 89) argue that greater attention to the supply side of local economies would yield important economic results. This approach would give attention to the supply side of labor market activity, capital accumulation and the importance of amenities, agglomeration economies, and spatial location in influencing the pattern of economic activity. The attachment of rural people to family, social, and cultural ties in their home areas exerts important influences on potential migrants. Research has shown that interpersonal relationships and regional cultural differences significantly affect the income differential required to induce migration from rural to urban areas (Deaton et al. 1982, Morgan and Deaton 1975).

Disequilibrium Theory

Kaldor (1970, 1985) has proposed a theory of disequilibrium economic change driven by low wages created by agglomeration economies. Patterns of uneven development so frequently observed among rural communities, and even within communities over time, tend to lend support to this theory of economic change. Recent work by Muesser and Graves (1990) appears to support earlier findings by Stevens (1980) that amenities play an important role in migration decisions. These factors make available a relatively higher quantity of quality labor at lower *efficiency* wage rates than a simple general equilibrium model would indicate, resulting in a lower wage for local manufacturing and service industries. An *efficiency* wage is the money wage divided by the productivity index. It is low when high quality labor contributes relatively more to output than is reflected by the money wage alone. It occurs when workers are willing to accept a lower money wage because of other (non-monetary) benefits that are place specific. A low *efficiency* wage is a powerful attraction to mobile capital because of higher profit potential for business and industry.

Export-Base Theory

Perhaps the most well-known approach to understanding rural economic development issues is the export-base theory which simply suggests that the local economy is driven by income flows generated by goods and services exported from the local region. Local spending from these income flows creates further economic activity in the local economy.

The traditional view was that manufacturing and resource industries (goods-producing industries) were the most effective vehicle for creating jobs and providing long-term economic growth (i.e., were the principal source of community wealth) and that the production level of the service-producing industries (transportation, communication, public utilities, trade, finance, insurance, real estate, government and other services) was dependent upon the level of activity in the goods-producing industries.

More recent thinking suggests that many of the service-producing industries also represent major exports from the local community (Porterfield and Pulver 1991). Many local services are made available to consumers outside the local community and thereby serve as an export base in the same way that a manufacturing firm generates outside income.

Industry Life Cycles

The movement of capital abroad is also a continuation of the industrial-product cycle which usually has different effects, depending on industry mix, in rural areas. The development of new products moves from scientific laboratories into pilot plants that develop and market the new product. Once the market for a standard product is established, larger scale production requiring routinized technique can be undertaken using lower cost labor. Technological change generates a new range of products and the cycle is repeated over and over again. Geographic movement toward areas of lower costs of production is an inherent aspect of this process.

Recent studies have placed great emphasis on the importance of new firms to stimulate the creation of jobs. Small firms are major contributors to the creation of jobs through firm births and expansions. Small firms are also the major contributors to development of new technology. These factors suggest that rural development programs must incorporate comprehensive technology development and transfer strategies with entrepreneurial development, venture capital, and human resource development.

Policy Alternatives and Trade-offs

Achieving the rural development goals encompassed by our earlier definition of rural development poses inherent conflicts in policy. Choices have to be made. Reasonable access to income, education, health services, and a healthy and inviting natural and built environment can be achieved only with concerted public policy. An immediate conflict arises over the need for short-term gains in private sector efficiency versus improved labor and environmental conditions. The longer term economic and social goals of the nation must be drawn upon and public sector investments with less immediate payoff may be required.

Protectionism Versus Free Trade

Some prominent observers of rural, social, and economic change have argued that an effective rural development strategy "will likely need to hinge on *insulating* rural America from international competition through, for example, selective protectionism, support for worker or community enterprise ownership, international policies to encourage the flow of capital into rural areas" (Buttel and Gillespie 1991, p 32). Such an extreme strategy would need to be very carefully justified on the basis of long-term economic gains that would diminish the need for social welfare payments. In other words, a good investment today may eliminate or forestall large fiscal outlays to ameliorate problems of poverty, school dropouts, and crime in the future.

At the same time, such protectionist policies run counter to our national need for economic efficiency and a free trade environment. It is important that short-term policies avoid establishing local or regional economic inefficiencies that will simply choke and stagnate the rural community in a competitive global economy. Greater understanding of the dynamics of economic change is critical in order to avoid such fundamental economic errors. National policies must be crafted carefully to ease transitions in rural and urban areas.

As the U.S. economy becomes more international and open to technology and cyclical changes on a global basis, more-powerful market forces will tend to favor centralization of economic activity within the United States. State governments, small towns and development organizations in rural areas need to form coalitions with public and private international institutions in order to become active participants in the global market. This will require substantial local leadership.

Investment Incentives Versus Tax Needs

The new dimensions of international capital mobility mean increased competition among localities and states of the United States (as well as among nations) for capital investment. Large local tax revenues are foregone in order to attract new investments. For example, concessions worth millions of dollars were given by the city of Memphis and the state of Tennessee to Federal Express to persuade it to locate its $75- to $100-million maintenance facility there. McKenzie and Lee (1991, p 17) see this competitive struggle for capital intensifying in the future as (1) "capital continues to become more mobile," and (2) "past government concessions alert other businesses of the prospects of making money by shifting portions of their costs to governments."

This new reality places a tremendous burden on less-mobile capital in the form of homes, farms, and small businesses that must pick up the additional revenue demands of state and local governments. These revenue demands are growing as a condition of generating the concessionary funds provided to more-mobile capital investments. National, state, and local policies that provide a socially acceptable framework of private capital investments must be carefully developed. We see this issue as one of the biggest challenges facing our country in the next century.

Spatial Efficiency Versus Preserving Community

Is space important to the quality of life in our nation? The questions raised recently in the *Economist* (1991, p 22) were "Should the nation seek to prevent rural depopulation and the collapse of rural society? And, if so, how does it go about it?"

The social consequences of failing to prevent rural depopulation remain hidden to the national consciousness, but there is little question that related social costs include increased crime, reduced health care, and school dropout problems, as well as other social ills. These costs could be reduced if we understood more clearly the desired optimal mix of public services at various levels of community size, and innovative ways of providing service delivery through new technologies.

The importance of agricultural policy and macroeconomic policies for shaping population patterns in the countryside has been talked about a great deal, but analyzed only very lightly. We do not really have a good understanding of the interdependence between the farm and nonfarm sectors of rural America, particularly as it affects self-perception, community pride, and leadership development. It may be that these factors are more important than capital investment.

Competition for Local Tax Revenues

A strong case can be made that the root cause of the relative decline in rural regions is that they become disadvantaged in attracting contemporary growth industries. Communities more-distant from population centers tend to have weaker economies, partly due to chronic underinvestment in public infrastructure. Highways and bridges have deteriorated and water and sewage systems have failed to keep pace with the needs of the community. Yet, these investment needs compete for the tax dollars needed for health and education improvements. The historic failure to adequately invest in public education has resulted in a work force ill-prepared to meet current and future job requirements.

Economic Development and the Environment

Farm and nonfarm developments will increasingly be scrutinized for their effects on the natural environment. Small towns and rural areas may be able to compete effectively with urban centers on the basis of environmental standards because of greater absorptive and regenerative capacity. On the other hand, distance costs for key services will continually place remote areas at a competitive disadvantage in the absence of new communications and technological improvements in telecommunications.

Present Policy Approaches

An effective farm policy may contribute to improved economic conditions in some rural areas, but as we have argued, more-targeted policies should encompass those factors that determine the responsiveness of local economies to national and international market forces. The national debate on the North American Free Trade Agreement (NAFTA) has crystallized the arguments on this issue. Wisdom and leadership will be required to resolve the inevitable conflicts.

Traditional agricultural policies are being broadened to encompass environmental sustainability and community infrastructure provisions. Farming once dominated rural economies, but is now only one of a number of basic rural income sources. The bulk of agricultural industrial activity takes place off the farm (Smith 1992, p 3). Nonfarm employment in manufacturing and services is much more important to overall rural economic wellbeing. The future of rural America is increasingly dependent on its capacity to stimulate the growth of high-technology manufacturing, producer services, tourism, and retirement-based industries.

Current programs of infrastructure finance, manpower training, education finance, communications technology, and health care improvements are vital foundations for future improvements in rural communities. These issues will be briefly addressed.

Financing Quality Education

The 1989 decision of the Kentucky Supreme Court, declaring unconstitutional the funding mechanism for the public schools in that state, is only the tip of the school-funding iceberg (Goetz and Debertin 1991). International competition, ongoing technological change, and continued mobility of human and financial capital ensures that problems of distribution efficiency, equity across districts, and funding will continue to plague many school districts in rural and urban areas of the country. Yet, quality education remains an important priority for the citizenry.

Small rural school districts, without distance-learning technologies, are unable to provide the range of language offerings and science laboratories needed to educate our future citizenry. At the same time, small schools provide leadership opportunities and instill self esteem in a large proportion of the student body. These strengths have, to a large degree, been lost in larger consolidated school districts.

Rural education needs to be reassessed to identify strengths and weaknesses and financing alternatives. Competing demands for limited revenues from an often-diminished tax base, plague many struggling communities. In a global economy these conditions will become more severe for weaker local economies. As implied earlier, each local jurisdiction has to assess the trade-off between investments in their children via school financing as compared to tax subsidies for new job creation. This trade-off is becoming more difficult.

The local educational system is an important aspect of economic development strategy as well, and studies have shown that the quality of education has an important bearing on the quality and number of new jobs created (McNamara et al. 1988). McNamara's (1986) research in Virginia revealed that a 10 percent increase in per pupil expenditures was associated with a 9-point improvement in reading scores and a 6-point improvement in math scores at the eighth-grade level. Related subsequent research revealed a clear relationship between higher standardized achievement scores and initiatives to improve public education. The same was true between higher scores and improvements in the local economy.

Poverty continues to be a major deterrent to school achievement, and one cannot address the educational issue without simultaneously being concerned about the local economic base and tax structure. Continuing assessments of these relationships will be critical to the future of rural America. We can expect distinctive changes in the funding responsibilities of localities, state governments, federal government, and the private sector.

Leadership

A few good leaders can make all the difference in the future of a community, a nation, or the welfare of the world. Traditionally, 4-H Clubs, homemakers' clubs, civic clubs, cooperatives, Chambers of Commerce, churches, and local governments formed a broad partnership that built effective leadership in rural communities across the country. This critical mass of support for leadership development is eroding rapidly.

The contemporary decision environment is now much more complex. Rural areas are at a great disadvantage. First, because of their smaller populations and less-diverse industrial bases, rural areas tend to possess a narrower knowledge base. Second, access to knowledge is further hampered by sheer distance from centers of specialized knowledge and technical assistance. Third, rural areas simply have less financial capacity to hire technical specialists and are more dependent on local volunteers (Pulver and Dodson 1992). Rural communities are faced with the need to shore up the foundations of leadership development and to take rural leadership training into a global dimension.

The economic and social forces identified earlier in this discussion are overpowering. They drain the talent pools of rural areas. Consequently, small and rural areas face the continual challenge of capturing the benefits of their own investments (in schooling) through job-creating strategies and social improvements.

Research has found that a principal difference between thriving communities and deteriorating communities in rural areas is the quality of leadership. More-progressive communities have more capable leaders (O'Brien et al. 1992, John et al. 1988). More knowledge is needed about the origins of leadership, supportive infrastructure for cultivating leadership, and opportunities for nurturing it.

Rural communities contain within themselves seeds of growth. These can be nurtured to create new entrepreneurial opportunities (i.e., newspapers, craft shops, tourist attractions) that can make a significant difference in the quality of rural life. It is important that the agenda include creating and nurturing local pride and taking advantage of the economic potential contained therein. This psychological and cultural commitment is probably our most unexplored resource.

Today, many argue that the level of local economic success is greatly affected by the quality of community leadership. Proponents of this position suggest that rural economic wellbeing is possible in spite of other detrimental conditions if community leaders are willing to take risks, are committed to sustained effort over time, seek external technical assistance, and make wise investments in their future. The great variation in income among rural communities with otherwise similar resource conditions offer some support to this argument (Pulver 1990, p 104).

Efficiency

Neoclassical economics suggests that the nation's efficiency is improved by shifting human resources to where they may be best used. Thus, attempting to preserve all rural communities will almost certainly lead to national inefficiency. National economic wellbeing will be reduced if people and industries are kept in places where the costs of production are too high for them to be competitive, nationally or internationally. Reich (1988) points out that a crucial point missed in this assessment is the social costs of crowding. He argues that there is a social value to dispersing our population across the land to reduce urban and coastal area pollution, inadequate housing, overloaded disposal facilities, traffic congestion, and unsafe places for children to play (Reich 1988, p 5). His recognition that policies must be aimed to enhance the transition of our rural economy into competitive manufacturing and service products is consistent with that of most observers of the rural United States.

A commitment to equal opportunity and balanced economic growth in the nation will require targeting federal aid and state assistance to the most-distressed rural areas. This targeting may be needed simply to ease the transition of some communities which cannot compete in the emerging international economy. On the other hand, research has shown that subsidy programs to upgrade water and sewer systems, improve access roads, airports, and other public infrastructure may play an important role in stimulating new job creation (Smith et al. 1978). Public policies must incorporate a strong research and development component in order to effectively target funds for maximum leverage. Creative local leaders dedicated to improving the quality of rural life and insuring greater opportunity for local citizens, both the young and the elderly, need to feel that they are supported by state and federal agencies with similar goals. There are a variety of approaches to achieve such supportive public policies.

Creative public- and private-sector partnerships (which include the educational institutions) will be a vital ingredient of future rural development policies. Continued exploration of innovative development approaches based on enterprise zones, tax-increment financing, and community development corporations will add to our knowledge of innovative policy approaches.

State-level Efficiencies

State initiatives to encourage and develop local entrepreneurship should be a focused strategy for economic development programs. Small firms historically have a low success rate. Programs such as small business development centers could help lower the loss ratio of local entrepreneurs. State programs could be developed to meet problems of local entrepreneurs—e.g., shortages of technical and professional workers, adequate financing for initial startups or venture capital, access to markets, and local business services.

Small- and medium-sized enterprises could initiate programs of industrial competitiveness based on networking. Through effective leadership and transfer of technology from Land-Grant Colleges and Universities, rural communities could gain competitive advantages in important areas of social and technological innovation. Industrial networks are basically an approach to competing on the basis of cooperation. Cooperation in this context is based on "strategic alliances, joint ventures and shared knowledge" (Rosenfeld 1993, p 1).

The industrial networking model has been based on extensive European experience, particularly Northern Italy and the recent innovations of Denmark. The Emilia-Romagna region of Northern Italy contains a number of small artisan firms which work together to build a growing and thriving manufacturing economy. Much of their success depends on devising sector-based strategies and cooperative relationships among firms, even as intense competition drives the innovations (Rosenfeld 1993, p 6). Substantial technical assistance and cooperation among federal, state and local governments as well as colleges, universities, trade and business associations, workers and managers are necessary to achieve these innovations.

Need for Broad Policy

Evidence suggests that communities, in promoting social and economic improvements, should not be locked into a single-minded approach. Self development is an important but not exclusive strategy for successful job development. Attracting outside industry is a critical strategy, but again is not an adequate approach. Pragmatic approaches that ensure that rural communities take advantage of their assets and work to the betterment of their citizenries must be taken. Effective local leadership is a critical ingredient for enabling communities to respond to a dynamic international economy. Improvements in telecommunications and other technologies will be required in order for small towns and rural areas to compete effectively. Equal opportunity and a sense of justice among rural citizens depends on continued innovations in health-care delivery, education, and manpower training as well as new, adaptive forms of infrastructure provision.

Rural communities need to take the lead in exploring interrelationships with similar communities so that community networking can match the pace of industrial networking. The combination of local initiatives, innovative state and federal policies, concerted outreach by Land-Grant

Colleges and Universities, and a national commitment has the potential to create an acceptable rural settlement pattern.

Summary

The central message of this chapter can be summarized in the following points:

1. Most indicators of economic wellbeing (e.g., per capita income, unemployment, education, health care) indicate that in general, rural areas are losing ground. This generalization calls for special rural-oriented public policy.

2. Small shares of rural regions are heavily dependent on farming and agriculture, but even in these regions most farm families are heavily dependent on off-farm income sources. Farm-dependent regions are doing quite well.

3. Rural regions are economically diverse with heavy dependence on manufacturing (not necessarily connected to farming), services production, tourism, and retirement. If rural economies are to improve, policies must address this diversity.

4. It is hypothesized that the rural economic lag is due to the friction of distance, inability to capitalize on economies of scale, shifts in macrovariables (e.g., fuel crisis, interest rates, foreign exchange rates, federal financial assistance), regulatory shifts, reduced access to capital, poor-quality leadership and lack of knowledge. Of these variables, there is strong evidence that the latter two are the most critical.

5. There is substantial theoretical and practical debate about the advisability of place-oriented policy (i.e., policy that supports the maintenance of a population in rural areas). The debate is influenced by arguments about the diseconomies of concentration (e.g., crowded housing, air pollution, declining educational levels, crime) plus the costs rural people bear when exporting their educated children *versus* the general economic inefficiency associated with maintaining resources in their current location.

6. Rural development requires a multifaceted policy including physical infrastructure (highways, telecommunications, schools) and social infrastructure (capital availability, education, leadership development, knowledge access). If rural areas are to be maintained with any sort of equity to urban areas, the policy initiative must be focused on the human infrastructure (i.e., formal education, leadership development, and continuous access to knowledge) necessary for ongoing economic adjustments.

7. Any turnaround in the deteriorating economic conditions in rural areas will require a major shift in policy perspective. Academics and policymakers in agriculture and rural life must recognize that agriculture is only one (albeit often a very important) sector of a diverse rural economy. A comprehensive rural policy that relates to manufacturing, services, tourism, leadership development, knowledge access, etc., is as important to agricultural interests as it is to others who live and work in the rural United States.

References

Bernat, Jr., G. Andrew. 1992. "Manufacturers' Restructuring in Non-Metro Areas Contributes to Lagging Pay, Job Instability." *Rural Development Perspectives* VIII(1):32-33, USDA, ERS.

Buttel, Frederick H., and Gilbert W. Gillespie, Jr. 1991. "Rural Policy and Perspective: The Rise, Fall and Uncertain Future of the American Welfare-State." Chapter 2 in *Future of Rural America*, Kenneth E. Pigg (ed.), Westview Press, pp 15-40.

Castle, Emery N. 1991. "The Benefits of Space and the Cost of Distance." In *Future of Rural America*, Kenneth E. Pigg (ed.), Westview Press, pp 41-55.

Deaton, Anne S., and Mary Simon Leuci. 1991. "Missouri Rural Health: A Community Challenge." Paper presented at 7th Annual Conference of Texas Rural Health Association.

Deaton, B. J. and Glenn L. Nelson. 1992. "Conceptual Underpinnings of Policy Analysis for Rural Development." *Southern Journal of Agricultural Economics* 24(1):87-99, July.

Deaton, B. J., L. C. Morgan, and K. R. Anschel. 1982. "The Influence of Psychic Costs on the Geographic Allocation of Human Resources." *American Journal of Agricultural Economics* 64(2)177-187.

Deaton, Brady J., and Bruce A. Weber. 1988. "The Economics of Rural Areas." In R. J. Hildreth, Kathryn L. Lipton, Kenneth C. Clayton, and Carol C. O'Connor (eds.) *Agriculture in Rural Areas Approaching the 21st Century*, Iowa State University Press: Ames, pp 403-39.

Deavers, Kenneth L. 1989. "Rural America: Lagging Growth and High Poverty... Do We Care?" *Choices*, 2nd Quarter.

______. 1992. "What is Rural?" *Policies Studies Journal* 20(2):184-189.

Economist. 1991. "Where Breakdown and Bankruptcy Play." November 2.

Goetz, Stephen J., and D. L. Debertin. 1991. "Rural Education and the 1990 Kentucky Educational Reform Act: Funding, Implementation and Research Issues." Agricultural Economics Research Report 54, University of Kentucky, College of Agriculture, Lexington, Kentucky.

Hansen, Niles. 1988. "Economic Development and Regional Heterogeneity: A Reconsideration of Regional Policy for the United States." *Economic Development Quarterly* II(2):107-118.

Harris, Thomas R. 1988. "Economic Development Strategies for State Government: Suggestions and Examination of National Trends." Paper presented at the 1988 National Conference of State Legislature Annual Meeting, Reno, Nevada.

John, DeWitt, Sandra S. Batie, and Kim Norris. 1988. *A Brighter Future for Rural America.* National Governors' Association, Washington, D.C.

Kaldor, Nicholas. 1970. "The Case for Regional Policies." *Scottish Journal of Political Economy* 17(1):337-348.

______. 1985. *Economics Without Equilibrium,* Amonk, New York: M.E. Sharpe Incorporated.

Markley, Deborah M. 1984. "The Impact of Institutional Change in the Financial Services Industry on Capital Markets in Rural Virginia." *American Journal of Agricultural Economics* 66(5):686-693.

McKenzie, Richard B., and Dwight R. Lee. 1991. "Capital Mobility: Challenges for Business and Government." *Contemporary Issue,* Series 47, Center for the Study of American Business, Washington University, St. Louis, Missouri.

McNamara, Kevin T. 1986. "A Theoretical Model for Education Production and Empirical Test of the Relative Importance of School and Non-School Inputs." Ph.D. Dissertation, Virginia Tech, Blacksburg, Virginia.

McNamara, Kevin T., W. P. Kriesel, and B. J. Deaton. 1988. "Manufacturing Location: The Impact of Human Capital Stocks and Flows." *The Review of Regional Studies* 18(1):42-48.

Morgan, L. C., and B. J. Deaton. 1975. "Psychic Costs and Factor Price Equalization." *Southern Journal of Agricultural Economics* 7(1):233-238.

Muesser, Peter R., and Philip E. Graves. 1990. "Examining the Role of Economic Opportunity and Amenities in Explaining Population Redistribution." Working Paper #90-4, Department of Economics, University of Missouri, Columbia, Missouri.

O'Brien, D. J., and E. W. Hassinger. 1992. "Community Attachment Among Leaders in Five Rural Communities." *Rural Sociology* 57(4):521-534.

Peterson, Wallace C. 1992. "What is to be Done?" *Journal of Economic Issues* XXVI(2):337-348.

Porterfield, Shirley L., and Glen C. Pulver. 1991. "Exports, Imports, and Locations of Services Producers." *International Regional Science Review* 14:(1):41-59.

Pulver, Glen C. 1990. "The Response of Public Institutions to the Changing Educational Needs of Rural Areas." *National Rural Studies Committee: A Proceedings,* Cedar Falls, Iowa, pp 103-198.

Pulver, Glen, and David Dodson. 1992. *Designing Development Strategies in Small Towns,* The Aspen Institute, Washington, D.C., pp 6-7.

Reich, Robert D. 1988. "The Rural Crises and What to Do About It." *Economic Development Quarterly* II(1):3-8.

Richardson, Harry W. 1973. *Regional Growth Theory.* London: MacMillan Press.

Rosenfeld, Stuart. 1993. "Staying Small, Acting Big." *Firm Connections* 1(1):1,6 and 8.

Smith, E. D., B. J. Deaton, and D. R. Kelch. 1978. "Location Determinants of Manufacturing Industries in Rural Areas." *Southern Journal of Agricultural Economics* 10(1):23-32.

Smith, Steward. 1992. "Farming Activities and Family Farms: Getting the Concepts Right." Presentation at Joint Economic Committee Symposium, Washington, D.C.

Stevens. J. B. 1980. "The Demand for Public Goods as a Factor in the Non-Metropolitan Migration Turnaround." *New Directions in Urban-Rural Migrations*, D. L. Brown and J. M. Wardwell (eds.), New York: Academic Press.

U.S. Department of Agriculture. 1990. *Financial Market Intervention As A Rural Development Strategy*, Economic Research Service Staff Report No. AG ES 9070.

20

Rural Credit Programs

John R. Brake and David A. Lins

U.S. food and agricultural policies have generally attempted to balance public concerns about adequate and stable supplies of high quality, reasonably priced food with concerns about incomes of agricultural producers. Policy decisions also reflect a desire to develop and maintain a viable farm and rural economy, with minimal barriers to entry into agriculture and rural businesses.

Economic theory and analysis come into credit issues primarily in two ways. First, it is assumed that businesses will purchase inputs to expand production if they can make a profit. If credit is not available, their profitability and economic contribution are restricted. Second, financial markets are expected to intermediate the flow of funds from savers to users. If savers (users) of funds in farm or rural areas are isolated from markets (or competition), then returns (costs) of funds will not be commensurate with risks and returns, creating economic inefficiency.

Credit relates to policy and economic concerns directly. Many firms have insufficient equity capital to purchase otherwise profitable inputs. They borrow the funds and repay the debt from profits. Hence, an important policy issue is the availability of credit to agricultural producers and rural businesses at competitive rates and terms. Federal policy responses have included four approaches: (1) establishment of the Cooperative Farm Credit System (FCS) to promote competition, (2) development of the Farmers Home Administration (FmHA) to fill credit gaps, (3) passage of regulatory legislation to promote soundness and to level the "playing field" among all agricultural lending competitors, and (4) development of lending programs and agencies to encourage rural development.

The FCS was initiated in 1916. In signing the Federal Farm Loan Act, President Woodrow Wilson said, "the farmers . . . have not had the same freedom to get credit on their real estate that others have had . . ." (Farm Credit Administration 1967). A non-real estate lending component was added in 1923 with establishment of the Federal Intermediate Credit Banks. Then, in 1933, Banks for Cooperatives and local Production Credit Associations were added to the FCS.

In establishing the Farm Credit Administration as part of the FCS in 1933, President Franklin Roosevelt indicated a purpose to ". . . maintain and strengthen a sound and permanent system of cooperative agricultural credit . . . for the purpose of meeting the credit needs of agriculture at minimum cost" (Arnold 1958).

The roots of *direct* government lending to agriculture began in 1918 when the federal government began lending to farmers who had suffered flood damage (Brake 1974). Additional purposes were added in ensuing years: Rural Rehabilitation Corporation loans to distressed farmers in 1933; Resettlement Administration supervised farm and home loans beginning in 1935; and tenant farm purchase loans beginning in 1937. Then, in 1937, the Farm Security Administration consolidated those purposes and began making loans to farmers unable to borrow from the usual credit sources.

In 1946, all of these purposes were consolidated into the Farmers Home Administration (FmHA), an agency of USDA, under the three categories of farm ownership loans, farm operating loans, and emergency loans. Broadened authorities, including community loans for facilities, water, sewer treatment, etc., were added to FmHA over the years, based upon reasoning that, for many rural communities, adequate credit at appropriate terms was lacking.

In 1963, D. Gale Johnson summarized the reasons for originating the various federal farm credit programs as follows: (1) there was not enough credit available to farmers, (2) change in the supply or availability of credit was unrelated to the needs of farmers, (3) credit was not available on terms suited to the particular needs of farmers, (4) farm credit provided by private sources was too costly, and (5) certain categories of farmers—especially low income farmers and tenant farmers who desire to become farm owners—either could not obtain credit at all, or could obtain it only in inadequate amounts (Johnson 1963).

The FCS and the FmHA operate in a very different manner. The FCS is a system of privately-owned, federally chartered, farmer cooperatives who compete directly with other lenders, such as commercial banks and insurance companies. FCS funds for lending come from sales of bonds and discount notes in the nation's money markets. FCS interest rates to farm borrowers are typically close to, and competitive with, interest rates from other sources. Farmer (borrower) boards of directors set policy for, and oversee management of, each institution in the system. The Farm Credit Administration regulates the system in line with congressional mandates and sound financial management principles.

The Farmers Home Administration is a government lender, and its personnel are federal employees. In its early years, funding came from government-borrowing in the money markets and by direct congressional appropriation; but in recent years, an increased proportion of its funding has come from private lenders who make loans that are guaranteed by FmHA. Studies of FmHA prior to the mid-1970s tended to verify that their loans had a public-purpose orientation: FmHA farm borrowers tended to be younger, with lower equity, farming smaller acreages, and renting more of their land than non-FmHA farm borrowers. In addition, loan terms to FmHA borrowers tended to be more favorable: lower interest rates, longer terms, and smaller down payments. In the late 1970s, however, FmHA lending (economic emergency loans, in particular) deviated from the earlier orientation.

For major institutional lenders, the federal government has specified regulators and "rules of the game" by which they must operate. The FCS has the Farm Credit Administration as its regulator. Banks, depending on their charters, are subject to the Federal Deposit Insurance Corporation, the Federal Reserve System, and state and national banking examiners. In setting regulations and requirements, Congress has emphasized safety and soundness, but equitability has not always been achieved due to conflicting Congressional committee jurisdiction.

Credit programs have also been used in attempts to enhance the economic viability of rural communities and businesses. For example, the Rural Development Act of 1972 was enacted "to provide an effective program to enable rural America to offer living conditions and employment opportunities adequate to impede the steady flow of rural Americans to our nation's large population centers . . . to make it desirable for Americans to actually return to our rural areas, thereby lessening the burdens and problems of the modern big city." The main provisions of this act were carried out through various nonfarm loan and grant programs of the Farmers Home Administration.

In the past, there have also been extensive efforts to encourage and accelerate rural electric power and telephone development, primarily through loans to local power and telephone cooperatives as well as to private companies. Often these programs provided very favorable terms and subsidized interest rates. The overall policy objective was to promote and accelerate development of farm and rural communities and businesses.

The Issues of Farm Credit Availability and Competitive Terms

Prior to establishment of the FCS, farm loan interest rates were generally high relative to nonfarm rates, the length of term on real estate loans was only 3 to 5 years, and the loan renewal decision was subject to economic conditions in the locality rather than to a borrower's specific situation. When caught in economic downturns, commercial banks were unable or unwilling to renew loans. Without alternative sources of credit, the farm purchaser had no place to get money to pay off the bank loan and many were forced to give up their properties.

With the establishment of nationwide real estate lending by the Federal Land Banks in 1916, and later a nonreal estate counterpart in 1933, farmers in all parts of the country had an alternative source of credit for their farm operations. Evidence suggests that appropriate terms of credit to farmers and agricultural producers were achieved. The Federal Land Banks pioneered a long term loan on farm real estate so that 25-year loans, for example, were readily available to qualified borrowers. Interest rates to borrowers were, and still are, closely tied to rates in those money markets where the FCS raises its funds. Today, rates to borrowers are typically tied closely to "prime" rates—often 1.5 to 2.5 percentage points over the cost of FCS's money. The FCS's cost of money on the bonds it issues is typically only slightly above US Treasury Bonds of the same maturity. Hence, because of the FCS, farmers have, to some degree, been insulated from regional economic downturns or market factors causing artificially high interest rates.

Evaluating availability of credit is more difficult. Credit extension depends on a number of factors including availability of funds and risks of lending. The FCS and other commercial lenders to agriculture have often been conservative in line with stringent loan limits set by law. Their borrowers tended to be relatively safer risks—low debt in relation to assets, or high cash flow relative to debt service requirements. While such policies are typically viewed as good business practice, a higher risk segment of the farm community was not being served. It was that "credit gap" which led to a number of FmHA lending programs. FmHA authorities provided for lending high ratios of farm asset values to low resource or beginning farmers if budgets showed repayment potential. FmHA programs also provided loans to farmers in need because of natural or economic disasters. These programs provided credit when other lenders would not do so, and their purpose was to help disadvantaged or distressed farmers.

While the FCS has direct access to money and capital markets, alternative sources of farm loans sometimes have had difficulties supplying the desired amounts of credit to the sector. Commercial banks typically have emphasized short term loans rather than real estate loans due to their need for liquidity. Their borrowers were required to go elsewhere, often to the bank's FCS competitor, in order to get real estate financing. Also, in the late 1970s, many agricultural banks were "loaned up," or nearly so. That is, they had loaned as much money as they could, given the deposit base of their bank. Many actively sought alternative ways of getting funds to lend to their farm customers: participating with other lenders, starting new farm lending institutions with outside capital, and even discounting with FCS institutions. However, given the deregulation emphasis from the Depository Institution Deregulation and Monetary Control Act, the economic recession of the early 1980s, and the general farm financial stresses of the mid-1980s, loan-to-deposit ratios at agricultural banks have decreased substantially and loan funds have again become available from deposit sources.

Primarily in reaction to their need for better access to long-term funds, banks lobbied for establishment of a new institution. The Federal Agricultural Mortgage Corporation (Farmer Mac) was authorized in the Agricultural Credit Act of 1987. The purpose was to establish a secondary market for agricultural real estate loans for all farm lenders. This secondary market is available to lenders who can originate farm mortgage loans for their customers, sell them through Farmer Mac, and thereby become competitive with the FCS in farm real estate lending.

Market shares over time for the various agricultural lenders suggest a high degree of competition. In the last fifty years, FCS's share of the farm real estate debt market has varied from a low of 26 percent to a high of 61 percent. Commercial bank share has varied from 11 to 29 percent, life insurance companies from 14 to 42 percent, and FmHA from 7 to 15 percent. For nonreal estate loans, the FCS share has varied from 14 to 33 percent; commercial banks, 47 to 77 percent; and FmHA, 3 to 29 percent.

The appropriate role for FmHA has frequently been an issue. As a government lender, FmHA has sometimes been accused of bending to political influence in its decisions. The substantial rise in FmHA lending in the late 1970s and the early 1980s was one instance of such criticism. And because of high rates of delinquencies and loan charge-offs, there has often been concern over whether FmHA is really more a welfare agency than an agricultural credit agency for disadvantaged farmers.

Recent Legislative Actions

While credit issues may be addressed in periodic food and agricultural legislation, most major farm credit legislation is enacted outside "omnibus" food and agricultural legislation. The major farm credit legislation of the 1980s came in reaction to the mid-1980s farm financial crisis and took the form of the Farm Credit Amendments Act of 1985 and the Agricultural Credit Act of 1987 (Peoples et al. 1992). Recall that in the early to mid 1980s, farm income fell, market values of farm assets fell, and sector debt load became excessive. USDA's Economic Research Service estimated that 200,000 to 300,000 farmers became bankrupt, were foreclosed, or restructured between 1980 and 1988 (Stam et al. 1991). While the overall numbers and proportions of farms leaving the sector were consistent with historical experience of preceding decades, involuntary exits and loss of farms from the sector in the 1980s led to considerable public concern.

As the farm financial situation worsened in the early 1980s, federal initiatives attempted to address the problems. Federal credit programs urged debt restructuring and adjustment and provided credit subsidies estimated at $7.2 billion for 1986-88 (Stam et al. 1991). Liberalized loan policies and lending limits were designed to assist FmHA borrowers. At first, loan payments were consolidated or rescheduled. Next, FmHA implemented a Debt Set-Aside Program and Debt Adjustment Program to reduce the debt service costs of borrowers. New rights of borrowers were prescribed covering information about loans and appeals processes.

By 1985, FCS problems gained legislative attention. The Farm Credit Amendments Act of 1985 restructured the Farm Credit Administration to strengthen its regulatory role and remove it to "arms length" from FCS. The thirteen-member FCA Board, with representation from each of twelve Farm Credit districts, was replaced by a three-member Board, appointed by the President. The Act also strengthened rights of FCS borrowers concerning loan terms and access to loan documents.

In November 1986, a new Chapter 12 farm bankruptcy, designed specifically for farmers, was added to the bankruptcy code (Harl 1992). It was given an October 1, 1993 "sunset" provision. Restrictions limited its use to family-sized farms with their major income from farming. Among the important provisions of Chapter 12 compared to Chapter 11, were its less complicated borrower plan for debt restructuring, the write-down of secured debt to market value, shorter time limits on bankruptcy court procedures, and a substantially reduced requirement for creditor approval of the plan. One of the likely benefits of Chapter 12 was that many lenders voluntarily restructured farm debts rather than risk their borrowers filing Chapter 12.

By 1987, losses by farm lenders and the continuing farm crisis required a more substantive response. The Agricultural Credit Act of 1987 followed with significant institutional and agency changes: (1) debt restructuring policies were mandated; and while FmHA's restructuring procedures are more complex and detailed, a procedure was specified for both FmHA and FCS, forcing them to restructure loans when restructuring would be no more costly than foreclosure; (2) in the FmHA and FCS, rights of borrowers were further strengthened with respect to time deadlines and rights of appeal; (3) it became possible for FmHA borrowers to lease or repurchase their lost farms and they were given right of first refusal on sales or lease offers by others; (4) former FmHA borrowers were to have both repurchase rights at a calculated government net recovery value and homestead protection if the residence was on the property; (5) up to $4 billion of special bonding with Treasury guarantee and cost sharing was authorized to cover anticipated FCS losses and to restore investor and farmer confidence in FCS; (6) FCS borrowers were protected from losses in stock value; (7) new procedures were specified for capitalizing FCS; (8) the minimum stock purchase requirement for FCS membership was reduced, but new stock is now at risk; (9) regional bank mergers into Farm Credit Banks were mandated for the twelve Federal Intermediate Credit Banks and the twelve Federal Land Banks; (10) the 13 Banks for Cooperatives were given authority to vote on merger, and local PCAs and Federal Land Bank Associations (FLBAs) were given authority to merge if they served similar territories; (11) the Farm Credit banks in the twelve districts were required to study possible consolidation into no fewer than six. Changes resulting from these provisions are listed in Table 20.1 (Collender 1992).

The 1987 Act created two new institutions: the Federal Agricultural Mortgage Corporation (Farmer Mac) and the Farm Credit System Insurance Corporation (FCSIC). The Farmer Mac secondary market was to

TABLE 20.1 Farm Credit System Changes After the 1987 Act.

Characteristic	*Before*	*After*[a]
Stock purchase	5 to 10% of loan amount; stock not at risk	Minimum of 2% of loan amount, stock at risk
Number of institutions	12 Federal Land Banks 12 Federal Intermediate Credit Banks	11 Farm Credit Banks 1 FICB
	13 Banks for Cooperatives	3 Banks for Cooperatives
	708 local credit associations	259 local associations

[a] Table data reflect January 1992. As of July 1, 1993, there were 240 local credit associations, 10 FCBs and an FICB. The FICB was scheduled for merger into existing FCBs by October, 1993. Additional mergers and consolidations are in process at both local and regional levels.

facilitate borrower access to long-term, fixed-rate loans. Loan originators such as commercial banks, life insurance companies, and FCS lenders could then sell the loans to poolers. The money would be available for relending, and the originator could gain fee income from servicing the loans. Farmer Mac has FCA as its regulator. Farmer Mac has a 15-member board: five members selected from FCS, five from other participating lenders, and five public members appointed by the President. The Act also authorized a secondary market for FmHA guaranteed loans to expand credit availability to high risk farmers. In 1990, Farmer Mac was given permission to pool FmHA guaranteed loans, and later to issue its own securities to raise funds.

FCSIC, the second new institution, was established to insure securities issued by FCS, to provide aid to troubled FCS institutions, and to reduce the likelihood for any future public assistance to the FCS. Premiums, based on loan volume and loan risk, are paid into the fund by FCS entities until a secure base amount of 2 percent (or such amount as determined by the FCSIC Board) of loans outstanding is reached. The FCA board also administers FCSIC.

While the mid-1980s legislation provided much help to borrowers and lenders, the Food, Agriculture, Conservation and Trade Act of 1990 refined and modified some of the earlier legislation. The required FmHA holding period on acquired property was reduced. Beginning farmers were given rights of first refusal on FmHA inventory property. Lifetime caps of $300,000 were imposed on FmHA write-downs and write-offs. Debt restructuring rights were curtailed for those possessing nonpledged assets. A 10-year recapture was imposed on debt buy-out property to reduce or eliminate financial windfalls to former borrowers. FmHA was also encouraged to emphasize guaranteed loan programs and to graduate borrowers from direct loans to guaranteed loans.

In retrospect, however, while the 1980s legislation provided much help to both borrowers and lenders, perhaps the greatest help to the sector came from commodity price support and income enhancement programs which totaled more than $150 billion in the 1980s (Stam et al. 1991).

Concern over the viability of rural communities and businesses has been a roller coaster issue—very high on the agenda of policy debate and concern at some times, and very low on the policy agenda at other times. The 1990 Act established a Rural Development Administration (RDA) within USDA, but there is continuing discussion and argument over its role and funding. The basic thrust of many of the efforts to promote rural development has continued to focus on credit enhancements through government loan guarantee programs; and, to that end, a number of former FmHA activities dealing with rural business and infrastructure development have been split off and put under the RDA.

Likely Issues for Future Consideration

Competitive Issues

1. Farmer Mac was designed to improve the ability of farm lenders to provide long term real estate financing to farm customers. However, Farmer Mac and the Farm Credit Banks have an element of competition in that each raises its funds by selling bonds in the financial markets. Many FCS officials have shown little interest in using Farmer Mac. When FCS might have used Farmer Mac for rural housing loans, it has often chosen instead to use Fannie Mae (FNMA). In late 1992, however, the Farm Credit Bank of Columbia, South Carolina, was approved as a Farmer Mac pooler, ostensibly for rural housing loans. Hence, the Columbia FCB may become an important pooler of rural housing loans for local, outside Columbia, FCS associations who are restrained by the 15 percent statutory limits on rural housing segment of their loan portfolios. Also, with the merger of most associations from the FICB of Jackson, Mississippi, into the Columbia FCB, there may exist potential for the Columbia-based PCAs in Alabama, Louisiana, and Mississippi, with access to Farmer Mac, to compete directly with the Texas-based FCS real estate lenders in those same states.

The continued unwillingness of some FCS officials to utilize Farmer Mac raises questions about whether FCS nonusers of Farmer Mac should sit on the Farmer Mac board. The Board was set up to include representation of the various users; it is less clear whether nonusers need representation.

2. Competition within the FCS became an issue when Congress authorized and encouraged merger and consolidation of local lending associations. In many instances, the real estate and nonreal estate associations with similar, but not identical, local geographic territories have combined. This has led to instances of over-chartering (competition between two FCS associations for similar loans). Such over-chartering has been approved by FCA on an ad hoc rather than a policy basis. System officials are concerned about the application of joint and several liability when (say) two such institutions are in competition. FCA is studying the issue, but Congress may wish to consider if a more- directed policy on competition within FCS is needed, or, alternatively, if the previous policy (exclusive lending territories, no over-chartering of local associations, and no intrasystem competition) should be reimplemented.

3. The FCS is currently studying possible changes in its operations and structure to better serve rural America. As a result, expanded lending authorities will likely be sought. With a declining customer base, one strategy is development of a more rural, as opposed to farm, orientation. In addition to improving their competitive position *vis à vis* other lenders, authority to expand lending to include farm service or farm-related businesses could spread lending risks and reduce costs per dollar of loan by increasing lending volume. FCS is currently restricted: (1) in rural home lending to communities of less than 2,500 population and to no more than 15% of its portfolio; and (2) in lending to businesses providing on-farm services. Increasing the population limits to communities of up to 10,000 or larger and increasing the portfolio percentage limit, would be more in line with FmHA restrictions and would expand potential FCS loan volume. Broadened authority for lending to farm-related businesses could also increase potential loan volume. The issue really concerns whether FCS should be released from its single industry restriction in recognition of the attendant higher risks of single-industry lending. These risks might be reduced by expanding lending authorities. Or, the risks could be addressed by raising capital and insurance standards.

4. Equity among lending institutions, sometimes referred to as establishing a "level playing field," is often an issue. Two items of concern are the degree of regulation and taxation of FCS institutions relative to other commercial lenders. Legislation affecting banks typically comes from banking bills rather than from food and agricultural bills. Nevertheless, Farm Credit legislation may change the competitive advantage by establishing policies or procedures for FCS institutions or for FmHA unlike those now in place for banks or insurance companies. The assessment level required to replenish the bank insurance fund—by one estimate, 23-31 basis points on loans—may not be commensurate with the 15 to 25 basis points plus assistance repayment set-aside that FCS institutions must pay.

Commercial bank regulators and examiners often have limited knowledge of agriculture, and may inadvertently, if not deliberately, discourage farm lending.

A second issue concerns the tax status of FCS real estate lenders. FCS associations with the former FLB charters are exempt from state taxes, whereas merged agricultural credit associations are not. Hence, FCS entities that continue to operate under an FLB charter have a tax advantage not available to commercial banks. This tax aspect has also been a factor keeping some FCS local lending associations from becoming agricultural credit associations that can offer both long- and short-term lending services from the same office. There will likely be discussion of this tax issue in the near future.

Efficiency/Structure Issues

1. Benefits of being a Government Sponsored Enterprise (GSE) are at issue (Lins and Barry 1984). FCS, as a GSE given its original public funding, has enjoyed agency status. However, initial government capital has been repaid and continuing agency status has been questioned. Competitors have argued that agency status gives the FCS an unfair advantage in issuing securities in financial markets. FCS has responded that benefits of agency status are out-weighed by restrictions in FCS lending authority and competitor depository functions and benefits of deposit insurance programs. In essence, agency status is an offset to single industry lending risk. Yet, if agency status reduces the cost of issuing FCS securities, then the lower cost and subsequent lower interest rates to farm borrowers subsidizes, and may even over-allocate, funds to the farm sector.

The 1987 Act permitted any FCS institution to terminate its status as a member of FCS. At the time, however, there were no regulations on the process for doing so. New FCA regulations were adopted in February 1993. Hence, exit from FCS is not now an issue, but it could become so in the future if large numbers of institutions were to attempt to exit from FCS.

2. FCS efficiency is directly related to FCA regulatory costs and structure. FCS continues to press for limiting the budget of FCA, its regulator, and for diminishing FCA's influence on day-to-day management of FCS entities. The 1987 Act requires FCA to examine local associations, and given FCA budget increases of less than the rate of inflation, the increased examination load will further remove FCA from direct influence over day-to-day business management of the various entities. The issue concerns an appropriate relationship between a regulator and a regulated entity. FCA should not be involved in day-to-day management of FCS institutions nor should FCS have major influence on its regulator's budget.

3. The Farm Credit System Insurance Corporation (FCSIC) was established in the 1987 Act when existing loss-sharing procedures were found to be ineffective. FCSIC was set up to accumulate reserves to cover possible future loan losses and to reduce the likelihood of needing future government bailouts of FCS. It is authorized, as specified by Congress, to assess premiums against FCS entities until the fund reaches a secure base amount (SBA), initially set at 2 percent of outstanding FCS loans. With expiration of the FCS Financial Assistance Corporation,

FCSIC is responsible for financial assistance to FCS entities, and if needed, for closing or putting failing institutions into receivership. It is already understood that FCSIC will redeem the Financial Assistance Corporation bonds that covered the failure of the Federal Land Bank of Jackson. Those bonds become due in 2005.

Several issues revolve around FCSIC: (1) with its purpose different than that of FCA, what is an appropriate board of directors? FCSIC was set up with the same board as FCA, but recent action specifies a different board makeup in 1996; (2) should FCSIC depend upon FCA examiners for its information?, and (3) should FCSIC have its own special examiners?

Also, the insurance fund (SBA) grows in two ways—from premiums and from earnings on the fund. So when the fund reaches the 2 percent (or other specified level), presumably premiums can cease. Should interest earnings beyond that needed to maintain the SBA remain in the fund or be returned to FCS institutions? FCSIC and FCS officials have different views on that issue. Further, is 2 percent the appropriate amount for the SBA? These issues deserve action before the fund reaches the secure base amount, expected in the latter 1990s.

4. Government-lending reorganization/efficiency issues will likely receive continued emphasis. There are, and will likely be, further efforts to consolidate local USDA programs. One proposal is to combine several of USDA's farm service agencies, including FmHA, into a Farm Services Administration. This combination, it is argued, would require fewer local offices and personnel to carry out the mandates of these farm programs. The continuing decline in farm numbers, it is argued, require fewer local offices and personnel to carry out agriculturally related programs.

FmHA may suffer from this reshuffling. Restructuring may not adequately consider the credit skills and competence needed for FmHA lending programs. The appropriate mission of FmHA and its structure, in terms of its mission, requires redefinition. In line with FmHA's pre-1970s mission—to help low-resource, disadvantaged, or disaster-struck farmers—the Agricultural Improvement Act of 1992 refocused programs more toward beginning, low-resource, and minority farmers. That contrasts with FmHA's late 1970s and 1980s emphasis on rescue of established but financially weak farm borrowers.

Other FmHA issues include guaranteed versus direct loans. In general, policymakers have pushed FmHA more toward guaranteed loans and away from direct loans as a way of reducing subsidies and moving loan decisions into the hands of private lenders. If, however, FmHA's public purpose mission is to help disadvantaged, low resource, low equity, and/or beginning farmers, then it may be that direct loans are best for this purpose. FmHA needs to assess the effectiveness of its various programs.

Further, attention should focus on changing the complex procedures for holding and selling farm properties in the FmHA inventory. FmHA incurs large costs for taxes, upkeep, and lost income on acquired properties. Regulations are being modified. These substantial costs could be reduced significantly by early auction/bid procedures to move the acquired properties out of inventory. And, while beginning farmers must receive first chance to buy such properties, often these "taken back" properties are in dilapidated, deteriorated condition. FmHA should be enabled to assess the value of each property taken back in terms of its potential as a start-up property for beginning farmers. A property found to be otherwise, i.e., a "failure mill," should be sold immediately to the highest bidder.

Barriers to Capital and Credit Flows

In general, it appears that capital and credit are available to the agricultural sector and that financial markets work effectively to allocate funds to the sector and to businesses within the sector (U.S. General Accounting Office 1992). With several competitor systems (commercial banks, Farm Credit System, insurance companies, and input suppliers) serving the sector, interest rates and loan terms tend to be competitive among credit sources and subject to influences of financial markets.

A variety of loan terms is generally available, although the volatile interest rates of the 1970s and 1980s tended to force real estate lenders into providing loans with adjustable rates rather than long-term fixed-rate loans. Farmer Mac was designed to provide such long term financing; but, with the pressure to rapidly build volume and little demand from farmers for fixed-rate loans during a period of declining interest rates, it has not been very successful in attracting pools of long-term, fixed-rate loans.

While there are few barriers to the general movement of funds, farm lenders claim that environmental liability legislation and EPA interpretations both restrict the availability of credit to farmers and make it more costly. Since this is an issue affecting all real estate lenders, jurisdiction is not likely to fall under the agricultural or agricultural credit legislation committees. Nevertheless, to limit their liability on environmental damages, farm real estate lenders have begun to require environmental audits as a part of the loan approval process, and the costs of the audit, often substantial, are passed on to the borrower. In instances where there are questions or concerns about prior environmental damage, the loan is usually not approved and the prospective buyer of the property may not be able to obtain real estate credit. One issue, of course, is whether the lender is the appropriate party to be held accountable for environmental damage by a current or earlier owner. Can environmental responsibility be brought about in a more rational manner?

Another possible concern over barriers to credit flow is the availability of credit to different risk segments of the farm sector. General lending policies and past experience suggest that low resource, low equity farmers and beginning farmers would seldom be served by conventional lenders. Without borrower equity or farming experience, repayment risks are high, and credit is typically unavailable. FmHA programs, in particular, have attempted to provide credit to this particular segment of the market. Occasionally other lenders have offered credit to this group through special programs or by requiring co-signers of notes or guarantees from parents or relatives. Is it in the public interest to support loans with such high risk?

Farm policy makers and the farm community often express concern about the relatively small number of young people entering farming. They conclude that there must be barriers to entry into the farm sector and that the most likely barrier is credit. It is difficult, as mentioned above, for young people with low equity to get access to credit. Yet, with a relatively constant or slowly decreasing land base in agriculture and both technology and economics together pushing towards larger average sizes of farms, there is not an opportunity for a new entrant each time an operator retires or leaves the sector.

Availability of Credit to Rural Businesses and People

Similar to the issue of availability of credit to the farm sector, there are concerns over the availability of credit to those living and working in rural areas. These include: (1) a concern that rural areas are somewhat isolated, so that financial markets may not be accessible to savers or borrowers who live in rural areas, (2) a concern over lack of competition in rural areas such that rural inhabitants and businesses have either no, one, or at best a few, financial institutions nearby, and (3) a concern that stable or shrinking rural communities will be unable to obtain funding to maintain or improve their infrastructure—roads, water systems, sewage facilities, etc. Without doubt, rural areas do not have a wide array of choices when it comes to financial institutions. However, problems may be more severe for lending than for saving. Rural inhabitants have access to a variety of mutual funds that are easily available by phone or mail. Loans may be more difficult to obtain. The low volume of deposits and loans in rural institutions due to market size make it uneconomic to have very many competitors. Further, with the low volume of business, loan rates are likely to be higher than in more competitive situations.

As a result of these concerns, special government programs have sometimes been developed. One approach was to give FmHA authority to make, and later to guarantee, rural housing loans. This authority was based on a perception that rural housing credit on a par with urban areas was not available. Later, the FCS was authorized to make housing loans in narrowly defined rural areas. Limitations were put on both FmHA and FCS to prevent full-scale competition with lenders serving larger, more heavily populated areas and providing a wider range of services.

While the lack of credit available for housing in rural areas was relatively easy to document, availability of credit for rural businesses—especially new or start-up businesses—has been much more difficult to evaluate. Certainly one public concern is that the recent financial crisis in the banking sector and consolidation taking place within that sector may have adversely affected credit availability in rural areas. Concerns continue that large holding companies are taking over small, rural banks, then withdrawing from lending to local businesses and siphoning funds from rural to metropolitan areas.

A fundamental problem is that we are not really sure how to promote rural development. Will investments in rural infrastructure induce movement of businesses to rural communities? Are there some existing businesses in rural communities stifled in growth due to the lack of credit? Are there potential new credit worthy businesses that are not being established due to lack of credit, but which would be funded if federal guarantees or initiatives were available? In short, it is not clear whether the real problem of rural development is a shortage of ideas and economic opportunities as much as it is a shortage of credit.

Efforts to accomplish more in the area of rural development may come through the establishment of the Rural Development Administration in USDA. That agency will take over FmHA's rural community, business, and industry programs. Yet, the issue remains of whether there are unmet productive uses of credit in rural areas. If so, RDA may be a useful approach. If not, it is questionable that throwing dollars at rural areas will create rural development. Again, the real issue is RDA's appropriate role.

In short, credit markets in rural areas are probably lacking in some respects. The relatively low volume of financial transactions limits the number of competitors providing financial services, and the low volume likely leads to higher costs in servicing the loans. Hence, given relatively few alternatives and the economics of the market, rural borrowers may pay higher rates. Whether there is a segment of the rural-business market or entrepreneurial market for which credit is unavailable is less clear. It is a question that needs assessment and research.

Likely Policy Approaches

It seems likely that near term policy approaches and legislative consideration will emphasize organizational/structural issues. Major attention will focus on the FmHA organization, downsizing and targeting of its efforts under a revised mission. Similarly, organizational/ structure issues in the Farm Credit System and FCA will receive consideration. Farmer Mac, a new struggling institution, will receive legislative attention intended to strengthen its influence, change its organization, refocus its mission, and/or perhaps, even to terminate its charter—depending upon its perceived effectiveness. Rural development issues will likely continue to gain attention, and public policies directed toward assistance through credit programs can be expected.

References

Arnold, C. R. 1958. "1933-1958: Farmers Build Their Own Production Credit System." Farm Credit Administration Circular E45.

Brake, John R. 1974. "A Perspective on Federal Involvement in Agricultural Credit Programs." *South Dakota Law Review* XIX.

Collender, Robert N. 1992. "Changes in Farm Credit System Structure." In *Agricultural Income and Finance Situation and Outlook Report* AFO-44, ERS, USDA.

Farm Credit Administration. 1967. *The Federal Land Bank System, 1917-1967* Farm Credit Administration Circular E43.

Harl, Neil. 1992. "Chapter 12 Bankruptcy: A Review and Evaluation," *Agricultural Finance Review* 52.

Johnson, D. Gale. 1963. "Agricultural Credit, Capital and Credit Policy in the United States." In *Federal Credit Agencies—A Series of Research Studies Prepared for the Commission on Money and Credit.* Englewood Cliffs, NJ: Prentice Hall.

Lins, D., and P. Barry. 1984. "Agency Status of the Farm Credit System," *American Journal of Agricultural Economics* 66:601-606.

Peoples, K. L., D. Freshwater, G. D. Hanson, P. T. Prentice and E. P. Thor. 1992. *Anatomy of an American Agricultural Credit Crisis*, Landham, MD: Bowman and Littlefield Publishers.

Stam, Jerome, Steven R. Koenig, Susan E. Bentley, and H. Frederick Gale, Jr. 1991. "Farm Financial Stress, Farm Exits, and Public Sector Assistance to the Farm Sector in the 1980s." ERS, USDA, Agricultural Economic Report No. 645.

U.S. General Accounting Office. 1992. "Rural Credit: Availability of Credit for Agriculture, Rural Development, and Infrastructure." Washington, D.C.: GAO/RCED-93-27.

21

Issues in Research and Education

Robert L. Christensen and Ronald C. Wimberley

Title XIV of the Food and Agriculture Act of 1977 and of the Food Security Act of 1985, and Title XVI of the Food, Agriculture, Conservation, and Trade Act of 1990, provide the basic federal funding authorization for the Land-Grant College and University system's agricultural research and extension programs. Among other things, Title XVI of the 1990 Act identifies six primary purposes:

1. Satisfaction of human food and fiber needs.
2. Global viability and competitiveness of U.S. agriculture.
3. Expanded economic opportunities and quality of life for farmers and citizens.
4. Improved productivity, development of new crops, and new uses for agricultural products.
5. New information and systems that enhance the environment and a sustainable agriculture.
6. Improved human health through provision of a safe, wholesome, and nutritious food supply.

This chapter reviews national food and agriculture legislation as it affects the Land-Grant College and University research and Cooperative Extension system. Some of the issues to be examined are:

1. Should other public and private research, education, and for-profit entities (i.e., non-land grant institutions) be permitted to compete for funding authorized under Title XVI of the 1990 Act? What are the arguments for, and against, open competition for these public dollars?

2. Should research and extension programs be limited to agriculture and rural areas, or is the mandate of larger scope? Is potential alienation of traditional political support offset by the larger benefits to society from these expanded programs?
3. Is interdisciplinary and multidisciplinary research (the current emphasis) more productive than is research within the independent disciplines?
4. Special project and initiatives funding of agricultural research and extension is increasingly preferred to formula funding. Will specific designation of topics limit the creative quest for knowledge and limit the capacity of both research and extension to respond to the larger range of public needs?
5. Is the debate over revision of formula funding equally applicable to research and extension. What are the factors to be considered in this debate?
6. Is the National Research Initiative (NRI)—a competitive grants program—an appropriate model for identifying social and economic research needs?

The Land-Grant System

External and Internal Changes

In 1870, nearly half of all U.S. employment was in agriculture and about three-fourths of the U.S. population lived in rural areas (Morris and Wimberley 1992). The 1862 and 1890 Morrill Acts created the Land-Grant Colleges, largely out of concern for the social and economic well-being of rural Americans. By 1990 the farm population (i.e., persons living on working farms) had fallen to less than 2 percent and the rural population to slightly less than 25 percent of the total. The magnitude of these changes and their significance has not been lost on some observers.

> The United States the land-grant universities were created to aid and assist no longer exists. . . . I have serious questions about the political viability of the land-grant university . . . If it cannot justify its existence, in these times of declining resources and middle class taxpayer unrest, Land-Grant universities may cease to exist in their current structure and number. (Stauber 1992)

Today's Land-Grant Colleges and Universities are indeed quite different from the institutions created by the 1862 and 1890 Morrill Acts, the Hatch Act of 1887, and the Smith-Lever Act of 1914. In many Land-Grant

Colleges and Universities, the removal of the agricultural research and extension components would hardly be noticed by the majority of students and the general public. This fact, coupled with an increasing ignorance of the historic Land-Grant mission among administrators and faculty within the institutions, contributes to the concern for viability. The conditions outlined above have led to calls for a new focus for land-grant research and extension programs (Stauber 1992; Morris and Wimberley 1992; Just and Huffman 1992).

Contemporary Land-Grant Colleges and Universities almost universally aspire to be "world class," a term implying recognition of institutional excellence by the scholarly community. Only a few Land-Grant College and University presidents would cite their experiment stations and extension services as a basis for claiming such distinction. Yet in the arena of agricultural research and extension, U.S. Land-Grant Colleges and Universities would dominate the field of "world class" institutions.

At some Land-Grant Colleges and Universities, administrators seem to place a low priority on the missions of applied agricultural research and educational outreach. It also seems that applied research and extension education are becoming less valued in peer-group recognition processes. The erosion of internal support for these traditional missions is a matter of concern.

An excellent paper by Just and Huffman (1992) on the structure, management, and funding of agricultural research should be required reading for members of the Land-Grant system and for policymakers at the federal level. Just and Huffman outline some of the significant problems facing public agricultural research in the United States and offer propositions for addressing these problems.

Why Have Public Research Funding Anyway?

Nearly ten years ago, Johnson (1984) projected, on the basis of world population trends, the need to increase average crop yields by 70 percent by the year 2030. Johnson pointed out the need for academic activity in problem and subject-matter research, and the development and dissemination of disciplinary knowledge. He concluded that while privately supported research will generate much new technology, in many instances the public interest will best be served by funding of research and the dissemination of new knowledge that is non-proprietary.

The reason why continued public support of agricultural research is necessary is simple. Agricultural research is a public good because the benefits realized in productivity and efficiency ultimately accrue to the public. There are several compelling reasons for continuing publicly funded research. While public benefits also result from proprietary research, they remain subordinate to the primary goals and interests (including profits) of private firms. This means that research having less potential for success is not likely to be pursued by the private sector. Proprietary research must consider the risks of a noncommercial result and consequent inability to recover the research investment. It may not be possible for the sponsor to protect and control certain research findings, reducing the work's potential profitability and further endangering its investment. Certain types of research require facilities and staff that a private firm would find difficult to assemble or willing to risk.

Seaver (1989) reviewed a number of studies of returns on investments in publicly funded agricultural research. These studies found internal rates of return ranging from 45 to 202 percent per year. Seaver also cites a substantial body of evidence to show that, while early adopter farmers capture returns from increased efficiency, consumers are the primary beneficiaries of the agricultural research investment. One study (Evanson 1979) found the rate of return to agricultural extension during the period 1948-1970 of 110 percent). According to Eddleman (1982):

> Even with all the evidence about the large and positive contributions to productivity, governments (and particularly the federal government) continue to underinvest in agricultural research and education. This factor is a constraint on the development and application of new knowledge for agriculture.

In the Public Interest

In his recent book, Hansen (1991) described agricultural policymaking in three time periods. His main concern is the influence of farm interests on agricultural legislation. He argues that an interest group has the greatest influence on a policy when the group appears better able than competing groups to help a legislator get re-elected, and when it appears the group will continue to have that competitive advantage. His analysis concluded that interest groups can be highly influential when they provide legislators and staff with information about their constituents and their needs.

McGuire (1992), says ". . . in the present condition of our political process, the focused concerns of a minority will always prevail over the unfocused concerns of the majority." He notes the influence of some environmental groups in public opinion and policy. However, the point extends to the broader context of the Land-Grant mission for research and education. Even in rural America, farmers today are a minority in our population. Yet, through their organizations, farmers are potent lobbying groups with the Administration and Congress. In addition, the National

Agricultural Research and Extension Users Advisory Board is weighted toward representation of agricultural producers.

What does this mean with respect to the future for agricultural research and extension? Moseley (1992), an Indiana farmer and former Assistant Secretary of Agriculture, recently spoke to this question. Although addressing extension, his remarks may also apply to research. He urged that extension be reoriented to problems and issues of a broad spectrum of the public and not just those in rural America. He said, ". . . when the public believes enough that extension is the mechanism for change, when it is believed that the system is the best at serving the present need, then the system can leverage itself into a position of influence and the public will pay."

He urged more attention and resources for programs addressing public policy, consumer and family issues, youth leadership training, and rural development. He implied that this would require shifting of resources away from production agriculture and suggested basic changes in the way extension agricultural agents and specialists approach their jobs. He emphasized his view that extension personnel should function as "knowledge brokers" rather than as "answer persons."

The Institutional Context

To reverse the decline in political and budget support for agricultural research and extension will require that disciplines and institutions address the causes of the decline. Several years ago, Shuh (1986) expressed the view that the agricultural research agenda in the Land-Grant Colleges and Universities has been captured by the disciplines and their internal professional goals. As a consequence, research has become more heavily weighted to methodological development and less oriented to the solving practical problems.

McDowell (1992) states, "When taken together with the other problems in the University, extension . . . will not make the 20 more years that some of its most severe critics give it. Unless there is profound change, I believe it will be virtually dead in 10 years." McDowell charges that agricultural interests, as primary advocates and supporters, have captured the Land-Grant agricultural research and extension agendas. At the same time the political clout of agriculture and rural America continues to decline.

Researchers, because of a somewhat greater degree of insulation from service demands of agricultural interests, would be more able to address the larger issues of rural America. Here McDowell echoes Shuh and says, "It is the scholarly communities and individual views and values of the scientists, as much as anything else, that controls that agenda." And, ". . . the research agenda of the university [is] captured by the professors . . .." Finally, with respect to the agricultural community, McDowell asks whether it is getting what it *wants* or what it *needs* from the system? He then argues that it gets what it wants but not what it needs.

Uncertainty of Funding

While there are always some uncertainties with respect to federal funding of Land-Grant research and extension, the record shows steady growth in Hatch and Smith-Lever funding over the past decade. There is, of course, more variability in special projects and competitive grants funding. Federal support for base funding of agricultural research and extension has been strong in spite of occasional challenges by critics.

State funding for extension has, in some states, been increasingly problematic in recent years. There have been those who have questioned the rationale of continuing support for an agency that has, in their view, become obsolete. Others recognize the obsolescence charge in regard to certain programs and activities, while strongly supporting the value of extension. This has led to initiatives, supported by many legislators, that broaden the traditional clientele of extension to include suburban and urban people. They have focused on issues like solid waste management, alcohol and drug abuse by young people, and drinking water quality.

In all of this, extension at the state and local levels has maintained a commitment to its grassroots clientele through user advisory committees, and there has been limited shifting of resources among priorities. Traditionally, extension's strongest political supporters come from its agricultural and youth constituencies in rural areas. The agricultural lobby has strongly resisted any loss of staff support serving its interests and has been lukewarm to new extension programs serving other constituents, even when new funding was appropriated to support such programs.

Who Sets the Priorities?

The Food and Agriculture Act of 1977 established two advisory groups to assist the federal government in establishing research and extension needs and program priorities. The Joint Council on Food and Agricultural Sciences and the National Agricultural Research and Extension Users Advisory Board have been continued in subsequent legislation. The Users Advisory Board reviews policies, goals, plans, needs, issues, and accomplishments of the agricultural research and extension system and provides policy, priority, and strategy recommendations to the Secretary of Agriculture. The mission of the Joint Council is to improve the planning and coordination of research, extension, and teaching activities relating to the food and agricultural sciences.

The national Land-Grant system has created three policy committees relating to agriculture: the Experiment Station Committee on Organization and Policy (ESCOP), the Extension Committee on Organization and Policy (ECOP), and the Resident Instruction Committee on Organization and Policy (RICOP). These committees develop and propose policies relating to their respective areas and interact with the Administration and the Congress on issues relating to legislation and appropriations.

In addition, virtually every state has an assortment of advisory committees representing the many constituencies that provide counsel and support to the colleges of agriculture, and their experiment stations, and extension services. It is here that state-specific agendas for research and extension are formed. Research and extension programs in the states must be responsive to the needs and priorities of their state in order to enjoy continued public support.

Congress influences the agricultural research agenda and extension's program priorities through specification of topics and funding. The first level of influence is provided by Title XVI of the 1990 Act. Further instruction is provided in other sections of the 1990 Act, where specific programs are mandated and responsibility given to the agencies. Perhaps most prescriptive are the appropriations to carry out programs.

In the end, scientists in the Land-Grant system still exercise considerable latitude in identifying their research agendas. They individually select the topics that stimulate their intellectual curiosity and satisfy disciplinary norms. Grant funding availability is certainly one important factor, but is not the only motivation for a researcher. Similar considerations are increasingly affecting the extension educator.

Competition

Program funding historically directed to the agricultural experiment stations and the extension system is increasingly becoming available to other public and private entities. Examples from the 1990 Act are the competitive grant program funding authorizations contained in the National Competitive Research Initiative (NCRI) [Section 1615(a)(6)], Best Utilization of Biological Applications [Section 1621(b)]; Alternative Agricultural Research and Commercialization [Section 1660(a)]; and Research Regarding the Production, Preparation, Processing, Handling, and Storage of Agricultural Products [Section 1644(b)]. It should be noted that not all authorizations actually receive funding.

It is important to maintain the agricultural research programs conducted by the Land-Grant institutions. These programs represent the nation's base of agricultural research and our capability of addressing national agricultural needs across the states. In this manner, Land-Grant College and University research and extension objectives serve the public interest in the localities where the problems exist. The Land-Grants tend to be more attuned to ongoing pure and applied scientific issues and less subject to passing political philosophies and initiatives that affect federal agencies.

Funding from private industry primarily serves technological development, as broadly defined. Funding from foundation sources—whether federal [as from the National Science Foundation (NSF)] or private—tends to reflect relatively focused topics identified by the donor organization. Driven by the need to establish an external funding record, some research and extension faculty members may pay little attention to the university's mission or state needs. Greater support through the federal formula funding sources (e.g., Hatch and Smith-Lever) would help to ensure that research and extension program priorities are not diverted from the mission and goals of the experiment station and extension systems. The executive and legislative branches of government need to understand the public research and education system in order to foster and promote activities consistent with needs of the public constituencies these systems are designed to serve.

Disciplinary and Multidisciplinary Programs

One traditional image of the scientist is a solitary figure in the laboratory who mixes chemicals and occasionally shout, "Eureka." In the real world, "Eureka" does occasionally echo from labs. However, there now seems to be more emphasis on team research than in the past.

Team efforts within a discipline are not the same as multidisciplinary or interdisciplinary approaches. Multidisciplinary and interdisciplinary research is characterized as research where a problem is identified that requires the attention of more than one discipline for discovering solutions. The multidisciplinary approach requires joint planning and coordination of research by administrators and scientists from the disciplines involved. Scientists from each discipline would then independently pursue their research according to the comprehensive plan. However, unless frequent communication and conscious coordination occurs, this approach often results in gaps in knowledge and non-comparability in findings.

The interdisciplinary approach goes a step further. This approach forms a working team of individuals from the separate disciplines and works within a common research plan so that an integrated model is generated. When this approach is used, alternative solutions may be tested with more complete understanding of complex interrelationships.

Realizing that agricultural production is really the management of a complex biotechnical system has led to the popularity of systems approaches. Further, the systems approach tends to require that more than one discipline is necessary to address problems. As a result, the 1985 and 1990 food and agricultural legislation contains language that calls for research on production systems. In fact, two of the six areas identified in the National Research Initiative RFP (CSRS 1992) are simply titled "Animal Systems" and "Plant Systems."

National Research Initiative (NRI) proposals of three types are accepted: (1) fundamental, (2) mission-linked, and (3) multidisciplinary. The NRI specifies that ". . . no less than 30 percent ($27.690 M) . . . shall be made available for grants for research to be conducted by multidisciplinary teams . . . " (NRI 1993). Perhaps the most direct incentives for system-oriented, interdisciplinary activity were exemplified in the "Low Input, Sustainable Agriculture (LISA)" program in which the evaluation criteria included:

> Most LISA projects have goals that are best attained through an interdisciplinary approach, involvement of both research and educational activities, or cooperation of farms or other private organizations with universities or government agencies. Proposals of this type will be judged according to the efficacy of the plan for assembling an appropriate team of participants and devising an effective team strategy. (U.S. Department of Agriculture 1989).

Such efforts to encourage multidisciplinary, interdisciplinary, and collaborative activity are laudable. At the same time, it is important that the pendulum not swing so far in that direction that solitary researchers or single discipline teams find it difficult to get funding. Each can make significant discoveries that may be integrated into systems.

Funding for Experiment Station and Extension Programs

Competitive Grants—The Rationale

One way Congress and the Administration ensure that state experiment stations address issues of national concern is to earmark research funding through a system of competitive grants. Examples include the National Competitive Research Initiative (NCRI) and the Sustainable Agriculture Research and Education (SARE) programs. The SARE program replaced the LISA program previously cited.

The National Science Foundation (NSF) has administered competitive grant programs for many years. At this writing, however, there is debate within NSF over the need to make programs more responsive to the needs of industry (Park 1992). There is a perceived need to focus and speed the transfer of research results from the academic community to the marketplace that would, in turn, result in stronger public support for research in the long run. A major issue is the degree to which industry would control the academic research agenda. A companion concern is the potential loss of support for long-term curiosity-driven basic research lacking obvious immediate application; such research has frequently led to important scientific breakthroughs. There is an obvious parallel set of questions that apply to competitive grants for agricultural research.

The justification for open access to the competitive grant programs of the 1990 Act is that it helps ensure that the scientists with highest qualifications, regardless of affiliation, will have opportunity to compete for funding. Another feature of competitive grants is that, since funding would not be available to all states, grants can be made that are large enough to assemble the critical mass of knowledge and technology necessary to problem solution. In response to the alleged bias toward the larger institutions, it is argued that the purpose of federal funding via competitive grants is not to strengthen institutions, but to enhance the creation of new knowledge by assisting the most-qualified scientists, regardless of institutional affiliation.

Land-Grant College and University scientists challenge the competitive grants approach on a number of different grounds. Some of these are noted here.

Competition Versus Equity Among States

Although an assumption of the competitive grants strategy is that the process will select the highest qualified researchers and institutions, there are further considerations. For example, the federal government is actually in partnership with the states through the Land-Grant Colleges and Universities. Indeed, research and extension funding provided by the states often considerably exceeds the federal contribution. Therefore, the states should have weight in assigning priorities, and local university scientists and administrators should play a major role in determining needs and resource allocations within their state institutions. By their nature, competitive grants cannot be expected to distribute equitably across the states. It is the largest, most prestigious, and well-financed universities that are in the best positions to compete. Their faculties provide disciplinary breadth and depth. These universities, because of reputation and funding, have been able to attract many of the best and brightest scientists. They generally have superior research facilities. In short, they have an inherent advantage over most smaller universities. Smaller institutions will become even less able to compete unless they have developed special niches of disciplinary expertise.

There is concern in the Land-Grant Colleges of Agriculture over the deterioration of research facilities resulting from years of underinvestment. This problem has been especially troublesome due to the limits included in appropriations language in recent years, on indirect cost charges allowed for agricultural research.

Competitive Grants and Research Programs

Much scientific discovery is evolutionary, i.e., new research is based on knowledge resulting from past research. Competitive grants for research projects, in comparison with formula funding, appear less compatible with evolutionary programs. Competitive funding is toward discrete projects, more narrowly focused, and with limited time frames. Once the project is completed, a new focus must be identified and pursued. Unfortunately, this approach is discontinuous. In contrast, the continuity of research program funding through a formula is more compatible with the long-term evolutionary process of developing knowledge.

There is a need to distinguish between research projects and research programs. A project is of limited scope and duration and aims to resolve a particular, ad hoc research problem. A research program is an ongoing effort to coordinate research projects addressing larger research issues. A similar distinction may be made between extension projects and programs. For example, a farm management program will consist of a number of individual projects or educational activities such as farm records, tax management, enterprise analysis, investment decisionmaking, etc. The priority given to each of these topics may change from year to year, based on perceived need.

Competitive Funding and Extension Programs

For extension, the merits of competitive grant funding seem somewhat questionable. A basic assumption is that new knowledge or approach, program materials, and/or curriculum developed by a successful extension competitor will be adopted by other state extension services. However, several factors work against that assumption. First, it needs to be recognized that substantial funding was made available to the state receiving the grant to make it possible to develop and implement the program. Even though a smaller level of funding may be required for adoption by another state, it may still exceed the smaller state's funding capability. Second, a program appropriate to the needs of clientele in one agricultural, social, or local area may not be transferred easily to another agricultural, social, or local area. Third, the priorities of other state systems may not coincide with those of the grant funding. Fourth, transferability depends on the size, skills, and interests of staffs in other states.

Do Competitive Grants Match the Land-Grant Model?

At issue is whether competitive grants are a desirable mode for funding the Land-Grant system in view of the total welfare of the country and the citizens of those states in which the less favored institutions are located. It is in the best interests of the public that *all* universities in the Land-Grant system be strengthened, as originally legislated. Funding for competitive research grants should not be such a large share of total funding that formula funding cannot maintain long-term research programs. Similar limits should be placed on competitive and special grants for extension in order that the individual states may have the discretion to design and implement formula-funded programs attuned to their priorities and needs.

Non-Land-Grant Agencies

Should other public or private educational institutions, or research or consulting firms compete with the Land-Grant system for federal agricultural appropriations? The problem is that this competition can jeopardize the strength and scope of research and extension programs at the Land-Grant Colleges and Universities and damage the continuity of programs serving agricultural and rural needs. Of particular concern is the loss of the vital linkage for technology and knowledge transfer that exists between research and extension within the Land-Grant system. It would appear to be more rational to support on-going, long-term research programs rather than a series of stop-and-go projects.

Priorities for the 1995 Legislation

Expanding the National Research Initiative

Support for the National Research Initiative in the 1990 Act and subsequent appropriations bills represent a signal success by the Land-Grant Colleges and Universities and federal agencies in putting together and marketing to policymakers a comprehensive plan for six priority areas.

However, the shares of funding given to these priorities seems woefully out of balance. The Plant Systems and Animal Systems titles account for two-thirds of the Initiative's appropriations in 1993 ($61.532 million of the total $92.298 million). The percentage allocated to the three areas titled "Nutrition, Food Quality, and Health," "Markets, Trade, and Policy," and "Processing, Added Value, and New Products" amounts to only 11 percent of the total.

Unfortunately, the allocation of larger sums of funding totals to animal and plant research perpetuates the traditional favoritism given to technology while neglecting economic and social research on the viability of farms, the agricultural production and marketing system, global competitiveness in agricultural trade, rural communities, conservation of land and water resources, safety and wholesomeness of food and water, and the relationships between agriculture and the economic well-being of consumers. At the same time, funding must be sufficient to maintain and improve U.S. competitiveness in plant and animal agriculture in world markets.

The Social Science Agricultural Agenda Project (1991) summarized the situation as follows:

> The widespread criticism of the USDA/land-grant college system for being too technocratic and neglecting alleged adverse social impacts of the technologies it has created on family farms, the environment, food chains, depopulated rural communities, food quality, rural health, agribusinesses, farm laborers, the structure of the agricultural sector, etc. *These criticisms need to be, can be, and should be channeled into support* for funding multidisciplinary problem-solving and issue-oriented work on the social and humanistic dimensions of current technical-change problems and issues by social scientists and humanists in our agricultural institutions. (*Emphasis in the original.*)

Wimberley (1992), in his presidential address to the Rural Sociological Society emphasized the need for rural policy initiatives for both economic development and human resource development. He points out that both must be taken into account if rural viability is to be sustained. Success in dealing with rural problems and achieving rural viability requires that both economic and human resource development policies and programs operate in concert.

A major effort should be made by the Land-Grant system to place before the Administration and Congress evidence of need and priorities for legislation that will significantly increase funding for economic development, social, and nutrition initiatives.

Among the least supported areas of research and extension is rural development, in both its economic and human resource dimensions. While verbally acknowledged as important, appropriations for rural development have been abysmal. By any measure, funding of the regional rural development centers is low. It seems desirable to provide additional funding for a "National Research and Extension Initiative for Rural Development" in the 1995 legislation. Perhaps the newly created Rural Development Administration in the U.S. Department of Agriculture can provide national leadership and support the creation of expanded rural development funding support for the Land-Grant Colleges and Universities.

Caution should be advised in more aggressive adoption of the competitive grants approach to funding research and extension. Concerns about the long-run implications of the several dimensions of competitive grant funding were raised earlier in this chapter. Competitive grants are a useful mechanism to encourage attention to problems of high priority and urgency. The National Agricultural Research Initiative, by giving focus to the research agenda, appears to have been instrumental in stimulating additional appropriations. At the same time, care needs to be taken that directed competitive grant programs not stifle the opportunity for the exercise of intellectual curiosity or to fail to provide support for research and extension where program continuity is important.

Unfortunately, the national extension system's set of national initiatives has been less successful in generating increased funding. In 1992, only three of the seven national priorities (Water Quality, Food Safety and Quality, and Youth at Risk) received funding. Among the unfunded priorities of extension were Nutrition Education, Rural Economic Development, Sustainable Agricultural Systems, and Waste Management.

There appears to be a paradox, or "catch 22," for cooperative extension. Among the National priorities proposed by extension, but for which no appropriations were made, are included areas that critics charge have been inadequately addressed by extension. The fact that the system identified these priority issue areas clearly indicates a willingness, given additional funding, to conduct pertinent educational programs. The fact that appropriations were not provided would suggest that the congress did not share these perceptions of priorities for national educational programming. A major assignment for the national extension system for the new food and agricultural legislation and annual appropriations process will be to persuade the Congress and the Administration of the need and urgency for educational programs directed to these issues. Perhaps most important in this process will be the extension system's ability to provide evidence that it can provide cost-effective programming that assists in solving problems and creating new opportunities.

Policymakers will be challenged as the new food and agricultural legislation is formed. It will take courage to define policies that will chart new directions for research and education into the next century. There is opportunity for aggressive Land-Grant leadership in a range of issues, from those of family farm viability to biotechnology, and from environmental protection to rural economic development. A proven system for research and education is now in place in the Land-Grant Colleges and Universities. With this existing system, initiating an expanded agenda of research and public education can be accomplished at less cost than would be needed to create new agencies and bureaucracy.

References

Cordes, C. 1992. "Science Foundation Under Intense Pressure to Lead Movement to Transfer Academic Research to Industry." *The Chronicle of Higher Education.* Page A24. October 7.

Eddleman, B. R. 1982. *Impacts of Reduced Federal Expenditures for Agricultural Research and Education.* IR-6 Information Report No. 60. Interregional Cooperative Publication. Mississippi State, MS: Mississippi State University.

Evanson, R. E., P. E. Waggoner, and V. W. Ruttan. 1979. "Economic Benefits from Research: An Example from Agriculture." *Science* 205:1101-1107.

Hansen, J. M. 1991. *Gaining Access: Congress and the Farm Lobby, 1919-1981.* Chicago, IL: The University of Chicago Press.

Johnson, G. 1984. "Academia Needs a New Covenant for Serving Agriculture." Special Publication. Mississippi State, MS: Mississippi State University Agricultural and Forestry Experiment Station.

Just, R. E., and W. E. Huffman. 1992. "Economic Principles and Incentives: Structure, Management, and Funding of Agricultural Research in the United States." *American Journal of Agricultural Economics* 74(5):1101-1108.

McDowell, G. R. 1992. "The New Political Economy of Extension Education for Agriculture and Rural Communities." Paper presented at the Annual Meeting of the American Agricultural Economics Association. Baltimore, Maryland. August 9-12.

McGuire, R. T. 1992. "A Capital Dilemma: Making the Right Decisions About Using Our Most Limited Resources." *Choices* (Third Quarter):12-15.

Morris, L. V., and R. C. Wimberley. 1992. "Old Missions and New Directions for Land-Grant Universities." Presentation to the 50th annual Professional Agricultural Workers Conference. Tuskegee University, Tuskegee, AL. December 7.

Moseley, J. 1992. "Cooperative Extension: What They Should be Doing and What They Should Stop Doing." Paper presented at the Annual Meeting of the American Agricultural Economics Association. Baltimore, MD. August 9-12.

Park, R. L. 1992. "Public Confidence in Science is Eroding." *The Chronicle of Higher Education.* Page A48. September.

Seaver, S. K. 1989. "Who Says Public Investment in Agricultural Research Doesn't Pay?." Storrs, CT: University of Connecticut. AES/CES Research Report 89-901.

Shuh, G. E. 1986. "Revitalizing Land Grant Universities—It's Time to Regain Relevance." *Choices* (Second Quarter):16-10.

Social Science Agricultural Agenda Project. 1991. *Social Science Agricultural Agendas and Issues.* Executive Summary. East Lansing, MI: Michigan State University Press.

Stauber, K. 1992. Speech given at the Spring Symposium of the Board on Agriculture and the Professional Scientific Society's National Academy of Sciences. Irvine, CA. April 3.

U.S. Department of Agriculture. 1993. Program Description and Guidelines for Proposal Preparation and Submission. Washington, D.C.: Cooperative State Research Service.

______. 1992. Program Description and Guidelines for Proposal Preparation and Submission. Washington, D.C.: Cooperative State Research Service.

______. 1992. "1990 Annual Report on the Food and Agricultural Sciences." Washington, D.C.: Cooperative State Research Service, Research and Education Committee. February.

______. 1989. "Call for Proposals—North Central Region 1989 LISA Program." Washington, D.C.: USDA Research and Education Grants Program. Cooperative State Research Service.

______. 1993. "Policy Perspectives on Social, Agricultural, and Rural Sustainability." *Rural Sociology* 58:1-29.

Subject Index